Recent Research in Psychology

Jeannette P. Ward
William D. Hopkins
Editors

Primate Laterality

Current Behavioral Evidence of Primate Asymmetries

With 31 Illustrations

Springer-Verlag
New York Berlin Heidelberg London Paris
Tokyo Hong Kong Barcelona Budapest

Jeannette P. Ward, PhD
Department of Psychology
Memphis State University
Memphis, TN 38152
USA

William D. Hopkins, PhD
Division of Behavioral Biology
Yerkes Regional Primate
Research Center
Emory University
Atlanta, GA 30322
USA

Library of Congress Cataloging-in-Publication Data
Primate laterality: current behavioral evidence of primate
asymmetries / [edited by] Jeannette P. Ward, William D. Hopkins.
p. cm. -- (Recent Research in Psychology)
Includes bibliographical references and indexes.
ISBN 0-387-97961-1. -- ISBN 3-540-97961-1
1. Laterality. 2. Left- and right-handedness. 3. Primates--Behavior. I. Ward, Jeannette P. II. Hopkins, William D. III. Series.
[DNLM: 1. Dominance, Cerebral--physiology. 2. Laterality--physiology. 3. Primates--physiology. WL 335 P952]
QP385.5.P77 1993
152.3'35--dc20
DNLM/DLC
for Library of Congress 92-484856

Printed on acid-free paper.

Production managed by Dimitry L. Loseff; manufacturing supervised by Vincent Scelta.
Camera-ready copy prepared by the editors using WordPerfect 5.1.
Printed and bound by Edwards Brothers, Inc., Ann Arbor, MI.
Printed in the United States of America.

9 8 7 6 5 4 3 2 1

ISBN 0-387-97961-1 Springer-Verlag New York Berlin Heidelberg
ISBN 3-540-97961-1 Springer-Verlag Berlin Heidelberg New York

This book is dedicated to all those,
past, present and future,
who recognize the power of the comparative method
to expand and alter our perspectives
and enhance our understanding
of the behavior of all organisms,
including ourselves.

Acknowledgments

Both editors would like to express especial and sincere appreciation to the gifted young artist, Charles Alexander, for the contribution of his original graphic interpretation of primate laterality for the frontispiece of this book. He may be contacted at 3797 Tutwiler, Memphis, TN 38122, or by phone at (901) 452-5105.

I would like to express appreciation to Peter F. MacNeilage who provided the initial inspiration for a book on the topic of primate laterality, to Lauren Julius Harris for his many helpful comments and suggestions on the development of the book, and to all the contributing authors whose patience with a long and difficult process has sustained this editor to the work's completion. Conversations with many colleagues, other than those directly involved, have contributed to the development of the book by influencing my ideas about laterality; for this I am grateful to Guy Mittleman, Robert L. Pusakulich, Lesley J. Rogers, Ian Q. Whishaw, and the talented students that have worked with me in the study of laterality: Geri R. Alvis, Allyson Bennett, Deanna L. Dodson, Carl F. Larson, Garrett W. Milliken, Caroline G. Sanford, and Donna K. Stafford. I would also like to express my deepest gratitude to R. Bruce Masterton, who first introduced me to the strategy of species comparison in the study of brain and behavior. Support for this work was furnished in part by National Science Foundation Grant BNS-8707754 awarded to Jeannette P. Ward and by a Centers of Excellence grant awarded to the Department of Psychology, Memphis State University, by the State of Tennessee. **J.P.W.**

Assistance in the completion of this book was provided in several forms. Partial financial support was provided by NINDS grant 29574. Long hours and many discussions with J. P. Ward, J. Fagot, J. Vauclair and R. Morris have provided many insights with respect to my evolving thinking on the topic of laterality and evolution. Over the course of my career, I have received help and assistance from many people, but several stand out, including Dr. Steve J. Suomi, Dr. Robin D. Morris, Kim A. Bard, and Dr. Duane M. Rumbaugh. **W.D.H.**

Contents

Contributors

Leah E. Adams-Curtis
Department of Psychology
Washington State University
Pullman, WA 99164-4830

Kim A. Bard
Department of Reproductive Biology
Yerkes Regional Primate Research Center
Emory University
Atlanta, GA 30322

Linda Bolser Gilmore
Language Research Center
Georgia State University
Atlanta, GA 30303

Douglas F. Carlson
Department of Psychology
Michigan State University
East Lansing, MI 48824-1117

Joël Fagot
C.N.R.S.- L.N.F.
Unité de Neurosciences Cognitives
31, ch. Joseph Aiguier
13402 Marseille cedex 09
France

Dorothy M. Fragaszy
Department of Psychology
University of Georgia
Athens, GA 30602

Stanley D. Glick
Department of Pharmacology and Toxicology
Albany Medical College
47 New Scotland Avenue
Albany, NY 12208

Lauren Julius Harris
Department of Psychology
Michigan State University
East Lansing, MI 48824-1117

William D. Hopkins
Division of Behavioral Biology
Yerkes Regional Primate Research Center
Emory University
Atlanta, GA 30322

Masao Kawai
Institute and Museum of Primatology
Japan Monkey Centre
Inuyama, Aichi, 484
Japan

James E. King
Department of Psychology
University of Arizona
Tucson, AZ 85745

Virginia I. Landau
Department of Psychology
University of Arizona
Tucson, AZ 85745

Ralph A. W. Lehman
Department of Surgery
Division of Neurosurgery
College of Medicine, University Hospital
The Milton S. Hershey Medical Center
P. O. Box 850
Hershey, PA 17033

Peter F. MacNeilage
Departments of Linguistics and Psychology
Calhoun 501
University of Texas at Austin
Austin, TX 78712

Garrett W. Milliken
Department of Psychology
Memphis State University
Memphis, TN 38152

Robin D. Morris
Department of Psychology
Georgia State University
University Plaza
Atlanta, GA 30303

Bruno Preilowski
Weissenau-Fieldstation
Department of Psychology
University of Tübingen
Rasthalde 3, D-7980
Ravensburg-Weissenau
Germany

Dulce D. Shafer
119 Duran Drive
San Rafael, CA 94903

Donna K. Stafford
Department of Psychology
Memphis State University
Memphis, TN 38152

Jacques Vauclair
C.N.R.S.- L.N.F.
Unité de Neurosciences Cognitives
31, ch. Joseph Aiguier
13402 Marseille cedex 09
France

Jeannette P. Ward
Department of Psychology
Memphis State University
Memphis, TN 38152

David A. Washburn
Language Research Center
Georgia State University
Atlanta, GA 30303

Kunio Watanabe
Koshima Field Laboratory
Primate Research Institute
Kyoto University
Ichiki 16-1, Kushima, Miyazaki 889-33
Japan

1
Handedness in Apes and Monkeys: Some Views From the Past

Lauren Julius Harris

Not too long ago, it was commonly assumed that cerebral functional asymmetry was a uniquely human characteristic. Today, however, many researchers in the field are coming to believe that it may well be a fundamental feature of mammalian and perhaps of all vertebrate brains. This new view has been spurred by demonstrations of lateralization on perceptual and motor tasks in a variety of species, including birds, rodents, cats, monkeys, and great apes (see Andrew, 1983; Corballis, 1983; Denenberg, 1981, 1983; Glick, 1985; Nottebohm, 1977; Rogers & Anson, 1979; Walker, 1980). It also has raised a great deal of interest in the question of whether animals show lateralization in limb use in a way that approximates human handedness. Again, a variety of species have been studied, but apes and monkeys have received special attention because of their biologically close relationship to man as well as their human-like hands and great dexterity. On this issue the picture is far less clear. Although there is broad agreement that individual animals may show reliable hand preference, the question of whether handedness is a species characteristic remains controversial (e.g., see MacNeilage, Studdert-Kennedy, & Lindblom, 1987, and commentaries). Some of the controversy will perhaps be resolved by the new research reports in the current volume.

This presentation of new studies of handedness in apes and monkeys marks an appropriate occasion to note that interest in primate handedness is not so much *new* as it is *renewed*. That is, almost from the outset of the scientific study of *human* handedness, interest began to turn to this different but related question. Unfortunately, except for a few studies from the 1930s and 1940s, this early work appears to be largely unknown today. If so, the reasons are not hard to understand. The early literature on primate handedness is scattered and often difficult to find, and even in its own time, it was the subject of very few reviews and analyses to which researchers might refer today.[1] The result is that we have only the sketchiest historical context for current work. My aim in this chapter is to enlarge the picture by bringing together some, but by no means all, of the older material. Although the early literature, like the current literature, was not confined to studies of apes and monkeys, these were the primary subjects of interest for many of the same reasons that make them so today. Our focus will therefore be on primates. I also shall concentrate on the period from the 1860s through the early twentieth century — the peak period of early study.

As we shall see, early writers were not in agreement about handedness in apes and monkeys. The main reason, which will be clear enough as we proceed with our review, is that the empirical evidence was often inconsistent and even wildly contradictory. Disagreements about evidence, however, were widened by further disagreements on two other, related issues.

Origin of Human Handedness

The first issue had to do with the origins and reasons for *human* handedness. At least since the Hellenistic Age, there seems to have been universal agreement that handedness was a human trait, with right-handedness as the modal form. It was the explanations of handedness that provoked debate. Some saw handedness as a natural, or native, attribute, with social factors playing only a secondary or supporting role. For example, Aristotle, perhaps the most prominent exponent of this view in ancient times, regarded the right side of the body as the source of all motion, called laterality a *natural* state, and, therefore, thought it natural for man to use his right limbs in preference to his left (*De Partibus Animalium*, 684a, lines 27 ff.). For his teacher Plato, however, symmetry was the natural state, and any deviation was the result of education and training (*The dialogues of Plato*, Laws, 7, line 795). Plato also saw an advantage for symmetrical development on the grounds that two well-trained limbs were better than one.

Aristotle's "nativist" and Plato's "nurturist" views were to find their counterparts in later historical periods. For example, in the early 19th century, a prominent champion of the nativist position was the great British

---

[1]Among works in English, the best source is Yerkes and Yerkes' book on the great apes (1929). Among works in German, there are Bardeleben's (1909) and Ludwig's (1932) monographs on laterality in animals and man, which include brief sections on non-human primates. In the French literature, there is Guldberg's (1897) essay on "dyssymetries in morphology and function in man and the higher vertebrates." These, along with other earlier works on laterality, were helpful in my search for original data papers and other reports. In nearly all cases, I have been able to consult the original documents. Where secondary citations had to be used, the sources are given in the text. Unless otherwise noted, translations and paraphrases from non-English works are my own.

anatomist Sir Charles Bell, who called right-hand preference "a natural provision" (1833, p. 125). On the nurturist side was the French physiologist Marie François Xavier Bichat, who declared that "this discordance [between the left and right sides] is seldom or never in Nature, but is the manifest consequence of our social habits.."(1805/1809, p. 22). Bichat had in mind the near universal training of the right hand for writing and for other acts of skill — practices that had been the rule in European society for over a millenium (see Harris, 1980a, 1989b, 1990).

Structural Explanations of Human Handedness

Disputes about human handedness frequently turned on opinion about the existence and significance of structural (physiological or anatomical) features, with nativists asserting their existence and relevance, and with nurturists questioning or denying them. Because debate about handedness in apes and monkeys drew on many of the same kinds of arguments, let us briefly describe these structural theories .

For many nativist theorists, human handedness was the result of blood supply (Harris, 1980a). Aristotle believed that the blood flowing to the right side was purer and hotter than that flowing to the left (*De Partibus Animalium*, 670b). The reason was that the aorta with its branches supplies the left side of the body, while the larger vena cava supplies the right side. Many variations on this theory were to follow. In the sixteenth century, the anatomist Gabriello Fallopio proposed that the crucial feature was the rightward location of the Azygos vein (cited in Browne, 1646), while others thought it was the rightward position of the liver, which was still regarded by many to be the source of blood. In the early nineteenth century, the crucial feature was said to be the subclavian arteries because the right subclavian is closer than the left subclavian to the heart (the explanation is cited in Bell, 1833, pp. 123-124, who rejected it; a later advocate was Hyrtl, 1860). Finally, when the brain became the focus, handedness began to be seen as a product of the physiological superiority of the controlling cerebral hemisphere. Now, to explain right-handedness, it was supposed that the left carotid artery's normal emergence directly off the aorta caused more blood to flow to the left hemisphere; where this anatomical asymmetry was reversed, left-handedness was the result (e.g., Ogle, 1871).

The nineteenth century also saw the creation of an ingenious theory that shifted the emphasis from blood to balance. According to the physiologist Andrew Buchanan (1862), the right lung, being larger, receives more air on inspiration, making the right side of the chest bulge out more than the left, carrying with it the lower ribs and liver. As the liver swings to the right, so does the center of gravity, causing "a shift of balance to the left side," leaving the right limbs "freer for action" (1862, p. 152). Buchanan assumed that this development typically would await the end of the first year, or at least until the child had begun to walk, because only then would the child be able to apply sufficient force to realize the superior power of the muscles on the right side. Buchanan himself did not measure the internal organs, but others did, and the results appeared to support his analysis. For example, Moorhead (1902) weighed the viscera of human fetuses and concluded that the child begins its extra-uterine life with "a marked right-sided bias" (1902, p. 403).

Buchanan seems to have assumed that the right arm was inherently more powerful than the left but that the advantage could be realized only after an adjustment in the body's "dynamic balance". Other nativist theorists emphasized the musculature directly. Of course, the larger muscles of the dominant arm could be said to ensue from that arm's greater use, but the premise was that muscularity and use jointly reflected a predisposition for superior development of that side. Examination of the arm muscles of the newborn human infant was held to be the crucial test. Results, however, were mixed, some studies finding differences favoring the right arm (e.g., Biervliet, 1899), and others finding none (e.g., Harting, 1869). Comparison of the forearm bones gave the same mixed results (e.g., Bischoff, 1863; Debierre, 1887; Gaupp, 1889; Moorhead, 1902; Theile, 1884).

All such asymmetries of the body — stomach, intestines, spleen, liver, lung, and the heart and its large vessels — were acknowledged by Bichat (1805/1809, p. 9), but he dismissed them as irrelevant. For him, the crucial factor was what he believed to be the essential symmetry of the sense organs — eyes, ears, nostrils — as well as the brain itself, whose parts, he said, "resemble each other on every side" while single organs "are all symmetrically divided by the median line" (p. 8). Likewise, the limbs show "a perfect equality of volume, number of fibres and nerves" (p. 22). Inasmuch as "two parts essentially alike in their structure cannot be different in their mode of acting" (p. 14), Bichat saw the anatomical evidence as confirming his thesis that human handedness was a socially-determined trait.

The Link Between Man and Ape

Just as opinion was mixed about the reasons for human handedness, even among those who accepted it as a fundamentally biological phenomenon, so was opinion about the relation between man and ape, and thus about the significance of the question of primate handedness for understanding human handedness. Among those who believed that handedness existed in other species, there was agreement that this signified continuity across species. It was also agreed that this meant that human handedness itself must be 'natural', because handedness in other animals presumably had not been induced by education or social convention of the sort to which human beings were routinely subjected. In other words, human handedness must be innate because there could be no other explanation of its existence in other species. (Let us call this position # 1.)

Those, however, who averred that handedness did *not* exist in other species were not of one mind about what this meant for understanding human handedness. Some took it to mean that human handedness must be natural and that the underlying natural reasons, whatever they might be, must be absent in other species (position #2). The implication, then, was that handedness (laterality generally) appeared for the first time only with the evolution of man and that it was linked to some peculiarly human feature, either neurological (an asymmetrical cerebral cortex) or behavioral (e.g., upright stance). For others, however, the lesson was that the uniqueness of human handedness must represent the influence of uniquely human cultural institutions (position # 3). (In a variant on this last position, a critic noted evidence suggesting that where animal handedness *had* been found, it had been established adventitiously or through training, and concluded that training and social convention must be the fundamental reasons for human handedness as well.) Position # 2 thus assumes non-continuity across species, whereas position # 3 assumes continuity.

Knowledge of Apes and Monkeys in Ancient Times

With this background we can begin our review of the early literature on primate handedness. For the most part, I shall discuss this work in chronological order so that we can more easily track its development.

Accounts of the physical and behavioral characteristics of apes and monkeys date back at least to the ancient cultures of Europe, Asia, and Africa through the writings of such voyagers and chroniclers of natural history as Hanno, Aristotle, Agatharchides, and Pliny, and through the anatomical descriptions of Galen (for reviews, see Huxley, 1863; Morris & Morris, 1968; Reynolds, 1967; Yerkes & Yerkes, 1929). Unfortunately, it is often hard to tell whether the creatures being described are anthropoid apes, monkeys, or baboons, or even, in certain cases, human beings (for example, some may have been pilose humans). Most authorities, however, credit Aristotle's account in *Historia Animalium* for taking a step toward taxonomic adequacy of description and for indicating the ancients' familiarity with at least three groups of non-human primates: tailless apes, tailed monkeys, and baboons, if not with any of the anthropoid apes.

Aristotle described only physical characteristics and said nothing about use of the hands, much less handedness. But since he regarded the right side of the body to be the source of motion not only in man but in *all* species, the possibility occurs that he would have been disposed to ascribe handedness -- presumably right-handedness -- to apes and monkeys. As he said, it was "natural for every animal in active operations [i.e., in movements and manipulations] to use the parts on its right side in preference to those on its left" (*De Partibus Animalium*, 684a, l. 25). Nearly four centuries later,

Pliny the Elder (A.D. 23-79), perhaps in the spirit of Aristotle's maxim that the right side was the origin of motion, wrote that "all animals start walking with the right foot and lie down on the left side" (*Natural History*, 1940, p. 591, Book XI, CV. 253). In Pliny's brief passage on apes and monkeys (1942, Book VIII, LXXX), however, the question of preferential limb use in animals does not arise, though he alludes elsewhere to the prominence of right-handedness in human beings (1942, p. 557, Book VII, XVII). Likewise, other later accounts of apes and monkeys, while richer than Aristotle's in description of behavior, seem to have left aside the question of preferential limb use. We therefore shall have to seek our first account in a later period.

17th Through the Early 19th Century

As Yerkes and Yerkes (1929, p. 8) have pointed out, from the second century to the sixteenth, acquaintance with the primates, meaning the anthropoid apes, "seems chiefly to have perished with its possessors." Such records as there are (e.g., accounts by Marco Polo, in the thirteenth century) are of negligible scientific value and afford no additional information over the ancient sources.

In the sixteenth century, a new era began. Descriptions from the ancient texts were compiled, attempts were made to distinguish the true accounts from the false, and further new reports began to appear. By the seventeenth century, several extended descriptions of anthropoid apes had been published, and it is in a work from this period that I have found what perhaps is the first reference to primate handedness.

The author was the seventeenth century English physician Thomas Browne. The account occurs in his *Pseudodoxia Epidemica* (1646), commonly known as *Vulgar Errors*. This was not a study of primates. It was instead a work designed by Browne to combat what he regarded as the erroneous beliefs of the general public. One of these "vulgar errors," in his view, was the belief that right-handedness was natural. Browne objected on the grounds that "if there were a determinate prepotency in the right, and such as ariseth from a constant roote in nature" (1646, Book 4, Chapter 5, p.187), then right-sidedness might be expected in other animals. But four-footed animals, Browne said, showed no preference, and animals that used their upper limbs like hands sometimes even showed a preference for the *left* side:

> for we find not that Horses, Buls, or Mules, are generally stronger on this [right] side; and as for animals whose forelegs more sensibly supply the use of armes, they hold if not an equality in both, a prevalency ofttimes in the other, as Squirrels, Apes, and Monkeys (p. 187).

In his reference to "Monkeys," Browne could have been relying on personal observations. Monkeys were familiar creatures as pets and performers, having served in these roles from the time of their introduction to European life in the eleventh and twelveth centuries. The reference to "Apes" is harder to pin down. Browne may have had in mind not true apes at all but Barbary apes, or macaques, which, like other monkeys, had long been enjoyed as performing animals and pets in Europe. Here, Browne again could have been drawing on personal observations. If, on the other hand, he was referring to true apes, we can be far less sure of the species. They might have been gorillas or chimpanzees, which the Englishman Andrew Battell had described in a well-known report first published in 1613 in a volume edited by Samuel Purchas (1613/1625). The first anthropoid ape believed to have actually reached Europe was a chimpanzee from Angola, described in 1641 by the Dutch physician and anatomist Nicholas Tulp. The animal was presented to Frederick-Henry, the Prince of Orange, and was given the name Orang-outang by Tulp, although it was actually a chimpanzee (see Morris & Morris, 1968, pp. 144-145), and possibly a pygmy chimpanzee (see Reynolds, 1967, p. 44). As I have seen neither Battell's nor Tulp's reports, I cannot say whether either one gave an account of hand preference. No such account occurs in excerpts quoted in other sources (Huxley, 1863, Chapter 1; Morris & Morris, 1966, pp. 152-153; Reynolds, 1967, pp. 42-43; Yerkes & Yerkes, 1929, Chapter 2).

Whatever the source of his statement about handedness in monkeys and apes, Browne, given his views about handedness, was understandably skeptical of blood supply explanations, whether the focus was on the Azygos vein or the liver. In this connection, it is interesting to note how he used evidence on ape anatomy to contest the blood supply explanation: "we observe in Apes and other animals, whose Liver is on the right, no regular prevalence [of handedness] therein" (1646, p. 188). For this account, Browne may well have drawn on ancient texts, in particular, Galen's *On Anatomical Procedures*. Galen had dissected Barbary apes, baboons, and possibly even true apes, and had concluded that "of all living things the ape is likest man in viscera, muscles, arteries, veins, and nerves, as in the form of the bones" (Galen, 1956, Book I, p. 2). After Galen, no one else seems to have dissected monkeys until the 16th century, and, in the case of apes, not until 1699 when the English physician Edward Tyson published his epochal account of his dissection of a chimpanzee, which he called "orang-outang." [2]

After Browne's time, apes increasingly became the subject of great curiosity in Europe. In England, a chimpanzee arrived in London in 1738

---

[2]According to Montagu (1943, p. 245), Tyson was using the term in its then customary sense as a general, or family, name, and not as a specific name; the same presumably may be said of Tulp's use of the same name, as noted earlier in the text.

and was an immediate public success, judging from an excerpt quoted by Montagu (1943) from a magazine article of the time. The excerpt noted that the animal "feeds herself with her hands as a human Creature" (*The London Magazine*, 1738, p. 438; in Montagu, p. 247), so perhaps a reference to hand preference was included.

The first documented account of hand preference that I have found (although I have been able to examine only a few works of natural history in this period) is in a book published more than 100 years later, in 1850, by Rennie, the author of a popular and technically excellent series of books on nature. In a volume on the "cheiropeds" (animals whose feet are formed into hands, that is, monkeys, apes, opossum, and lemurs), Rennie wrote about the chimpanzee "Tommy," a popular and much-celebrated denizen of the London Zoological Gardens, who arrived from Africa in 1835 when he was about 18-20 months old. (An unsigned magazine article the next year ["The chimpanzee," 1836] is presumably about this animal.) Of Tommy, Rennie said, "We observed that he used the right hand in preference to the left, and had obviously greater power and facility of action with this than with the opposite member" (1850, p. 67). Rennie went on to note that in human beings, handedness "has generally been attributed to the effects of education," but "in Tommy, at least, it was a "natural action, since he was perfectly unsophisticated in this respect" (p. 67). Rennie then asked, would it not be "highly interesting" to find out whether the same preference was exhibited in other primates, and consequently, "how far it may depend upon some necessary and inherent principle of the animal conformation, rather than upon mere education" (p. 67).

Given the late appearance of chimpanzees in Europe and the rarity of their numbers, it is understandable that scientific study was so meager. The same can be said for the orang-utan; the first specimen did not arrive in Europe until 1776, when a single animal was brought to England and presented to William V (Reynolds, 1967, p. 53). As for the gorilla, it was another seventy years before the first scientific description of gorillas in Africa (Savage & Wyman, 1847) and 8 more years — 1855 — until the first living gorilla arrived in Europe.

Research in the Era of Darwin and Broca and Afterwards: 1860s — 1910

So far, all is preamble to our story, as neither primate study nor the study of human handedness had become established as major fields of scientific research. Until this happened, the study of *primate* handedness would not itself be regarded as a subject for investigation in its own right. As it happened, both areas of study were to see important advances within the same brief span of time.

In the case of primate study, the first major impetus came in 1859 with the publication of Charles Darwin's *On the Origin of Species*, a work that argued for continuity across species in physical as well as psychological characteristics. A few years later, Thomas H. Huxley (1863) published *Evidence as to Man's Place in Nature*, which advanced and documented the thesis that man was more closely related to the most highly developed apes than the higher apes were to the other primates. Huxley argued that man and ape were members of the same order, and he invoked Darwin's theory to explain their differences and similarities. Darwin's *The Descent of Man* (1871) elaborated the same theme of continuity. As for handedness, broader interest in the subject was launched in the 1860's when Paul Broca (1865) proposed that the control centers for manual dexterity (right-handedness) as well as for speech lay in the left cerebral hemisphere (see Harris, 1991).

Now, with this combination of crucial development in evolutionary biology and in the theory and evidence of localization of function, scientists began to see new significance in the old question of whether handedness was among the characteristics common to human beings and other animals. As mentioned earlier, some believed that it was a common feature, reaching its pinnacle of expression in man. Others were skeptical. We begin with the positive reports.

Positive Reports of Handedness in Apes or Monkeys

In Broca's time, possibly the first report of laterality data in apes or monkeys was by William Ogle (1871), a British physician and classical scholar (his English translation and annotation of Aristotle's *De Partibus Animalium* is still the standard). Ogle's interest in this question seems to have been kindled by Bichat's (1805/1809) proposition that human handedness was not the result of nature but of custom and education. Ogle objected on several grounds. For one, how could education explain the "peculiarity" of left-handedness, which manifests itself "before education begins" and often persists despite the parents' efforts to overcome it (1871, p. 286)? For another, how could education explain handedness in monkeys? Ogle's evidence consisted of observations of 23 monkeys (species unstated) at the Zoological Gardens in London.[3] The method and results were as

[3]Ogle explained that he had to make observations on his own because he had "hunted in vain" for information on this point in works on natural history (1871, p. 286). As I have already mentioned, my own examination of the early literature in natural history was likewise in vain (with the exception of the references to Browne, 1836, and Rennie, 1850). Because I presumably covered at least some of the same ground as Ogle examined, Ogle's statement makes me more confident that handedness in apes and monkeys had not yet become a subject of significant interest.

follows:

> If, standing close to a monkey, one offers it a nut or apple, the monkey takes it with the nearest, and so the most convenient, hand, be this the right or the left, and will proceed to use both or either indifferently in carrying it to its mouth. But if, instead of standing close to the monkey's cage, one stands, bait in hand, at some distance — at such a distance that is, that right and left hand are equally distant from the tempting morsal — the monkey will stretch out one of its arms as far as possible through the bars of the cage; and in the great majority of cases the arm thus extended will be the right one. Some few monkeys, it is true, will stretch out the left limb; but this is comparatively exceptional, and, as I have said, in most cases it is the right arm which will be used preferentially.(p. 287)

In all, 20 of the 23 monkeys consistently used their right hand for reaching for and grasping an object outside their cage; the remaining 3 used the left. Ogle also noted that on repetition of the experiment, with few exceptions, each monkey acted "in precisely the same manner as it did on the first trial" (p. 287). Ogle likened the few exceptions to the way "a right-handed man may occasionally extend his left arm instead of his right to grasp an object" (p. 287). Ogle therefore concluded that monkeys, "like men, are as a general rule right-handed" (p. 287) — a conclusion clearly at odds with Thomas Browne's. Ogle also drew the same conclusion about the reason for handedness in monkeys as Rennie (1850) had about the chimpanzee "Tommy" — that it could *not* be due to education. But unlike Rennie, Ogle saw the same implication for *human* handedness:

> Now, it will hardly be asserted that a monkey is disciplined by its parents to use one hand in preference to the other; and the only conclusion one can draw is that the similar dextral pre-eminence of man and monkey depends on some common fact in their anatomical structure.(p. 288)

Which anatomical structure? Ogle believed that it was the arrangement of the blood vessels to the brain, with the left hemisphere having the advantage in both species. [4]

Broca (1877) himself also admitted to the possibility, indeed the liklihood, of handedness in other species. He noted that the "dyssymetry" of the convolutions was "very pronounced in man," but that "there is a less but still

---

[4]Ogle (1871) offered the same explanation for what he reported to be foot preference in parrots, namely, their preference for holding food in the left foot while perching on the right. The question of "footedness" in parrots has a history of its own and became the subject of considerable controversy involving some of the same persons who were engaged in the dispute over handedness in monkeys and apes (Harris, 1989a).

very evident degree of dyssymetry in the great apes" (1877, p. 527). By one account, he is also reported to have said that not only were monkeys right-handed ("droitière"), so were the majority of mammals, including lions, tigers, leopards, cats, and dogs (cited by Delaunay, 1874, p. 74). Nowhere in any of Broca's writings, however, have I found any reports of actual observations of handedness in apes or monkeys or any other animals.

A different opinion about animal handedness was offered by the missionary-explorer David Livingstone. Whereas Ogle had found, and Broca had claimed, that monkeys were *right*-handed, Livingstone is said to have stated that, except for man, "all animals are *left*-handed ("'Alle Thiere,' sagt Livingstone, 'sind linkshändig'"; quoted in Vierordt, 1881, pp. 427-428). Livingstone's views on this matter were also cited by other writers including the physicians Louis Jobert (1885, p. 13) and Gustav Guldberg (1897, p. 24) and the criminologist Cesare Lombroso (1903, p. 441). Lombroso repeated Livingstone's statement without comment, but Jobert was suspicious. "Cette opinion," he said, "est évidemment exagerée" (1885, p. 13). Guldberg called it "exagerée et prématurée" (1897, p. 24).

Livingstone's statement presumably was based on observations of some sort, although there is no way to tell from these second-hand accounts. In 1882, however, a personal observation was reported by Jules Deniker, a member of the Société Anthropologique, the Paris-based society that Broca himself had founded. In that year, two apes — an orangutan and a chimpanzee — were presented by an M. Bidel at his zoological exhibition in Paris, and, great apes still being exotic creatures for both the scientific and general public, the laboratory of anthropology of the School of Advanced Studies ("l'Ecole des hautes études") in Paris established a commission to examine the two animals. The commission consisted of Deniker and two other anthropologists Topinard and Chudzinski, and it was Deniker who reported the results to his colleagues in the Société Anthropologique.

The orang-utan was an adult male, the chimpanzee a juvenile female. In his report, Deniker (1882a) mentioned only the animals' physical and certain behavioral characteristics, but, in the discussion that followed, added "one peculiarity ["une particularité"] that will interest you":

> The orang and the chimpanzee always used their left hand for seizing objects. Now, in all the drawings, one represents the anthropoids hanging to the branches of the trees with the right hand. It is possible, after that, that it is not the same arm that serves them in climbing and in the execution of acts that require more delicacy; it would be interesting to study the comparative force of the two arms.(1882b, p. 340)

If Deniker was correct in stating that in drawings, anthropoids are shown hanging to tree branches "with the right hand," one wonders whether this was merely an artistic convention based on the assumption that apes *are* right-handed, or whether it was based on real evidence. In any case, by the 1890s, it appears to have been widely assumed that most apes *were* right-handed, notwithstanding Deniker's observations and Livingstone's declaration to the contrary. This position is reflected in a statement by the German anatomist Rudolf Fick (1895, p. 6). Speaking of a captive adult orang-utan who, in grasping objects, seemed more skillful with his right hand and apparently preferred it, Fick concluded that the animal was right-handed, "as indeed most apes are" (quoted in translation in Yerkes & Yerkes, 1929, p. 118). (In 1891, Deniker published another report on zoo animals, this time on two young gorillas recently placed on exhibition in Paris, but said nothing about hand preference.)

In 1901, another, and evidently positive, statement was reported by Osawa, a Japanese scientist. According to the German abstract, which seems to be the only record of the research, Osawa stated that monkeys are sometimes right-handed, sometimes ambidextrous, and rarely left-handed (1901, p. 28). The abstract did not describe the species or testing method, or whether Osawa was recounting his own observations or merely repeating those of others. In 1902, a similar account was given by Bernard Langkavel (1902; cited in Seeligmüller, 1902, p. 51; also cited in Cunningham, 1902, p. 285). It is conceivable that Langkavel was merely drawing on Osawa's earlier statement, or perhaps on Ogle's work.

In 1907, another positive report was published, this time based on what we can be sure were new data. The author was the British physician Sir James Crichton-Browne (1907). Crichton-Browne cited Ogle's (1871) positive report as well as those of Osawa (1901) and "several [unnamed] German and American observers who have independently investigated the matter and have supported Dr. Ogle in his view" (1907, p. 629). But Crichton-Browne regarded the "tokens of right-handedness accepted by these observers" to be "somewhat faint and ambiguous" (p. 629). After making "a few observations" of his own that left him "inclined to agree" with Ogle, he arranged for further observations by the custodian of the monkey house at Regent's Park Zoo (London Zoological Gardens), where Ogle had made his observations more than 30 years earlier. His description of the custodian's qualifications suggests that Crichton-Browne had some doubts about his own qualifications as well as those of previous observers: the custodian was "adept at monkey manners and free from scientific predilections" (p. 629). Nonetheless, the custodian agreed with Ogle and the others: his "decided opinion is that anthropoid apes exhibit a preferential use of the right hand" (p. 630). Unfortunately, Crichton-Browne provided no details, not even about the particular species examined. Despite the custodian's report, Crichton-Browne also suggested why handedness in anthropoids would still be less than that found in man.

> It is to be borne in mind that the upper limbs in the monkey are still largely used for locomotion, in which both sides must equally participate if progress

> in a straight line is to be accomplished, and are only to a limited degree manipulatory and prehensile in gathering and shelling nuts or pods, in opening shell-fish, in pulling up roots, in picking thorns or burrs from its fur, or in hunting for parasites. And we should not, therefore, expect anything like such a division of labor as in man when they are emancipated from that more servile and automatic office.(1907, p. 630)

Crichton-Browne also supposed that right-handedness would be found in other animals:

> But whenever the upper limbs have functions assigned them other than locomotion, divergence will, I fancy, begin; and I am disposed to believe that the first dawnings of right-handedness are to be recognized not only in the monkey, but much lower down in the animal kingdom.(1907, p. 630)

Asymmetries in Length of Bones

Over the next few years the argument for primate handedess was supported by reports of anatomical asymmetries in the limbs themselves. Prominent contributors to this work were two German anatomists, Karl von Bardeleben and Theodore Mollison. Bardeleben (1909) found that anthropoids had larger and heavier right arms and therefore concluded that anthropoids were preponderatingly right-handed. Mollison (1908) made similar analyses of primate skeletons but reached a somewhat different conclusion. The gibbon, orang-utan, and man were, indeed, naturally right-handed, but the chimpanzee and gorilla were left-handed (1908, pp. 112-115; cited in Yerkes & Yerkes, 1929, p. 410). In a subsequent paper, Mollison concluded that:

> In the arm the right side is greatly preferred in the gibbon, orang, and man, the left, at least in the lower arm and also less plainly (there is more frequent equality) in the chimpanzee and gorilla. All three bones of the leg predominate easily on the right side in the orang and gibbon, on the left side in the gorilla (1910-1911, p. 196; quoted in translation in Yerkes & Yerkes, 1929, p. 410).

Negative Reports and Criticisms

Judged in its totality, the evidence over four decades — from 1871 to 1911 — would seem to make a case for handedness in primates, right-handedness in the case of monkeys, and either right- or left-handedness for anthropoid apes. But negative reports were no less frequent.

Some negative reports were simple statements, without elaboration, that handedness had *not* been observed in this or that case. One such occasion came in the discussion of Deniker's (1882) report to the Paris Société Anthropologique of left-handedness in the orang-utan and chimpanzee. One participant recounted that his own observation of two macaques showed that both used their two hands "indifferently for seizing objects without marked preference for one of them" (Coudereau, 1882, p. 341). No reply from Deniker was noted.

So far as can be told, the author of this particular negative report was not committed to any particular view about animal or human handedness. There were certain critics, however, who were convinced that handedness was an unnatural, as well as unfortunate, product of prejudice or social custom. Aside from challenging evidence that handedness was natural in human beings, these advocates of what came to be known as "ambidextral culture" looked to the evidence of other species for the reason that if other species showed no lateral preference, ambidexterity must be the more natural and thus more desirable state in man as well. One such advocate was the English writer and social critic Charles Reade, a widely celebrated author of what he called "novels with a purpose" — novels that attacked the evils of Victorian society. One of these evils, in Reade's estimation, was the "pagan prejudices against the left hand" (19 Jan 1878, p. 51). Reade wrote exaltingly of equal-handedness, or ambidexterity, and urged its blessings on every child. To attack the notion of some anatomists that right-handedness in man was due to an asymmetry of the viscera favoring the right side, Reade noted (as had Thomas Browne more than two centuries before) that "this slight preponderance of the viscera is found in the ape and other animals, yet they are either-handed. Droll logic, to cite as the cause of a bad habit in one animal a feature that animal has in common with a thousand animals not one of whom is nursed into that bad habit" (23 March 1878, p. 234). For Charles Reade, it was "all man's own doing that he is any more semiplegiac or lop-limbed than a lion, a raccoon, a fox, a tiger, or an ape" (19 Jan 1878, p. 51).

The movement for ambidextral culture advanced over the years, culminating, in the early 1900s, in the work of John Jackson (see Harris, 1985a, 1985b). Jackson was the iconclastic founder and president of the Ambidextral Culture Society, a society based in London for the promotion of even-handedness. Like Reade, Jackson regarded even-handedness to be the more natural and desirable state and saw the best proof of this in what he presumed to be the natural ambidexterity of animals, unencumbered as they were by culture and superstition. Any claims to the contrary, like Ogle's (1871), thus fundamentally challenged his thesis. Unlike Reade, however, Jackson decided to check things for himself and so made his own observations, evidently at the same London Zoological Gardens as Ogle had visited and that Crichton-Browne was to visit two years later. Unfortunately, Jackson, a stickler for details when criticizing others, presented none himself. He simply noted that the results failed to confirm Ogle's report. The monkeys "exhibited the most perfect two-handedness" (1905, p. 28).

Finally, negative reports, along with criticisms, came from persons with more conventional, or less ideologically-colored, views about *human* handedness (i.e., they accepted the evidence that human handedness was rooted, at least in part, in structural or physiological factors). Among the first was a British physician, W. A. Hollis (1875). Hollis conducted several experiments with a variety of species, including the rhesus monkey (*Macaca mulatta*), the bonnet monkey (*Macaca radiata*), the lion-tailed macaque (*Macaca silenus*), and the Java monkey (*Macaca cynomologus*). Finding no preference for use of the right limb, Hollis concluded that dexterity was "confined to the human race" and that "the monkey tribes, the present representatives of our Simian ancestry (if such they may be), use their right and left limbs indiscriminately to grasp any object offered to them" (1875, p. 266). Unfortunately, Hollis provided no details, so there is no way to reconcile his negative findings with Ogle's positive report (1871). Ogle, of course, might have asked whether Hollis had required full arm-extension when offering the monkeys objects to grasp, since it was only under those circumstances that Ogle himself had found any hand preference.

Like Ogle, Hollis also saw implications in his results for structural or physiological explanations of human handedness. The difference is that Ogle invoked his own positive results with monkeys in support of such an explanation of human handedness (namely, the blood supply theory), whereas Hollis saw his own negative findings as undermining a structural explanation. In Hollis' case, it was the visceral asymmetry theory. Upon inspecting specimens of monkeys (preserved at the Museum of the Royal College of Surgeons) and finding that they showed asymmetries in lung size as did human beings, Hollis concluded that the greater size of the right viscera, "although possibly determining in some way the primaeval selection of the right-hand in man, does not necessarily conduce to dexterity [in man] as shewn by the actions of monkeys and other animals" (p. 267).

Ogle's (1871) blood supply theory was also challenged directly by the physician George Sigerson (1884). Sigerson pointed out that many other animals, besides monkeys and man, had the same vascular arrangement (Sigerson cited the work of the anatomists Richard Owen, 1868, vol. 3, pp. 535 ff, and Carl Gegenbaur, 1878) but showed no indication of right-handedness (1884, pp. 42-43). Sigerson therefore thought it "quite possible" that the "alleged facts [about right-handedness in monkeys] were insufficiently, and therefore inaccurately, observed" and here cited "contradictory accounts" with which "my own investigations coincide" (p. 43). These were that the "Sphinx baboon" in the London Zoological Garden "has been seen to fling projectiles with both hands," and that the "Aye-Aye, according to Mr. Bartlett, used only the left hand in feeding," even though in this animal "both carotids arise from a common branch of the brachio-cephalic" (footnote # 17, p. 43).[5] Sigerson concluded that it would require "the most scrupulous vigilance to eliminate from such a test the influence which right-handed men may have had upon animals apt to imitate, and probably taught (more or less consciously) by their human attendant or proprietor" (p. 43).

Still another critic was the physician Daniel G. Brinton (1896), who wrote that anthropoids as well as other primates closest to man "are ambidextrous, displaying no preference for either hand" (1896, p. 181). Instead of documenting this statement, Brinton justified it on the grounds that the "final source" of handedness was "*the erect posture of the human species*," which was lacking in anthropoids and in other primates. Brinton reasoned that the erect posture:

> introduces a new distribution of force in the economy; it opposes the powerful retardation of gravity to the distribution of the arterial blood above the level of the heart. The great arteries arising from the aorta carry the blood by an appreciably shorter course and in less time to the left brain than to the right. Its nutrition, therefore, is the more abundant and its vitality the more active of the two hemispheres. Hence the right side of the body, which it controls, is more ready to respond to a stimulus on account of its higher innervation. (p. 181)

Note that Brinton's analysis thus incorporated the very same blood-supply theory that Sigerson (1884) had criticized.

Of all the many contributors to the debate about primate handedness, one of the most prominent was the anatomist Daniel J. Cunningham (1902). In an address to the Royal Anthropological Institute of Great Britain and Ireland, Cunningham ranged broadly and with scholarly attention to detail over the whole subject of handedness, including the question of handedness in other species. Cunningham noted that opinion on the matter of animal handedness was divided, with some reports (Langkavel, 1902; Ogle, 1871; Osawa, 1901; Seeligmüller, 1902) concluding that right-handedness was predominant, but with others (Hollis, 1874 [sic]) disputing this view.

---

[5]Sigerson cited "Duncan — 'Apes and Monkeys' — *Cassell's Natural History*, vol. i" for this report. The account in fact is not from this section of the book but from the section on lemurs, which was written by Murie and Duncan. The "Aye-Aye", presumably named after its cry, is a nocturnal lemur, *Daubentonia madagascariensis*. According to Murie and Duncan, Mr. Bartlett (the Superintendent of the Zoological Gardens) did say that in feeding, only the left hand is used, but he added that "the examination of the mode of taking her food requires careful attention, owing to the very rapid movement of the hand" (Murie & Duncan, 1877-1883, vol. I, p. 255). The word "her," furthermore, implies that only a single animal was observed. It is interesting that Murie and Duncan give the "Aye-Aye" a different genus name, *Cheiromys madagascariensis*, evidently no longer in use. According to the authors, *Cheiromys*, from the words for hand and rat, referred to the animal's skull and front teeth, which resemble those of rodents, and the unusual forelimbs, with fingers "of all kinds of lengths." Especially striking-looking are the long slender joints of the "skeleton-looking middle finger," which, according to the eighteenth century naturalist Sonnerat, the animal uses to draw worms out of holes in trees (pp. 251-252).

Cunningham named two reasons why, a priori, apes might be expected to show preference for the right upper limb. First, in ape and man, the viscera were disposed in a very similar manner, meaning that in the ape "the centre of gravity also lies to the right of the mesial plane" (1902, p. 285). Second, the ape's hand, rather than being devoted entirely to locomotion, is "endowed with many of the characteristics which distinguish the human hand..." (p. 285).

Despite these points in favor, Cunningham himself was inclined to take the negative side because, for many years he had had "an intimate experience of both the higher and lower apes in the Gardens of the Royal Zoological Society of Ireland, and I have never been able to satisfy myself that they show any decided preference for the use of one arm more than the other" (p. 285). In this, Cunningham also saw important clues for understanding human evolution because, if he was correct in supposing that the ape is truly ambidextrous, it could mean that in human evolution: "right-handedness did not assert itself until the upper limb had been set absolutely free from the office of locomotion, and had assumed the higher duties which are now assigned to it" (p. 286).

Cunningham also helped enlarge the scope of the analysis of primate handedness by focusing on neuroanatomical structures. Although most anatomists at the time were looking for the bases of human hemispheric specialization in cerebral blood supply or the weight of the cerebral hemispheres, interest had begun to turn to finer structures; in particular, to certain details of the fissures. Here, in contrast with the brain weight measures, substantial and reliable lateral differences had been found. For instance, Eberstaller (1884) had found that the Sylvian fissure in the adult human brain was longer on the left than on the right (cited in Bonin, 1962, p. 5). Cunningham (1892) himself had discovered further anatomical differences in the fetal human brain as well as in the brain of the gorilla, chimpanzee, and orang-utan. In all cases, the posterior end of the left Sylvian fissure was more depressed than the right, leading Cunningham to conclude that the entire region of cortical surface bounded below by the Sylvian fissure (i.e., the region of the planum temporale, part of Wernicke's area in the human brain) was more extensive on the left side.

Even though the human neuroanatomical evidence was clear and substantial, Cunningham's (1902) skeptical views about handedness in apes made him doubt its relevance for understanding lateral cortical specialization:

> That this [greater depression of the left Sylvian fissure] is in any way associated with right-handedness, or even with the localization of the active speech centre in the left cerebral hemisphere, I am not prepared to urge, because the same condition is also a characteristic of the ape. This it is true would offer no impediment to the acceptance of this explanation by those who believe that the ape is right-handed, but, as I have already stated, I cannot persuade myself that the ape possesses any superior power in either arm. (1902, p. 293)

Cunningham's was not the last skeptical voice to be heard. Another was that of the Philadelphia ophthalmologist George M. Gould (1908). Of David Livingstone's statement that all animals are left-handed (Vierordt, 1881), Gould said, "I suspect this is all error, because, as a rule, it would disadvantage rather than help them in the animalian struggle" (1908, p. 10).

Gould's objection seems to have been purely theoretical. John B. Huber (1910), however, appears to have had some observations in mind, although he provided no documentation. He noted that lower animals, at least those that have not been taught tricks, use their fore paws indiscriminately, and this was true even of monkeys or gorillas. Even among these: "which of all animals use the fore paws mostly as hands, there is no suggestion of preferential use or superior expertness in the left or the right hand" (1910, pp. 260-261).

The geneticist H. E. Jordan (1911) was suspicious as well. He regarded the view that certain animals, including monkeys, are predominantly right- or left-handed to be "probably without foundation in fact" (1911, p. 27). As for Bardeleben's (1909) conclusion that anthropoids are preponderantly right-handed on the basis of the greater size and weight of the right forelimb, Jordan said that it introduces "an element of confusion into the discussion" (pp. 26-27). The reason was that right-handedness, in the ordinarily restricted sense, "signifies essentially greater dexterity with that hand in executions requiring skill and delicate manipulations" (pp. 26-27), and that it therefore seems necessary to distinguish between functional (deftness) and structural (size and weight) bilateral asymmetry. Here, Jordan added, "Cf. Weber," a reference presumably to the psychophysicist Ernst Weber's *De Tactu* (1834/1846). In this work, Weber noted that, whereas in most men the right side is physically stronger than the left, the contrary is found for tactual sensitivity (see Mollon, 1978). The distinction was seen as relevant to a complete understanding of handedness, especially where superior performance was shown by the conventionally-defined 'non-dominant' hand (see Harris, 1980b, p. 308).

As noted earlier, the popular press had long shown an interest in primates in general, but especially in the exotic chimpanzees and other great apes; so as the debate over the question of animal handedness heated up in the scientific literature, the popular press also began to take notice. Writing in *McClure's Magazine*, the science writer Edwin Tenney Brewster (1913) said that "The report that certain animals," including monkeys, "are predominantly right or left-handed is probably without foundation in fact" (1913, p. 27). He also suggested that, at least where reports of left-handedness were concerned, it was a matter of training as well as biased observations:

> Whatever we do to them tends to drift to our right side; and they, facing us,

put out the left paw. But, as anybody can prove for himself, a record of the number of times any animal uses one hand or the other, when either would be equally convenient, shows that neither has any advantage over the other. One must, therefore, keep count. A left-handed act looks off, and so contributes more than its share to the general impression". (p. 168)

For Brewster, "Man, then, is the only 'dextrous animal'" (p. 168).

Into the Modern Era: 1911 — 1941

Through the first decade of the 20th century, no consensus seems to have been reached about primate handedness. For every positive account, there was a negative one. But as we have seen, most of the literature consisted of a motley collection of pure opinion, anecdotes, and informal uncontrolled observations of only a few animals (although "pure opinion" appears to have been more common among those on the negative side of the ledger). On the few occasions when a large number of animals were observed, we are told little or nothing about the number and species tested, the methods used, or even the actual results found. One exception would be the report by Ogle (1871), which at least described the task and reported the number of animals observed. Furthermore, most of the evidence, such as it was, came from observations of monkeys. On the great apes, there were hardly any data at all, but this is understandable given their scarcity and late arrival on the scientific scene.

All this started to change in the second decade of the new century. Although anecdotal accounts and informal uncontrolled observations continued, a new kind of study also began to appear — more sophisticated in scientific methodology and quantitative analysis, more enlightened about animal behavior, and, for the first time, supplementing observations with experimental manipulation. It is no coincidence that the authors of these new reports were different as well. Whereas the earlier debate was dominated by physicians, naturalists, neurologists, and anatomists — most with probably only modest training in research methodology and with minimal experience with primates — the new generation of investigators were accomplished in these very areas. The new work also was significantly aided by the establishment of laboratories and field stations for primate research. These included: the Anthropoid Research Station established by the Prussian Academy of Science on Tenerife in the Canary Islands, where, during World War I, the Gestalt psychologist Wolfgang Köhler carried out his field studies of problem-solving in the apes (Köhler, 1927); the primate center started by Robert M. Yerkes at Yale University in New Haven, Connecticut, later supplemented by a subtropical anthropoid breeding station near Jacksonville, Florida (Yerkes, 1961, p. 402; see also Yerkes &

Yerkes, 1929, pp. 581-591); and a primate laboratory established at the University of Wisconsin, in Madison, by Harry F. Harlow. Finally, both in Europe and in America, primate research, and animal research in general, were catalyzed by the new psychologies of the day — Gestalt psychology in Europe, functionalism and behaviorism in America. In America, one important sign of the new focus was the founding, in 1911, of *The Journal of Animal Behavior*, which published several of the new investigations. Coming as they did from different psychological schools, the research directions were not the same. For example, it was American behaviorists, not German gestaltists, who supplemented observational studies of handedness with experimental manipulations.

It may make for a clearer exposition of this new era of research to discuss the studies of monkeys separately from those of the great apes. This division, of course, reflects the possibility of important behavioral differences between the two groups, but it also reflects differences in methodology. The monkey studies were more extensive, and only in monkeys was hand preference assessed by means of experimental manipulations as well as by more conventional observational techniques.

Research on Monkeys

Observational studies

Of the new generation of studies of monkeys, perhaps the first was by the German psychologist Oscar Pfungst (1912), already famous for his exacting analysis of the "calculating" horse, Clever Hans (Pfungst, 1907). In fact, although I have assigned Pfungst's study to this new era, I make this judgment only on the basis of Pfungst's outstanding reputation, since the report itself provides no more details than did many of the earlier accounts. All we are told is that in observations at the Berlin Zoo on different genera of monkeys (as well as apes, the total number of individuals being over 60), Pfungst failed to find any evidence for the predominant use of either hand that could not always be traced to training or to some earlier trauma.

In Germany a few years later, Sarasin (1918; cited in Ludwig, 1932, p. 296) drew a similar conclusion but from a far smaller sample. He observed hand use in three monkeys (*Cebus*, *Cercopithecus*, and *Macaca rhesus*) for picking up food, and, finding no evidence of preference, concluded that monkeys are basically ambidextrous ("völlige Ambidextrie").

In America, two new studies both appeared in 1913 in the same issue of *The Journal of Animal Behavior*. In descriptive details, one of these reports was no improvement on Pfungst's, although it should be said that handedness was not the authors' primary concern at this time. The authors were Karl Lashley, then a 23-year-old doctoral student in zoology at Johns Hopkins, and his mentor in his minor area of psychology, John B. Watson.

In a study of a young *Macaca rhesus* monkey in the Johns Hopkins University Laboratory, Lashley and Watson (1913) reported being unable to detect any hand preference in a variety of tasks and situations.

The other report was expressly concerned with handedness and provided much more information. The author was Shepherd Ivory Franz, a physiological psychologist associated with the Government Hospital for the Insane (now St. Elizabeth's Hospital), in Washington, D.C. Franz (1913) kept records of the number of times rhesus monkeys used each hand in taking food presented to them. The monkeys were tested in cages, and the food was placed on a glass plate at one end of the cage. The monkeys typically sat at the other end of the cage and therefore had to move forward to take the food. On each trial, two pieces of food were presented side by side, with the relative positions counterbalanced over trials, and with the pieces of food at varying distances from each other but always far enough apart so that both pieces could be obtained only with difficulty in one operation.

Summing across several different methods of data analysis (e.g., trials on which only one piece of food was taken vs. trials on which both pieces were taken), the results indicated that of six animals tested, one preferred to use the right hand, two preferred the left, and three showed no preference. For at least two animals, there also was no indication that food position influenced hand preference, but Franz repeatedly did find relatively long periods when each animal would use one hand exclusively, followed by a period of use of the other hand. Franz supposed that, in a long series, these alternatives should balance if due to extraneous causes, although for two of the three monkeys who had shown an over-all hand preference, the series were not long enough. Franz acknowledged that more observations were needed before any definite conclusions could be reached, but, noting the relevance of the observations to the question of the origin of human handedness, offered his data "in the hope that other workers with monkeys may be tempted to make similar observations and publish their results" (1913, p. 144).

Among the first to answer Franz's call was Watson's friend and colleague Robert M. Yerkes. Yerkes' observations were carried out in the laboratory of his former student G. V. Hamilton at Montecito, California, and were published as a part of his study, "The mental life of monkeys and apes" (1916). Yerkes noticed that Skirrl, a *Macaca cynomolgus* monkey, when driving nails always held the hammer in his left hand and the nail in his right. His curiosity aroused, Yerkes tested Skirrl and 5 other monkeys (either *Macaca cynomolgus* or *Macaca rhesus*; 1916, p. 3) by standing outside each animal's cage and holding out a peanut, keeping it just barely within reach, and then later by placing the peanut on a board. Each animal received 20 to 40 trials in 10-trial blocks.

The results suggested that 3 of the 6 monkeys were left-handed, one right-handed, and two probably ambidextrous. The 3 left-handers were Sobke, Gertie, and Jimmie. Sobke, who received 40 trials over two days, reached for and obtained the food each time with his left hand and could be induced to use his right only when the bait was held well toward the right side of his body. Gertie consistently used her left hand, never the right. Jimmie, likewise, used his left, but Yerkes called the tests unsatisfactory because Jimmie was extremely aggressive during the testing and focused on the experimenter rather than the food. The one right-handed animal was a female rhesus, who used her right hand 14 times, her left hand six times.

The two "ambidextrous" animals were Scotty and Skirrl. Scotty, in the first series of 10 trials, used his right hand 8 times, his left hand twice, but then in the second series, used the right three times, the left seven. Skirrl showed something very different; exclusively left-handed when using the hammer, he proved to be right-handed in reaching for food (17 of 20 trials) and, like Sobke, could be induced to use his left only with difficulty. Yerkes therefore concluded that Skirrl was "right-handed in connection with certain movements and left-handed in others" (1916, p. 117).

Although Yerkes thought these data "too scanty to be of any considerable value," they seemed to make sense in terms of his own personal experience. He himself was "naturally left-handed" but, because of long training, could use his hands equally well for writing and "various other activities." Nevertheless, he always used his left hand for "[d]elicate, exact, and finely coordinated movements," such as those of writing and using surgical instruments, but was more likely to use his right hand for "grosser movements involving the whole hand or arm" (pp. 117-118). Yerkes went on to say that it therefore did not seem improbable that we shall find "some specialization among the lower animals with respect to preference for right and left hand or arm," and that he would not be surprised to find that "it is the rule for animals to possess or to develop readily definite preferences for one hand in connection with a given act of skill and to have quite as definite a preference for the other hand in connection with a radically different kind of act" (p. 118). The inspiration for this statement presumably was Skirrl.

Experimental Training Studies

So far, all of the new evidence consisted of observations of spontaneous preferences in situations designed not to favor one hand or another (except for Yerkes' attempt to test the strength of Sobke's left-hand preference by offering the bait toward his right side). In 1913, Franz reported what may have been the first deliberate attempts to change a spontaneous preference. As Franz tells it, some six years earlier, he had observed that among a dozen monkeys, certain ones would steady themselves by holding onto the wire netting of the cage with one hand while using the other hand to grasp food presented to them from outside the cage. Some animals almost always held onto the wire netting with the right hand and took the food with the left;

other animals did just the reverse. Franz discovered that such preferences could be easily reversed after 20 to 30 trials by refusing to give the food until the monkey switched hands. From this, he concluded that hand preference in these animals might be due to extraneous factors, such as the experimenter's position or his offering the food in his right hand.

Franz himself did not carry out any further experimental manipulations, but he was instrumental in having them done by others. The first was Lashley, who, in 1916, after finally deciding to make his career in psychology, had moved from Baltimore to Washington to work with Franz. Franz placed two young male rhesus monkeys at Lashley's disposal and suggested that he study hand preference "with a view to the permanent modification of the normal condition by training" (Lashley, 1917, p. 178).

Lashley first determined whether the monkeys showed hand preference without training. Monkey No. 1 preferred his left hand, and monkey No. 2 was "decidedly right handed" (p. 178), but in each case, the preference seemed to be associated with the positions assumed by the monkeys when kept alone in different cages, and there were marked variations in hand use from day to day.

Lashley then began training each animal to use only the left hand in taking food from the experimenter's right hand. This was done by withholding the food and grasping or striking at the monkey's right hand whenever it was extended toward the food. Monkey No. 2 "did not take kindly to this training" (p. 178), so training was continued only on No. 1, a gentler and less fearful animal. The training worked. After nine successive failed attempts with his right hand, the monkey began to use his left hand predominantly, and then exclusively after 34 additional "negative" trials. After the first 100 trials, the left hand was used 98 times. After 500 additional trials, transfer of training to new situations was assessed. The amount of transfer varied, being greatest in a situation (taking food from a table) most similar to the original training. Finally, the conditions of reinforcement were reversed, and right hand usage was swiftly re-established.

Lashley's conclusion was different from Franz's. Having confirmed Franz's demonstration that hand preference could be established through training, Lashley did *not* conclude that handedness (at least in monkeys) lacked an innate, hereditary basis. Instead, he remarked that the ease of altering hand preference by training "will make it very difficult to establish the existence of any hereditary predominance" (p. 185). The reason was that:

> We never know the complete history of an animal or can exclude the possibility of a severe trauma which might condition the use of one or the other hand. Only the observation of the predominant use of one hand in a variety of situations and in situations entirely new to the animal could furnish reliable evidence for an hereditary predominance.(p. 185)

Soon after Lashley's report appeared, similar results were reported by

Helen D. Clarke Kempf (1917). The animals, again supplied by Franz, were six *Macaca rhesus* monkeys, including three adult males (A, B, and F), two young males (E and D), and one young female (C).

Like Lashley's study, Kempf's was in two parts. First, spontaneous hand use was observed in three different situations, each requiring the animal to reach for pieces of food placed in a variety of positions and distances from the animal. Each animal received more than 400 separate trials. Across all trials, monkeys A and B used their right hand on 100 percent and 96.3 percent of the trials, respectively; C was left-handed, although by a somewhat weaker margin (83 percent); and D, E, and F were left-handed but by a still weaker margin (63-65 percent).

Kempf then tried to change each animal's preference by offering the food from the experimenter's hand but not relinquishing it if the monkey reached with its accustomed hand. In this way, A and B were shifted to left-hand preference; C, D, E, and F to right-hand preference. Although each monkey tried repeatedly to grasp the food with the accustomed hand before using the other, within a few days every one had been switched successfully. Retesting between 3 and 4 months later showed good retention of preference for the trained hand except for A, who reverted completely to his original right-handedness.

Analysis of Task Effects

Except for Skirrl's use of a hammer and nails (Yerkes, 1916), in all of the studies to date, the measure of handedness was the hand used for picking up or reaching for food. In 1933, at the University of Chicago, Heinrich Klüver made an observation that would have important implications for methodology. He found that in Java monkeys (*Macaca cynomolgos*), hand preference for reaching could change depending on target location. For example, when a food box could be reached by pulling on a string to which it was connected, one monkey, D.-L., used his right hand 1.76 times more often than his left (438 vs. 297 times), and another monkey, B.-R., used his right hand about twice as often as his left (198 vs. 97 times). On these tests the ends of the string lay in easy reach, either within the cage or just outside. D.-L., however, used his right hand whenever the end of the string was *outside* the cage, and used his left hand when it was *inside*, whereas B.-R. showed a strong right-hand preference for both locations. For both animals, the effect persisted in further trials.

Observations with other Java monkeys suggested other effective task variables. For example, for picking food out of a box after pulling in the box from outside his cage, one monkey used his right hand 157 times, his left only 28. However, for reaching for a raisin placed at the end of a board 15 cm. above the floor and 15 cm. from the cage, the same animal showed virtually no preference (29 right, 24 left). The effect was also different for different animals. On the same two tasks, another monkey showed only a marginal right-hand preference for picking food out of a box (73 vs. 53) but an overwhelming right-hand preference for the raisin task (197 vs. 12). Klüver thus suggested that tests for handedness should acknowledge the possibility that a change in distance could affect hand preference.

Klüver's observations were among the first signs of a new scientific sensitivity to the complexities involved in the assessment of primate handedness. A similar awareness was appearing among psychologists studying handedness in human beings. For example, in her study of hand preference in young children, Updegraff (1932) excluded the "hand used for picking up five small articles" and the "hand used for picking up a spoon from the floor" from her test battery because these two measures had low reliability. Likewise, Koch (1933) showed that on tests requiring little skill (e.g., tests of "holding" and "carrying"), college students tended to use their hands indiscriminately. Tests having the greatest value in diagnosis were those requiring more refined and elaborate manipulation. Such observations of human subjects were strongly supported by the results of an unpublished study of monkeys carried out at The University of Wisconsin Primate Laboratory by Harlow, Settlage, and Grether (1933; cited in Kounin, 1938). Eight monkeys (2 Java, 5 Rhesus, and 1 Pigtail) were tested in 3 situations, described here in descending order of their adjudged demand on a high degree of coordinated manual skill: In the pipe test, two short pipes were placed through a board that was hung in front of the monkey's cage. Food was placed in front of each pipe, in irregular order, and the monkey had to retrieve the food by reaching through the pipe; in the cup test, a shallow container was placed about 15 cm. in front of the cage from which the monkey had to remove a piece of food; in the string test, the monkey secured food by pulling in a string placed about 8 cm. from the front of the cage. The results indicated that, on average, the tests most effective in eliciting hand preference as measured by strength of preference were the pipe tests (88-89 percent) compared with the cup test (76 percent) and the string test (73 percent). As another indication, all 8 animals showed preference in the same direction across both pipe tests, with the margin of preference being 90 to 100 percent for 5 animals. That is, if an animal preferred its left arm when the pipe was on the left, it used the same arm when the pipe was on the right. By contrast, only 4 animals showed consistency of direction across all four tests. Although the results indicate hand preference in individual animals, there was no consistency of direction across the 8 individuals. On the pipe test, 4 animals (1 Java and 3 Rhesus) used the right hand, and 4 (1 Java, 2 Rhesus, and the pigtail) used the left. Similar ratios appeared on the other tests.

The Harlow et al. study served as the model for a new study by Jacob Kounin, at the time a research assistant in the same institution. Kounin (1938) tested 8 monkeys (4 Rhesus, 3 Cebus, and 1 Spider), all either pubescent or pre-pubescent. Three different tests of handedness were used:

in Test I, the monkey had to extend its arm through a hole in the floor of its cage (which for this purpose had been placed on a table above the floor) to secure a piece of banana from a container. The banana was placed at a distance approximating the animal's maximum reach. In Test II, the banana piece was placed in a food box attached to the front of the cage, and the animal, in order to secure the food, had to put one hand through a small hole in the food box, lift the cover of the box, hold it open, and remove the food from a narrow slanting depression in the box — all by using only one hand. Finally, in Test III, the food reward was beyond reach, and the monkey could secure it only by drawing it in with a rake, the nearest end of which was placed 20 to 25 cm. from the animal's cage.

As in the Harlow et al. study, the results showed task effects, the greatest degree of hand preference being expressed in Tests I and II. Test III had the lowest diagnostic value, the lowest percent of hand preference, and the lowest reliability. The three Cebus monkeys also were the most consistent across tasks and, indeed, were the only ones to show definite handedness (2 preferred the right hand, the other the left). Kounin suggested that this might be related to the Cebus monkey's greater degree of manipulatory and investigatory behavior compared to the Rhesus. He noted that the Cebus monkeys frequently reached for and searched in the experimenter's pocket, whereas the Rhesus monkeys never did, and that true tool-using has been observed in Cebus but not in Rhesus monkeys. Unfortunately, because only 3 Cebus monkeys were tested, we cannot say whether the Cebus also show a species effect with respect to direction of hand preference.

Research on Apes

Of the research on the great apes, possibly the first report in this new era was that of Pfungst (1912). Recall that Pfungst said that he had found no evidence of predominant hand preference in apes or monkeys that could not always be traced to training or to some earlier trauma.

The next report, perhaps, was by Yerkes (1916). Along with the 6 monkeys mentioned earlier, Yerkes reported observations on a young orang-utan (not more than 5 years old) named Julius. Julius showed a strong left-hand preference. Indeed, "in difficult situations he almost invariably reached for food or manipulated objects in connection with food getting with the left hand" (1916, p. 117).

Yerkes reported no further tests with Julius, so we do not know whether the preference persisted or whether it could have been changed with training, as it was for the monkeys tested by Franz, Kempf, and Lashley. In any case, later anecdotal reports supported suggestions that the great apes showed handedness, although they disagreed about its direction. For example, W. Henry Sheak, an American animal trainer, flatly stated that "most chimpanzees are decidedly right-handed" but that Joe, one of the chimpanzees in Sheak's animal show, was "left-handed" (1924, p. 127). No documentation was provided for either claim. Information was somewhat better in the case of John Daniel, a young lowland gorilla who came to England from Gabon in 1918 and was kept in London by Alyse Cunningham until 1921 (Cunningham, 1921). One of the reports about John Daniel, or "Johnny Gorilla," the name Cunningham preferred, came from the German anatomist Rudolf Fick (1926), by way of a communication from Major Rupert Penny, Cunningham's nephew:

> I will mention here what Major Penny told me at my request, that the gorilla may be exclusively a right-hander; for example, both in eating and in throwing sand or stones, which he likes to do, as I myself observed. Major Penny has never seen him carry out these movements with his left hand. (1926, p. 432; quoted in translation in Yerkes & Yerkes, 1929, p. 410)

Fick himself, however, observed chimpanzees at the Zoological Garden in Berlin and at the Anthropoid Research Station on Tenerife and concluded that his observations did not reveal handedness in the animals, contrary to what had been claimed in the earlier statements by Mollison (1908, 1910-1911), as cited in the previous section:

> Mollison calls the orang the greatest right-hander after man, the chimpanzee and gorilla left-handers. With the chimpanzees from Tenerife and the other chimpanzees in the Zoological Garden in Berlin I found no *lasting* preference for one hand in grasping. Here and there I believed I had established in one or another an exclusive use of the right hand, and then a grasping with the left hand would throw this idea over. I should designate the chimpanzees as 'both handers' (ambidextrous). (p. 432; quoted in translation in Yerkes & Yerkes, 1929, p. 410)

A somewhat different impression of the work on Tenerife is suggested in Köhler's book, *The Mentality of Apes* (1927), although, in fairness, the evidence is so meager that no general conclusion can be drawn. In one of the many tests of the chimpanzee Sultan's ability to use implements, some food, tied to a string, was placed on the ground outside the barred windows of Sultan's house. Besides the 'right' string, two or three additional strings ran from the approximate direction of the food, crossing the 'right' string and each other, with their ends all near the bars of the window. Köhler noted that in trying to obtain the food, Sultan tended to prefer strings lying to his right and that, on all occasions that "require the slightest degree of skill, uses his *right* hand" (1927, p. 28; emphasis in original). In fitting together two tubular sticks, Sultan also would hold the larger one in his left hand while trying to insert the smaller one with his right hand (pp. 128-129; see also Plate III, facing page 128).

In the 1920s there were at least two other reports of handedness in great apes. One was a second lowland gorilla obtained by Alyse Cunningham after

John Daniel died in 1921 and whom Cunningham named Sultan — evidently a popular name — although he was known as John Daniel II when exhibited in the United States). According to a report by Le Gros Clark (1927), Sultan, like his predecessor, "was evidently right-handed, though sometimes when told to shake hands, did so with his left hand" (1927, p. 467). Witnesses said that Sultan also "showed a distinct preference for the use of the right hand in movements which required skill of any kind"(p. 475).

In America, Yerkes (1927) made observations of Congo, a 5-year-old mountain gorilla (*Gorilla beringei*). On a variety of reaching and tool-using tests, Congo repeatedly showed right-handedness, and sometimes also in conjunction with left-footedness.

> Right-handedness was early manifested in the various forms of stick problems, and by a systematic test in which the gorilla was made to reach for food I discovered that she used the right hand and arm approximately two-thirds of the time (1927, pp. 69-70).
>
> I set the problem in six different locations and despite the variation of conditions Congo consistently reached with her right arm and hand, if the object was near, and with her left leg and foot, if the object was beyond arm reach (pp. 71-72).

In their 1929 book on the great apes, Yerkes and Yerkes called this the "sole record of experimental test of handedness in the gorilla" (1929, p. 410). Unable to find any other observations, they simply noted that the behavioral evidence for John Daniel and Congo agree in indicating *right*-handedness and therefore contradict Mollison's (1910-1911) inference from physical measurements of bone length that the gorilla should naturally be *left*-handed. Yerkes and Yerkes, however, were reluctant to draw a strong conclusion, pointing out that, although there was "ample ground" for assuming that there often is limb preference for morphological or functional reasons, there was no evidence that "all or any one of the anthropoid types is definitely right-handed or left-handed..." (p. 538). They also suggested that the many instances of hand preference that have been discovered appeared to be "environmental accidents" (p. 217).

In the 1930s, at least one further observation of handedness in a great ape was reported — this one of a 6-year-old female orang-utan (*Pongo pygmaeus*) (Finan, 1939). The animal was made to sit on a chair before a table level with her elbows. There she was shown a rack containing two shelves, one at table-top level, the other raised 30.5 cm. above. Three targets were used, placed at a variety of distances, levels, and left-right positions: a 5-cm. wood cube, a ball 6.35 cm. diameter, and a pellet (shelled peanut). The results showed that the animal systematically adapted her style of prehension to the type and location of the target (e.g., thumb-forefinger opposition was preponderant when the pellet was distant and in the top shelf), but no tendency toward hand preference was observed. Instead, hand use was "conditioned" by stimulus proximity to a particular hand.

Throughout this new period, anatomical studies continued. The subject of one such examination was the gorilla Sultan (John Daniel II), who, recall, had been called right-handed according to the report by Le Gros Clark (1927). When Sultan died, Le Gros Clark examined the brain and found cortical asymmetries in the parietal and occipital areas, including left occipital petalia (protrusion of the left occipital lobe beyond the extension of the right). Like Cunningham (1902) before him, Le Gros Clark played down the possibility that the asymmetries could be related to handedness.

The neurologist S. E. Henschen (1926a) was less conservative even though, unlike Cunningham and Le Gros Clark, he did not study the brain directly. Instead, he examined the *skull* of a gorilla in the museum of University College, London, observed that "the left half...was more prominent than the right half," and concluded, "This gorilla was probably right-handed" (1926a, p. 123; see also Henschen, 1926b). In support, Henschen cited a later examination of the skulls of six other gorillas, this time adding that the greater left-sided prominence was in the posterior region and that it was greater in older males than in younger specimens or females. Henschen then mentioned having received a photograph "of a Gabon-gorilla throwing sand on his keeper with his right hand" (1926a, footnote 1, p. 123). Henschen's readiness to rely on a single photograph evidently was conditioned by his acceptance of the evidence of earlier reports, cited in his paper, of right-handedness in monkeys (e.g., Ogle, 1871; Osawa, 1901).

So seems to have been the state of the evidence, both behavioral and anatomical, through the 1920s and 1930s. In 1941, a new investigator, Glen Finch, looking back at the behavioral evidence, delivered a negative verdict: previous studies were few in number, and the data came from "scattered, largely non-experimental observations of a few animals" (1941, p. 211). Finch did not cite any of the earlier studies, but this is a fair assessment of the work reviewed to this point, including even Yerkes'(1927) study of Congo. (Judgment must be reserved about Pfungst's [1912] report, for reasons already mentioned, as well as Fick's [1926], which I have not seen first-hand). Finch was in a good position to improve matters. His own work was carried out at Yerkes' own Laboratories of Primate Biology at Yale, where Yerkes was still Professor of Psychobiology.

Finch devised 4 tests, all requiring the subjects — chimpanzees — to procure pieces of fruit through the 5-cm.-square apertures of their wire-netting cage-walls, and in such a way as to allow the use of only one hand. In one test, pieces of fruit were attached to the ends of lengths of string, which the subject had to pull in; in another, 10 pieces of fruit, spaced at close intervals, were placed alone on a board 2.54 cm. from the cage; in another, pieces of fruit were placed individually under a small hinged metal box, requiring the subject to reach through a hole in a piece of plywood, thereby upsetting the metal box; and in the last test, pieces of fruit were presented, individually, on a board 2.54 cm. from the cage-netting. Each situation was presented until the subject had procured 100 pieces of fruit, for a total of 800 trials (one repetition of the entire series). Subjects received 100 trials in a single session in one day, and tests were not continued on immediately succeeding days. Thirty chimpanzees were tested, each in isolation from the others. The result was that 18 animals used one hand on more than 90 percent of the 800 trials, with 9 preferring the right hand, and 9 the left. When the criterion was lowered to 80 percent consistency, 25 animals showed hand preference, 11 for the right, 14 for the left. Examination of the protocols for the 5 animals who showed the least-pronounced hand preference revealed that their low handedness scores were largely a result of low inter-test agreement as well as low single-test reliability.

Finch modestly concluded that his work "does not pretend to explore more than a rather narrowly limited aspect of chimpanzee lateral organization" (p. 118). Yerkes (1943) was bolder. He saw important implications in Finch's results for understanding not only handedness in individual apes but also handedness as a trait in human beings. In both species, Yerkes suggested that the "initial bias in favor of one hand or the other" was due either to a structural asymmetry in the brain or neuromuscular system, or to "slight inequalities in developmental status, strength, or immediate readiness for use" (1943, p. 114), and he went on to say that this explained why individual apes were as often left- as right-handed. But the question remained, why is man usually right-handed? Yerkes' answer was that:

> the initial bias is either confirmed or suppressed by social tradition. The design of our implements, the nature of our customs of work and play, altogether favor right-handedness, whereas in the chimpanzee any initial handedness tendency either develops or disappears without any appreciable cultural or other species-determined influence.(p. 114)

Yerkes ended on a note that would have been music to the ears of John Jackson and the followers of ambidextral culture:

> the fact seems clearly established that in respect to handedness and what it signifies developmentally the chimpanzee is free to act according to its natural characteristics and immediate maturational and other functional influences, while we humans, by contrast, are slaves to convention, having surrendered our boasted freedom of individual choice, decision, and direction of effort by accepting a form of limitation which at best is inconvenient to a considerable portion of us (estimated variously at 2 to 30 percent) and at worst interferes more or less seriously with our psychobiological development. I exhibit this contrast between human and anthropoid life chiefly because when a mode of action has become established as customary, as has handedness, it tends to be

> unquestioningly accepted as necessary, inevitable, right, even though the cost to the individual in sacrifice or inconvenience may demonstrably outweigh the social advantages. My tentative conclusion is that the free natural development observable in the chimpanzee is preferable to the custom-imposed handedness development of civilized man.(pp. 114-115)

Discussion and Conclusions

Finch's report in 1941 seems to have been the last new study of primate handedness until the end of World War II. (Primate research generally came to a halt as well.) Then, in the post-war years, a number of new studies began to appear (e.g., Cole, 1957; Cole & Glees, 1951; Grzimek, 1949; Riess, Ross, Lyerly, & Birch, 1949; Warren, 1953, 1958). Ever since, the publication curve has proceeded on a sharp upward course, matching the great growth of post-war interest in human cortical lateralization and handedness.

What lessons can be drawn from the earlier work? I think there are several. First, at the very least we can appreciate that ours is not the first or even second generation to regard the question of primate handedness as worthy of serious attention. Like our predecessors, researchers today see implications in the study of primate handedness for addressing fundamental questions about the nature of human handedness and, more broadly, about ape and human evolution (e.g., MacNeilage, 1987).

Beyond its historical interest, the early literature may also be informative about matters of contemporary concern. For example, it suggests that individual animals do show hand preference, not only the great apes but monkeys as well. In this respect, the early and the contemporary literature are largely in agreement. But what about the primary question whether apes or monkeys show hand preference as a species characteristic? Here, as we have seen, opinion was mixed, much as it remains today. Whereas some of the best of the very early studies, such as Ogle's (1871), suggested that the answer was yes, the better studies of the later era, such as Finch's (1941) study of chimpanzees, and Kounin's (1938) study of monkeys, provided few signs of directional consistency within species.

Several of the early reports also included insightful suggestions or observations that, had they been duly noted, might have accelerated methodological improvements in the scientific study of hand preference in primates. One example is Deniker's (1882b) suggestion that the arm used for climbing might not be the same as the arm used in the execution of acts requiring more delicacy, the point presumably being that the same hand might not dominate in all skilled acts. Yerkes (1916) sharpened the point. Even though Yerkes later (1943) implied that handedness was largely a social creation in human beings (so that where it also appeared in lower primates, it would be so *a fortiori*), he raised the possibility that if lower primates did show handedness, its form might differ from that shown in man. Recall his earlier comment that he would not be surprised to find that "it is the rule for animals to possess or to develop readily definite preferences for one hand in connection with a given act of skill and to have quite as definite a preference for the other hand in connection with a radically different kind of act" (1916, p. 118). Both Deniker's and Yerkes' proposals perhaps anticipate the analysis recently put forward by MacNeilage, Studdert-Kennedy, and Lindblom (1987). As mentioned in the beginning of this chapter, MacNeilage et al. are among those championing the view that great apes and monkeys do show handedness at the population level. They suggest, however, that the pattern is different from human handedness. Whereas in human beings, the dominant hand takes the lead in most skilled acts, MacNeilage et al. propose that monkeys and apes split responsibility between the hands, using the left for visually-guided reaching, the right for object manipulation.

Several of the early researchers also showed an awareness of the importance of task variables of the sort now being examined by researchers. For example, Ogle (1871) was probably correct in supposing that on a reaching task, the food bait must be placed sufficiently far away that the animal must stretch out one arm. Ogle did not explain why distance should be a factor, but he might well have argued that when the food was within easy reach, the task was not sufficiently demanding. Köhler (1925, p. 28) may have been making a similar point in emphasizing that the chimpanzee Sultan used his right hand whenever some degree of skill was called for.

Aside from the greater difficulty in reaching for a distant object, far reaches might also require a shift in posture, which has been shown to be important for demonstrations of hand preference in the bushbaby (*Galago senegalensis*) (Sanford, Guin, & Ward, 1984) and ruffed lemur (*Varecia variegata variegata*) (Forsythe, Milliken, Stafford, & Ward, 1988). MacNeilage (1987, p. 299) has suggested that Finch's (1941) finding a bimodal distribution of left and right preference (with some animals lacking preference) might be related to his use of "undemanding laboratory reaching tasks." Finch's tasks seem to me to have been no less demanding than those used in studies where species effects were found (certainly they required more precise skillful manipulation than Ogle's task did). Nevertheless, it appears that none of Finch's reaching tasks required the animal to stretch and thus shift posture, so perhaps this feature played a role in the animals' performance. All such effects also might prove to be highly idiosyncratic, depending on the particular animal and the testing circumstances. For example, Yerkes' observation of the gorilla Congo suggested that different distances can be associated with different hand-preference. Recall that Congo showed a right-side preference for near objects, a left-side preference for far objects.

The question of task variables is also raised in Brewster's (1913) suggestion that a left-hand preference may stem from food having been offered more often on the animal's left than right side (by a right-handed person). Contemporary studies have established that such adventitious factors can indeed have a powerful effect (cf. Collins, 1975).

The method used by Franz (1913), Yerkes (1916), and Kempf (1917) of studying hand preference by experimental manipulation also deserves further attention by contemporary researchers. The question should not only be whether hand preference can be modified experimentally by, say, contingent reinforcement, but whether there is also some asymmetric 'response preparedness' (cf. Seligman, 1970) such that one forelimb is more easily trained than the other. At least one new study of this sort has been reported, with squirrel monkeys, and the results seem to be more consistent with the first outcome than the second (McGonigle & Flook, 1978). In this case, however, hand preference was manipulated for only one kind of relatively simple behavior — namely, pushing back the cover on a foodwell at the center of a test tray. Further experiments should include a variety of measures varying in task demand, difficulty, and complexity.

As a last example of an early observation deserving further attention, consider Kounin's (1938) suggestion that hand preference might be more likely in those primates (Cebus monkeys in his experiment) that show a greater degree of manipulatory and investigatory behavior. These and other motivational variables might be of great importance in the planning of further studies, as does the possible role of highly specialized forelimbs as in the "Aye-Aye" (*Daubentonia madagascariensis*, or *Cheiromys madagascariensis*), which uses its unusual middle finger to draw worms out of holes (see footnote 5).

Finally, consider the question of the significance of neuroanatomical asymmetries. Recall that Cunningham (1902) was reluctant to conclude that the greater depression of the left Sylvian fissure in the human brain was associated with right-handedness or even with left-hemisphere localization of speech because the same condition was found in the ape, and Cunningham could not persuade himself that the ape possesses "any superior power in either arm" (1902, p. 293). Although others were less reluctant, even when the anatomical evidence was far more indirect (e.g., Henschen, 1926a), it was Cunningham's conservative position about the meaning of such neuroanatomical asymmetries that became, and long remained, the dominant view. Thus, as recently as the 1960s, the consensus was still that the human brain was anatomically symmetrical (much as Bichat had asserted in the 1800s), or at least that such asymmetries as were known to exist (such as the longer left Sylvian fissure) were of no great significance. As the anatomist Gerhardt von Bonin said, in a 1962 review, all of the known morphological differences, "are, after all, quite small" (1962, p.6), and "How to correlate these with the astonishing differences in function, such as the speech function on the left side, is an entirely different question, and one that I am unable to answer" (p. 6). Smith, writing in 1966, saw things in much the same way:

> Like most paired organs, the two great human cerebral hemispheres are symmetrical in gross appearance. In size, shape, and anatomic location, naked eye inspection reveals uniform and parallel organization of large and small hemispheric structures. Neurohistologic study shows a similar correspondence in even the finest cytoarchitectural and myeloarchitectural details. Experiments with animals using modern fine electrical recording and stimulation techniques consistently confirm historical findings describing the two hemispheres as also practically identical in function (1966, p. 109).

By 1968, of course, the picture changed dramatically when Geschwind and Levitsky reported that in the majority of human brains the left temporal plane was appreciably and obviously larger than the right. A similar neuroanatomical asymmetry has now been confirmed in fetal and infant brains, as well as in the brain of the great ape (e.g., Witelson & Pallie, 1973; Yeni-Komshian & Benson, 1976), and theorists today seem far more sympathetic to the possibility that this and a host of other asymmetries have functional significance, not just in man but in other species (e.g., Geschwind, 1985).

Clearly, it is not only the new anatomical evidence that has changed the intellectual climate. It is also the recent demonstrations of motor and perceptual asymmetries in apes and monkeys, renewed interest in the possibility of hand preference in nonhuman primates, and finally, evidence suggesting that the great apes can learn rudimentary symbolic communicative systems. In light of what he knew at the time, we therefore can hardly fault Cunningham (1902) for discounting his neuroanatomical data. Still, one wonders what effects his conservative stance might have had on future developments. What if Cunningham had been a trace readier to credit the ape with "superior power in either arm"? Could it, directly or indirectly, have hastened the discovery of the neural substrates for speech?

These are some of the lessons and questions that might be suggested by the earlier studies of primate handedness. In ending this review, I can think of no better words than those used by Yerkes and Yerkes (1929) at the end of their majestic work on the great apes. They said that if their survey had helped to conserve the time of the investigator, to improve historical perspective, and to suggest new problems, then their regrets for the hours lost to research will have been dissipated and their labors abundantly rewarded. I can scarcely hope for more from my own infinitely smaller efforts.

References

Andrew, R.J. (1983). Lateralization of emotional and cognitive function in higher vertebrates, with special reference to the domestic chick. In J. P. Ewert, R. R. Capranica, & D. J. Ingle (Eds.), *Advances in vertebrate neuroethology* (pp.477-509). New York: Plenum.

Aristotle (1912). *De Partibus Animalium.* Trans. by Wm. Ogle. In J.A. Smith & W. D. Ross (Eds.), *The Works of Aristotle*, Vol. 5. Oxford, England: Clarendon Press.

Aristotle (1878). *Historia Animalium*. Trans. by R. Cresswell. London: George Bell & Sons.

Bardeleben, K.von (1909). Ueber bilaterale Asymmetrie beim Menschen und bei hoheren Tieren*Anatomischer Anzeiger fur die Gesamte Wissenschaftlich Anatomie.*Ergan zungsheft zum 34 Band (*Verhandlungen der Anatomischen Gesellschaft*,23 Versammlung,Giessen,April 1909, 2-72).

Bell, C. (1833). *The hand: Its mechanisms and vital endowments as evincing design.* London: William Pickering.

Bichat, X. (1805/1809). *Physiological researches upon life and death.* Trans., by T. Watkins, from the 2nd Paris edition of *Recherches physiologique sur la vie et la mort*. Philadelphia: Smith & Maxwell.

Biervliet, J.J. van (1899). L'homme droit et l'homme gauche. *Revue Philosophique de la France et de l'Etranger, 47*, 113-143; 276-296; 371-389.

Bischoff, E. (1863). Einige Gewichts- und Trocken-bestimmungen der Organe das Menschlichen Korpers. In Henle & Pfeiffer (Eds.), *Zeitschrift für rationelle Medicin, Series 3.* Leipzig & Heidelberg: C.F. Wintersche Verlagshandlung.

Bonin, G. von (1962). Anatomical asymmetries of the cerebral hemispheres. In V. Mountcastle (Ed.), *Interhemispheric relations and cerebral dominance* (pp. 1-6). Baltimore: Johns Hopkins University Press.

Brewster, E.T. (1913, June). The ways of the left hand. *McClure's Magazine*, 168-183.

Brinton, D. G. (1896). Left-handedness in North American aboriginal art. *American Anthropologist*, *9*, 175-181.

Broca, P. (1865). Sur le siège de la faculté du langage articulé. *Bulletins de la Société d'Anthropologie de Paris, 6*, 377-393.

Broca, P. (1877). De l'inegalité dynamique des deux hémisphères cérébraux. *Bulletin de L'Académie de Médecine, 2nd Series, 6,* 508-530 (and discussion by Bouillaud, Bouley, and Broca, pp. 530-539).

Browne, T. (1646). *Pseudodoxia epidemica, or, Enquiries into very many received tenents, and commonly presumed truths.* London, printed by T.H. for Edward Dod.

Buchanan, A. (1862). Mechanical theory of the predominance of the right hand over the left; or, more generally, of the limbs of the right side over those of the left side of the body. *Proceedings of the Philosophical Society of Glasgow, 5*, 142-167.

Cole, J. (1957). Laterality in the use of the hand, foot, and eye in monkeys. *Journal of Comparative and Physiological Psychology, 50*, 296-299.

Cole, J., & Glees, P. (1951). Handedness in monkeys. *Experientia, 8*, 224-225.

Collins, R. L. (1975). When left-handed mice live in right-handed worlds. *Science, 187*, 181-184.

Coolidge, H. J. (1929). *Revision of the genus Gorilla.* Cambridge, Mass.: Museum of Comparative Zoology, Harvard University, *50*, No. 4.

Corballis, M.C. (1983). *Human laterality.* New York: Academic Press.

Coudereau, [no initial]. (1882). Discussion sur un orang et un chimpanzé. *Bulletins de la Société d'Anthropologie de Paris, 5, 3rd Series,* 338-341.

Crichton-Browne, J. (1907). Dexterity and the bend sinister. *Proceedings of the Royal Institution of Great Britain, 18,* 623-652.

Cunningham, A. (1921). A gorilla's life in civilization. *Bulletin of the New York Zoological Society, 24,* 118-124.

Cunningham, D.J. (1892). *Contribution to the surface anatomy of the cerebral hemispheres, with a chapter upon cranio-cerebral topography by Victor Horsley.* Royal Irish Academy: Dublin (Cunningham Memorial).

Cunningham, D.J. (1902). Right-handedness and left-brainedness. The Huxley lecture for 1902. *Journal of the Royal Anthropological Institute of Great Britain and Ireland, 32,* 273-296.

Darwin, C. (1859). *On the origin of species by means of natural selection.* London: John Murray.

Darwin, C. (1871). *The descent of man.* London: John Murray.

Debierre, [no initial]. (1887). Le developpement des membres du coté droit l'emporte-t-il originairement sur celui des membres du coté gauche? *Comptes Rendus Hebdomadaires des Seances et Mémoires de la Société de Biologie, 4,* 28-31.

Delaunay, C.-G. (1874). *Biologie comparée du coté droit et du coté gauche chez l'homme et chez les êtres vivants.* Paris: N. Blanpain.

Denenberg, V.H. (1981). Hemispheric laterality in animals and the effects of early experience. *Behavioral and Brain Sciences, 4,* 1-21.

Denenberg, V.H. (1983). Animal studies of laterality. In K. H. Heilman & P. Satz (Eds.), *Neuropsychology of human emotion* (pp.65-84). New York: The Guilford Press.

Deniker, J. (1882a). Sur l'orang et le chimpanzé par M. Bidel a Paris. *Bulletins de la Société d'Anthropologie de Paris, 5, 3rd Series,* 333-338.

Deniker, J. (1882b). Discussion. *Bulletins de la Société d'Anthropologie de Paris, 5, 3rd Series,* 338-341.

Deniker, J. (1891). Les gorilles vivant a Paris. *La Nature: Revue des Sciences et Leurs Applications aux Arts et a l'Industrie, Journal Hebdomadaire Illustre,* No. 965, 14 Nov., 24-25.

Eberstaller, O. (1884). Oberflachen-anatomie der Grosshirn-hemispharen. *Weiner Medizinische Blaetter, 7,* 479-482; 542-582; 644-646.

Fick, R. (1895). Vergleichend anatomische Studien an einem erwachsenen Orang-Utang. *Archiv für Anatomie und physiologie: Anatomische abteilung* (pp. 1-100). Leipzig: Verlag von Veit & Comp.

Fick, R. (1926). Massverhaltnisse an den oberen Gliedmassen das Menschen und den Gliedmassen der Menschenaffen. *Sitzungsberichte der Preussischen Akademie der Wissenschaften,* 417-451.

Finan, J. L. (1939). A note on prehension in a captive orang-utang. *Journal of Genetic Psychology, 55,* 435-437.

Finch, G. (1941). Chimpanzee handedness. *Science, 94,* 117-118.

Forsythe, C., Milliken, G. W., Stafford, D. K., & Ward, J. P. (1988). Posturally related variations in the hand preferences of the ruffed lemur (*Varecia variegata variegata*). *Journal of Comparative Psychology, 102,* 248-250.

Franz, S.I. (1902). On the functions of the cerebrum: I. The frontal lobes in relation to the production and retention of simple sensory motor habits. *The American Journal of Physiology, 8*, 1-22.

Franz, S.I. (1913). Observations on the preferential use of the right and left hands by monkeys. *Journal of Animal Behavior, 3*, 140-144.

Franz, S.I. (1917). The effects of cerebral destructions upon habit-formation and retention in the albino rat. *Psychobiology, 1*, 71-139.

Galen. (1956). *On anatomical procedures*. Trans. by C. Singer. London: Oxford University Press.

Gaupp, E. (1889). Ueber die Maass und Gewichts-Differenzen Zwischen den Knochen der rechten und der linken Extremitaten des Menschen. *Inaug. Dissert.*, Breslau.

Gegenbaur, C. (1878). *Elements of comparative anatomy*. Trans. by F.J. Bell,revised by F.R. Lankester. London: Macmillan and Co.

Geschwind, N. (1985). Implications for evolution, genetics, and clinical syndromes. In S. D. Glick (Ed.), *Cerebral lateralization in nonhuman species* (pp.247-276). Orlando, Florida: Academic Press.

Geschwind, N., & Levitsky, W. (1968). Left/right asymmetries in temporal speech region. *Science, 161*, 186-187.

Glick, S. D. (Ed.), (1985). *Cerebral lateralization in nonhuman species*. Orlando, Florida: Academic Press.

Gould, G. M. (1908). *Right handedness and left handedness*. Philadelphia & London: J.P. Lippincott Co.

Grzimek, B. (1949). Rechts- und Linkshandigkeit bei Pferdan, Papageien und Affen. *Zeitschrift für Tierpsychologie, 6*, 406-432.

Guldberg, G. A. (1897). *Etudes sur la dyssymétrie morphologique et fonctionnelles chez l'homme et les vértébres supérieurs*. Christiana(Oslo, Norway): H. Aschehoug & Co.

Harlow, H. F., Settlage, P., & Grether, W. (1933). Handedness in monkeys. Unpublished studies, University of Wisconsin Primate Laboratory, Madison.

Harris, L. J. (1980a). Left-handedness: Early theories, facts and fancies. In J. Herron (Ed.), *Neuropsychology of left-handedness* (pp.3-78). New York: Academic Press.

Harris, L. J. (1980b). Which hand is the "eye" of the blind? — A new look at an old question. In J. Herron (Ed.), *Neuropsychology of left-handedness* (pp.303-329). New York: Academic Press.

Harris, L. J. (1985a). Teaching the right brain: Historical perspective on a contemporary educational fad. In C. T. Best (Ed.), *Developmental neuropsychology and education* (pp.231-273). New York: Academic Press.

Harris, L. J. (1985b). The Ambidextral Culture Society and the "duality of mind." *The Behavioral and Brain Sciences, 8*, 639-640.

Harris, L. J. (1989a). Footedness in parrots: Three centuries of research, theory, and mere surmise. *Canadian Journal of Psychology, 43*, 369-396.

Harris, L. J. (1989b). Hand preference in gestures and signs in the deaf and hearing: Some notes on early evidence and theory. *Brain and Cognition, 10*, 189-219.

Harris, L. J. (1990). Cultural influences on handedness: Historical and contemporary evidence. In S. Coren (Ed.), *Left-handedness: Behavioral implications and anomalies* (pp.195-258). Amsterdam: Elsevier.

Harris, L. J. (1991). Cerebral control for speech in right-handers and left-handers: an analysis of the views of Paul Broca, his contemporaries, and his successors. *Brain and Language*, *40*, 1-50.

Harting, P. (1869). Sur une asymétrie du squelette humain se transmettant hereditaitment. *Archives Néerlandaises des Sciences Exactes et Naturelles*, *4*, 44-54.

Henschen, S. E. (1926a). On the function of the right hemisphere of the brain in relation to the left in speech, music and calculation. *Brain*, *49*, 110-123.

Henschen, S. E. (1926b). Ist der Gorilla linkshirnig? *Deutsche Zeitschrift für Nervenheilkunde*, *92*.

Hollis, W. A. (1875). Lopsided generations. *Journal of Anatomy & Physiology*, *9*, 263-271.

Huber, J. B. (1910). Why are we right-handed? *Scientific American*, *102*, 260-261; 268-269.

Huxley, T.H. (1863). *Evidence as to man's place in nature*. London: William and Norgate.

Hyrtl, J. (1860). *Handbuch der topographischen Anatomie* (4th Ed.). Vienna: Braumüller.

Jackson, J. (1905). *Ambidexterity or two-handedness and two-brainedness*. London: Kegan Paul, Trench, Trubner & Co.

Jobert, L. (1885). *Les gauchers compares aux droitiers aux points de vue anthropologique et medico-legal*. Lyon, France.

Jordan, H. E. (1911). The inheritance of left-handedness. *American Breeders Magazine*, *2*, 19-28.

Kempf, H. D. C. (1917). The preferential use of the hands in monkeys with modification by training and retention of the new habit. *The Psychological Bulletin*, *14*, 297-301.

Klüver, H. (1933). *Behavior mechanisms in monkeys*. Chicago: The University of Chicago Press.

Koch, H. (1933). A study of the nature, measurement, and determination of hand preference. *Genetic Psychology Monographs*, *13*, 117-221.

Köhler, W. (1927). *The mentality of apes*. (Trans. from the Second Revised Edition by E. Winter; 1st ed., 1925). London: Routledge & Kegan Paul Ltd.

Kounin, J.S. (1938). Laterality in monkeys. *The Journal of Genetic Psychology*, *52*, 375-393.

Langkavel, (1902); *Deutsche Revue*, *27* Jahrgang, April.

Lashley, K. S. (1917). Modifiability of the preferential use of the hands in the rhesus monkey. *Journal of Animal Behavior*, *7*, 178-186.

Lashley, K. S., & Watson, J. B. (1913). Notes on the development of a young monkey. *The Journal of Animal Behavior*, *3*, 145-139.

LeGros Clark, W.E. (1927). Description of cerebral hemispheres of the brain of the gorilla. *Journal of Anatomy*, *61*, 467-475.

Lombroso, C. (1903). Left-handedness and left-sidedness. *North American Review*, *177*, 440-444.

Ludwig, W. (1932). *Das Rechts-Links-Problem im Tierrich und beim Menschen*. Berlin: Verlag von Julius Springer.

MacNeilage, P. F. (1987). The evolution of hemispheric specialization for manual function and language. In S. P. Wise (Ed.), *Higher brain functions* (pp. 285-309). New York: John Wiley & Sons.

MacNeilage, P. F., Studdert-Kennedy, M. G., & Lindblom, B. (1987). Primate handedness reconsidered. (Target article, commentaries, and authors' response). *Behavioral and Brain Sciences, 10*, 247-303.

McGonigle, B. O. & Flook, J. (1978). The learning of hand preferences by squirrel monkeys. *Psychological Research, 40*, 93-98.

Mollison, T. (1908). Rechts und links in der Primatenreihe. *Korrespondenzblatt der deutschen Gesellschaft Anthropologie*, Braunschweig, *39*, 112-115.

Mollison, T. (1910-1911). Die Korperproportionen der Primaten. *Gegenbaurs Morphologisches Jahrbuch*, Leipzig, *42*, 79-304.

Mollon, J. D. (1978). Weber on sensory asymmetry. In M. Kinsbourne (Ed.), *Asymmetrical function of the brain* (pp. 318-325). Cambridge, England: Cambridge University Press.

Montagu, M. F. A. (1943). *Edward Tyson, M.D., F.R.S., 1650-1708, and the rise of human and comparative anatomy*. Philadelphia: The American Philosophical Society.

Moorhead, T.G. (1902). The relative weights of the right and left sides of the body in the foetus. *Journal of Anatomy and Physiology, 36*, 400-404.

Morris, R., & Morris, D. (1968). *Men and apes*. New York: Bantam Books, Inc. (orig. publication, 1966, New York: McGraw-Hill.)

Murie, J. & Duncan, P. M. (1877-1883). Lemurs. In P.M. Duncan (Ed.), *Cassell's Natural History: Vol. I* (pp. 210-258). London, Paris, & New York: Cassell Petter & Galpin.

Nottebohm, F. (1977). Asymmetries in neural control of vocalization in the canary. In S. R. Harnad, R. W. Doty, L. Goldstein, J. Jaynes, & G. Krauthamer (Eds.), *Lateralization in the nervous system* (pp.23-44). New York: Academic Press.

Ogle, W. (1871). On dextral pre-eminence. *Transactions of the Royal Medical and Chirurgical Society of London, 54*, 279-301.

Osawa, K. (1901). Ueber Linkshändigkeit. *Archives Italiennes de Biologie, 36*, 28-29. (Abstract of communication to International Congress of Physiological Sciences, Turin, Italy, 1901). Turin: Hermann Loescher.

Owen, R. (1868). *On the anatomy of vertebrates: Vol. III. Mammals*. London: Longmans, Green, and Co.

Pfungst, O. (1907). *Das Pferd des Herrn von Osten (Der Kluge Hans)*. Leipzig: Joh. Ambrosius Barth.

Pfungst, O. (1912). Zur Psychologie der Affen. *Bericht uber den VI. Kongress der Deutschen Gesellschaft fur Experimentelle Psychologie* (Leipzig), 200-205.

Plato (1953). *The dialogues of Plato: Vol. 4 (4th ed.)*. Trans. by B. Jowett. London and New York: Oxford University Press.

Pliny (1940). *Natural history: Vol. II*. Trans. by H. Rackham. Cambridge, Mass.: Harvard University Press.

Pliny (1942). *Natural history: Vol. III*. Trans. by H. Rackham. Cambridge, Mass.: Harvard University Press.

Purchas, S. (1613/1625). *Hakluytus posthumus, or Purchas his pilgrimes. Contayning a history of the world, in sea voyages & lande travells: Vol. 2 (4th ed.)*. London.

Reade, C. (1878). The coming man. [Letters to the editor] *Harper's Weekly*, 19 Jan., 50-51; 26 Jan., 74; 2 Feb., 94-95; 2 March, 174-175; 23 March, 234-235; 18 May, 394-395.

Rennie, J. (1850). *The menageries. The natural history of monkeys, oppposums, and lemurs*. London: Nattali & Bond.

Reynolds, V. (1967). *The apes*. New York: E.P. Dutton & Co., Inc.

Riess, B. F., Ross, S., Lyerly, S. B., & Birch, H. G. (1949). The behavior of two captive specimens of the lowland gorilla, *Gorilla gorilla gorilla* (Savage & Wyman). *Zoologica, 34*, 111-117.

Rogers, L. J., & Anson, J. M. (1979). Lateralisation of function in the chicken forebrain. *Pharmacology, Biochemistry, and Behavior, 10*, 679-686.

Sanford, C., Guin, K., & Ward, J. P. (1984). Posture and laterality in the bushbaby (*Galago senegalensis*). *Brain, Behavior, and Evolution, 25*, 217-224.

Sarasin, P. (1918). Uber Recht-und Links-Händigkeit in der Praehistorie und die Rechts-händigkeit in der historischen Zeit. *Naturforschende Gesellschaft in Basel. Verhandlungen, 29*, 122-196.

Savage, T. S. (1847). Notice of the external characters and habits of Troglodytes Gorilla, a new species of Orang from the Gaboon River; Wyman, J., Osteology of the same. *Boston Journal of Natural History, 5*, 417-443.

Seeligmüller, A. (1902). Rechts und Links. *Deutsche Revue*, 27 Jahrgang, April, 50-63.

Seligman, M. E. R. (1970). On the generality of the laws of learning. *Psychological Review, 77*, 419-429.

Sheak, W.H. (1924). Some further observations on the chimpanzee. *Journal of Mammalogy, 5*, 122-129.

Sigerson, G. (1884). Considerations of the strutural and acquisitional elements in dextral pre-eminence, with conclusions as to the ambidexterity of primeval man. *Proceedings of the Royal Irish Academy*, Series 2, Vol. 4, Science, 38-51.

Smith, A. (1966). Certain hypothesized hemispheric differences in language and visual functions in human adults. *Cortex, 2*, 109-126.

"The chimpanzee." (1836). *The Penny Magazine of the Society for the Diffusion of Useful Knowledge, 5*, 13 Feb., 57-60.

The London Magazine (1738), Thursday, 21 September, [Report of the arrival of a chimpanzee], 464-465.

Theile, F. W. (1884). Gewichtsbestimmungen zur Entwicklung des Muskelsystems und des Skelets beim Menschen. *Nova Acta der kaiserlichen Leopoldino-Carolinae deutsche academie der naturforscher*, Band xlvi. No. 3, Halle. (Current *Chemical Abstracts* title: *Nova Acta Leopoldina*).

Tulp, N. (1641). *Nicoli Tulpii, Amstelredamensis, Observationum medicarum libri tres*. Amsterdam.

Tyson, E. (1699). *Orang-Outang, sive Homo Sylvestris: or, the anatomy of a Pygmie*. A facsimile of the original 1699 publication, with an introduction by A. Montagu. London: Dawsons of Pall Mall, 1966.

Updegraff, R. (1932). Preferential handedness in young children. *Journal of Experimental Education, 1*, 134-139.

Vierordt, K. von (1881). *Physiologie des Kindesalters*. Tubingen: Druck von Heinrich Laupp.

Walker, S. F. (1980). Lateralization of function in the vertebrate brain. *British Journal of Psychology, 71*, 329-367.

Warren, J. M. (1953). Handedness in the rhesus monkey. *Science, 118*, 622-623.

Warren, J. M. (1958). The development of paw preferences in cats and rhesus monkeys. *Journal of Genetic Psychology, 93*, 229-236.

Weber, E. H. (1834/1846). Der Tastinn und das Gemeingefuhl. In R. Wagner (Ed.), *Hand-Worterbuch der Physiologie: Vol. III, Part 2,* (pp. 481-588). Orig. publ. *De pulsu, resorptione, auditu et tactu*. Leipzig: C.F.Koehler, 1834.

Witelson, S. F., & Pallie, W. (1973). Left hemisphere specialization for language in the newborn: Neuroanatomical evidence of asymmetry. *Brain, 96*, 641-646.

Yeni-Komshian, G. H., & Benson, D. A. (1976). Anatomical study of cerebral asymmetry in the temporal lobe of humans, chimpanzees, and rhesus monkeys. *Science, 192*, 387-389.

Yerkes, R. M. (1916). The mental life of monkeys and apes: A study of ideational behavior. *Behavior Monographs, 3*, no. 12, iv. + 145 pp.

Yerkes, R. M. (1927). The mind of a gorilla. *Genetic Psychology Monographs, 2*, Part I, 1-193; Part II, Mental development, 375-551.

Yerkes, R. M. (1943). *Chimpanzees: A laboratory colony.* New Haven: Yale University Press.

Yerkes, R. M. (1961). Robert Mearns Yerkes, Psychobiologist. In C. Murchison (Ed.), *A history of psychology in autobiography: Vol. II* (pp. 381-407). New York: Russell & Russell. (Orig. publ., Clark University Press, 1930)

Yerkes, R.M., & Yerkes, A.W. (1929). *The great apes: A study of anthropoid life*. New Haven: Yale University Press.

2
Patterns of Lateralized Behavior in Prosimians

Jeannette P. Ward, Garrett W. Milliken, and
Donna K. Stafford

In the preceding chapter, Harris has thoroughly documented the enduring interest of the human species in the question of lateral asymmetry in the behavior of nonhuman primate species. The appearance of this book with reports of a variety of nonhuman primate laterality studies originating in many parts of the world testifies to the resurgence of interest in this topic. This chapter reports the incidence and patterning of laterality phenomena in species of prosimians, a major group of nonhuman primates much less studied in all research contexts than their anthropoid counterparts. The relative neglect of these most primitive of primates, which represent the stem from which the primate order originated, has been undergoing remediation and there now exist many sources of information about this group (see for example, Charles-Dominique,1977; Charles-Dominiqueet al., 1980; Doyle,1974; Doyle & Martin, 1979; Jolly, 1966; Martin, 1990; Martin, Doyle, & Walker, 1974; Tattersall, 1982).

Evidence of prosimian forms, not substantially different from some of the modern species, has been dated to at least two million years past (Walker, 1974). The general phyletic relationship of prosimian to anthropoid primates is illustrated in Figure 2.1. Modern prosimian species are diverse in geographic location, habitat, structure, and behavior. The largest number of individuals and species are found in Madagascar and Africa; however, prosimians also appear in the South Seas (tarsier of the Phillipines and Borneo) and in India (potto, slow and slender lorises). Although many species of prosimian are arboreal, nocturnal, and more or less insectivorous, the lemurs of Madagascar include diurnal species that favor a diet of plant products and some of these, such as the ring-tailed lemur, have a more terrestrial habit. It has been suggested that these ancient forms incorporate a greater degree of structural and behavioral specialization than their anthropoid relatives (Jouffroy, 1962, 1975); certainly, in the aggregate, the prosimians represent a range and variety of types that is at least comparable

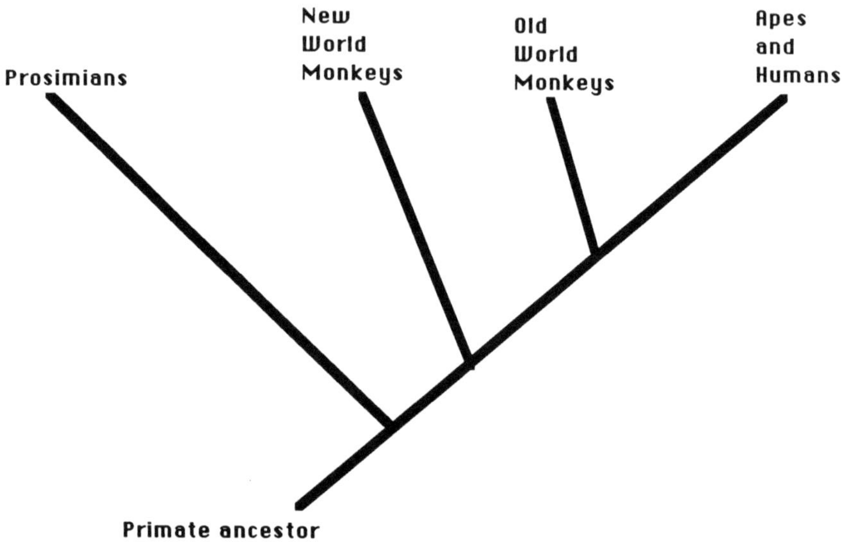

Figure 2.1. Phyletic relationship of prosimian and anthropoid primates.

to the anthropoid group.

In part, we chose to study laterality phenomena in prosimian species because of their taxonomic position in primate phylogeny. If behavioral lateralization occurs commonly in many extant species of prosimian, then it could be inferred that laterality is and has been a characteristic of the primate order, and likely of the common ancestor of primates. In addition, the greater simplicity of organization in prosimians (as compared to anthropoid species) in features such as brain structure and structure and use of hands, which, combined with a rich variety of adaptive types in prosimians, offers an opportunity to look at lateralized behavior freshly in a variety of primate species neurologically and behaviorally less complex than the anthropoids (Ward, 1991). The patterns of laterality that have emerged from our study to the present time, and the principles derived from them, compose the subject matter of this chapter.

Preference and Performance Measures

In the study of laterality as hand preference, the behavioral measures reflect a lateral bias in use and it is presumed that either hand could perform the task equally well. Reaching for food is a measure commonly employed. What is measured in preference tests is intrinsically different than what is measured in tests of performance. Hand preference measures reflect relative incidence of use. Hand performance measures are typically efficiency measures that make possible comparison of the manner of performance and/or outcome of use of the left or right hand. In this volume, this distinction is discussed as quantitative and qualitative handedness by Preilowski (see Chapter 5) and as manual preference and performance by Fragaszy and Adams-Curtis (see Chapter 3). These researchers have been principal contributors to the literature on performance laterality in anthropoid primates. As in the anthropoid work, most prosimian studies have been conducted to assess preference in hand use and these are discussed first in this chapter. The preliminary work that we have done to assess performance differences between the two hands in prosimians is treated subsequently.

Preference Research With Prosimians

The study of laterality in humans has included a large body of research into the functional specialization of the cerebral hemispheres, as well as research concerned more directly with the incidence of observable behaviors that have a consistent lateral bias. Although there are several important lines of inquiry into functional specialization in nonhuman primates (see for example, Hamilton & Vermeire, 1988, 1991; Heffner & Heffner, 1984, 1986; Hopkins & Morris, 1989; Hopkins, Washburn, & Rumbaugh, 1990; Petersen, Beecher, Zoloth, Moody, & Stebbins, 1978), most nonhuman primate studies have reported motoric lateral bias in a response system, for example that used to reach for food. Although we speculate in this chapter about functional specialization of neural systems, our research with prosimians has been principally concerned with the study of lateralization in response systems.

The study of lateralized behaviors in much of nonhuman primate research has been directed toward and interpreted in terms of the question of whether behavioral laterality as left or right directional bias occurs at the population level to the extent that right handedness does in the human population. We disagree that motoric lateralization should be conceptualized only in terms of a strong directional bias at the population level. We have rejected this homocentric orientation because lateralized behavior is a function of an individual organism and statements about population level lateralization are summaries derived from the study of individuals of a given taxonomic group. A species may be determined to have no lateralization *only* if no individual of that species is capable of demonstrating laterality in any response system. Fascination with the perceived strength of directionality in human population level laterality has seriously biased and limited nonhuman primate research and should be set aside in order to make way for more productive approaches. For example, our studies have consistently found important differences between strength and direction of lateral motoric biases. These results were foreshadowed by the important work of Collins (1969, 1975), who found that genetic selection for strength, but *not* direction, of lateralization was possible in mouse populations. In the prosimian research, different factors have been found related to strength of lateralization as compared to left or right directional bias.

Strength of Lateralization in State and Trait Factors

Factors that influence the strength of lateral bias in the behavior of an individual animal can be divided into two categories: *state factors* and *trait factors*. State factors are intrinsically more transient in their effects and are subject to changes in the test environment. The manipulation of posture to alter the strength of lateral hand bias in food reaching is a state effect that has been documented in several species. We chose to categorize this as a state effect because it involves temporary alteration of the postural state of the animal and also perhaps its general state of arousal. Trait factors in strength of lateral bias constitute relatively enduring characteristics of an individual animal and are ultimately based in the genetics of the individual. The heritability of strength of lateral bias in Collins' (1969, 1975) mice would be evidence of a trait factor. In the prosimian studies, the trait factors have been related to species membership and to age in ontogenetic development.

Whereas it is convenient to separate state and trait effects in order to think more clearly about the laterality phenomenon, it should be observed that they may be related, in that the same neural substrate may mediate both effects. For example, a member of a species with a typically more upright postural habit may be more strongly lateralized in hand use than another with quadrupedal habit (a trait effect), and yet both may vary in their manifested lateralization as a function of posture at the time of reaching (a state effect). The neural system involved in the species difference might well be the same one that mediates the more transient postural effects on reaching.

Posture as a State Effect

In the first systematic study of lateralized behavior in a prosimian species, the principal finding was that a colony of galagos (*Galago senegalensis*) housed in our laboratory had a population level left hand bias in executing their species-typical ballistic grab for a mealworm (*Tenebrio* larva) from a bipedal posture with vertical orientation of the head and trunk, but not when reaching to the floor for the same bait from a quadrupedal stance with horizontal orientation of the long axis of the body (Sanford, Guin, & Ward, 1984). Larson, Dodson, and Ward (1989) reported a replication and extension of the previous study with the same species. Ten galagos were tested in eight conditions designed to manipulate posture during reaching, visibility of the target mealworm, and angle of reaching. Only the manipulation of posture by high or low placement of the target mealworm affected the consistency of dominant hand use across the eight test conditions. Another replication of the target elevation test with a related species of galago (*Galago moholi*) failed to find an increase in strength of dominant hand use with target elevation (Dodson, Stafford, Forsythe, Seltzer & Ward, 1992). It is not possible to say whether this result represents a species difference or whether a modification to the test apparatus to accomodate the smaller stature of this species precluded measurement of a bipedal effect.

Generality of the postural effect across species was found in the study of five male sibling black and white ruffed lemurs (*Varecia variegata variegata*) housed on a moated island at a zoological park (Forsythe, Milliken, Stafford, & Ward, 1988). This Malagasy species is among the larger prosimians, heavy-bodied, and invariably quadrupedal in terrestrial locomotion and in much of arboreal movement. There is evidence of similarity to galagos in the manner of leaping from one vertical substrate to another (Pereira, Seeligson, & Macedonia, 1988). However, the postural orientation of the ruffed lemur is usually quite different than the galago, whose vertical clinging and leaping locomotion (Napier & Walker, 1967) is bipedally propelled by exceptionally long hind legs and who frequently regards the world from an upright posture, whether seated or standing.

The lemurs were evaluated for strength and direction of hand preference with a minimum of 100 responses in each of three conditions: 1) free foraging for seeds, etc., already present on the island; 2) retrieval of fruit pieces thrown across the moat to the surface of the island; and 3) retrieval of fruit thrown into the water at the edge of the island. Strength of lateral hand preference varied with condition and directly with the amount of postural preadjustment that accompanied reaching. In the free foraging condition the lemurs reached casually upon seeing a bit of food on the ground nearby, often from a seated or prone position, and rarely made a whole-body postural adjustment prior to reaching. In this condition only one lemur had a lateral bias and that was only 71% use of preferred hand. When fruit was thrown to the surface of the island, all lemurs manifested a dominant left hand with an average bias of 96.2%. In this condition the lemurs would watch the fruit as it sailed through the air and landed on the island, then locomote to the fruit and retrieve it. The condition in which fruit was thrown into the water of the moat was similar to the land presentation except that in this case the lemurs had to reach down into the water from a concrete ledge that required an extreme reach with one hand while propping on the sheer side of the ledge with the other. All lemurs had a left hand bias in this condition as well, and the average strength of hand preference was increased to 99.8%. In fact, in this condition only one of 515 reaching responses was executed with the nondominant hand. Not only did this study confirm the species generality of a postural effect on strength of lateral hand bias, it also showed that the effect was not exclusively related to bipedalism but could be demonstrated for other extreme postural adjustments that might accompany reaching. A similar effect of bipedal posture on strength of hand preference has been reported by Olson, Ellis, and Nadler (1990) for both gorillas and gibbons, although not for orangutans who defeated the bipedal reaching task by climbing the wire mesh and reaching from a hanging posture rather than a bipedal standing posture as did the gorillas and gibbons.

Arousal as a Possible Mediator of State Effects

From the time that the effect of posture on strength of lateral bias first came to our attention, we have sought to conceptualize the phenomenon and speculated about what type of neural mechanism might mediate this effect. Because our first observation was in relation to an elevation into bipedal posture by the galagos, we began to think in terms of what the relative instability of bipedal posture with an elevated center of gravity might require of the neural organization for posture and locomotion. Following Sherrington (1906), we believe that the nervous system has an essentially integrative function, and that many of its subsystems are globally interrelated, especially those action systems that support posture, locomotion, and other movement patterns.

Although reaching with the hand and arm is often spoken of as though the action of this member did not involve dynamic changes in other motoric systems, a simple demonstration suggested by Fortuyn (1982) will quickly convince the reader otherwise. One should assume a supine position on a solid surface such as the floor with all extremities relaxed and fully supported by the surface. The abdomen should be bare so that the navel can be observed. If one then lifts one arm slightly from the floor, the navel will be observed to shift toward the side of the lifted arm, so subtle and global are the distal effects of action systems. How much more interdependent these action systems must be when a bipedal organism is conducting its various interactions with the environment simultaneous to maintaining its ongoing antigravity dynamic with an elevated center of gravity and a maximum of two points of support.

Bipedalism, as a trait, constitutes a chronic condition of postural instability which we propose as a condition promoting increased strength of lateralization in action systems. The assumption of bipedal posture or of any other extreme posture that promotes instability of the organism in relation to the forces of gravity would impose a state effect with the same result. One of the most remarkable examples of this has been found in the marked increase of lateral hand bias in squirrel monkeys reaching up for food while balanced on their hind legs on a suspended rope as compared to reaching up while standing on a stable surface (James E. King, personal communication).

Although our evidence is largely in terms of postural effects, we have been tempted to speculate in terms of the broader concept of arousal. We have considered that the amplification of lateral bias with unstable postures may be mediated by a general arousal system. We were initially led to this thought because of the relationship of the vestibular system to the reticular activating system and the importance of these to postural reflexes. Talbott (1979) suggested that integration of overall postural mechanisms may be brought about by a system of visceral proprioception; the application of his neuraxial model to the phenomenon of the effect of bipedal posture on strength of lateralization has been explored elsewhere (Larson et al., 1989).

A neuraxial arousal system such as this could be specific to postural effects; however, we have been inspired by this activational model to speculate that other types of phenomena that have been deemed to have arousal effects might also intensify lateral biases. In pursuit of this question, studies should examine the influence of biological drives on strength of lateral bias. For example, the effect of degree of hunger on strength of hand preference should be studied. Most of our prosimian data on hand preference in simple food reaching have been collected at scheduled daily feeding times when the subjects may be presumed to be in at least a moderate state of hunger. We have followed this procedure because it requires only measurement of an established behavior and does not require that we train, and possibly influence, hand use. This procedure also makes possible the recording of a large number of food reaching responses in a period of time that is short relative to what would be required with satiated animals. About 80% of the prosimians that we have assessed have had a statistically significant lateral hand bias and most have been tested at scheduled feeding times.

One exception to this procedure was in the study of the black and white ruffed lemurs described above (Forsythe et al., 1988). In this study, data were collected at various times of day and not in any constant relationship to the regular feeding schedule. However, it was observed that in the discrete food presentation conditions the lemurs were very attentive to the researchers on the other side of the moat, watched the fruit intently as it approached the island through the air, and ran immediately to the spot where it fell. Every aspect of their behavior suggested a high level of interest and arousal. By contrast, the lemurs were relaxed while free foraging, with 76% of reaching responses made from a sitting or reclining posture. It is possible that arousal effects other than postural have contributed to laterality findings and these should be systematically examined in future studies.

The Effect of the Trait of Age on Strength of Hand Lateralization and Its Possible Relationship to the Mouth

It is common knowledge that human handedness increases in strength from infancy to adulthood, and there is at least one report of an increase in consistency of preferred hand usage with age in nonhuman anthropoids (Stafford, Milliken, & Ward, 1990). This ontogenetic effect has also been found in two prosimian studies with large and varied subject populations that permitted cross-sectional analysis of age effects.

The first study was a hand preference survey of 194 individuals of the *Lemur* genus resident at the Duke Primate Center in Durham, NC (Ward, Milliken, Dodson, Stafford, & Wallace, 1990). Six species of lemur and five subspecies of *Lemur fulvus* were assessed for strength and direction of hand bias in reaching for food during the scheduled daily feeding. Because these are breeding populations, the proportion of young animals was high; however, the range of ages represented was broad with all species groups having both adult and juvenile members. It was observed that juvenile lemurs had a strong tendency to pick up food with the mouth, rather than with the hand, and this behavior was added as a scoring category. When the use of the mouth in food retrieval by all juvenile lemurs (M = 29.4%) was compared to that of all adult lemurs (M = 11.8%), the difference in this tendency between the two age groups was substantial, $t(192) = 8.74$, $p < .001$. A similar analysis of strength of hand use between age groups with species pooled was not statistically significant because of the extreme variability between species (see Table 2.1). However, when adults of each species were compared to juveniles of the same species on strength of preferred hand use as shown in Table 2.1, the greater degree of lateral hand bias in the adults was apparent, $t(9) = 4.22$, $p < .002$.

Milliken, Stafford, Dodson, Pinger, and Ward (1991) found age effects in a study of a laboratory population of small-eared bushbabies (*Otolemur garnettii*). Individuals housed in small social groups of two to four animals were evaluated for lateralization in food reaching, duration of food holding, degree of manipulation of the food between the hands and mouth, and in a composite measure constructed from the three primary behavioral measures. Behaviors of the six adult males, nine adult females, and eight juvenile males were videotaped at the scheduled daily feeding. Both juvenile and adult bushbabies had a much greater tendency to pick up food from an uncovered bowl with the mouth than did the lemurs of the study cited above. Of the twenty-three bushbabies tested, seventeen had such a marked

Table 2.1. Strength of Hand Preference and Mouth Usage for All Lemurs as Percentage by Species and Age with Adults Including Ages 3-30 Years and Juveniles Ages 1-2 Years

	Hand		Mouth	
Species	Adult	Juvenile	Adult	Juvenile
L. rubriventer	99.8	82.0	23.7	49.5
L. coronatus	96.0	65.8	13.1	50.6
L. mongoz	87.2	77.6	7.2	18.1
L. f. collaris	85.5	70.1	8.6	26.0
L. catta	83.7	73.8	13.9	27.6
L. f. rufus	82.1	79.3	8.6	35.0
L. macaco	79.8	65.1	14.1	28.1
L. f. sanfordi	79.6	70.6	18.8	33.7
L. f. fulvus	75.1	69.7	4.8	20.5
L. f. albifrons	72.1	71.9	18.9	37.4

Note. Hand percentage for each species and age group is based on unimanual hand scores only and includes scores of left, right, and ambipreferent lemurs. Mouth use percentages calculated from hand and mouth food retrievals.

mouth preference that it was necessary to retest them using a covered bowl with a small opening (port) that required use of a hand to obtain the food. In the open bowl condition there was a trend toward greater use of the mouth by juveniles ($M = 76\%$) as compared to adults ($M = 60\%$) and none of the eight juvenile males preferred the hand to the mouth, whereas three of the six adult males used the hand preferentially. In the ported bowl condition, with food retrieval restricted to the hand, composite scores showed adults to be more consistent in use of preferred hand than juveniles ($t(21) = 2.25, p < 0.036$).

Throughout the course of the prosimian research we have become increasingly intrigued by the relationships between the mouth and hand. Infants initially take their nourishment exclusively by mouth from the mother, and during weaning solid food is taken by mouth first from the mother's hand. As the young become older, the mother increasingly declines to accomodate the young by sharing the food and the young prosimian then procures the available food directly, but still with the mouth. In lemurs the transition from mouth retrieval of food to primary use of hand lasts for about two years (Ward, et al., 1990). The slow development of hand use in food reaching in lemurs may reflect the maturational course of neuromuscular control for the hand. The sensory and motor development of the mouth so greatly precedes that of all other action systems in mammalian infants, and the hand displaces the mouth so slowly in lemur food retrieval, that we have been moved to speculate that the mouth may have an intrinsic lateral bias in use, and perhaps contribute to the ontogenetic development of hand preference.

Very early hand-mouth interaction in infant lemurs was also observed in a nonfeeding context during a pilot study (author's unpublished research). Seven infants of five different species were videotaped in daily 30 minute sessions at various times of day over a period of weeks. Lemur infants cling to their mothers continuously for the first month of life and slowly relinquish this security over the next two months. The infants ride with the long axis of their body and the tail wrapped crosswise around the mother's lower back and abdomen and cling to her with both feet and hands when she is moving about vigorously. When the mother is quieter, the infants spend a great deal of time in the oral and visual investigation of one of their own hands. Oral investigation includes licking, sucking, biting at, and what appears to be just mouthing. The behavior is more in the nature of "hand-mouth play" than purposeful grooming. During the first month or so of age, infants "played" with their hands in preference to investigating other body parts. As the infant began to leave the mother and move about independently, the pattern of grooming approximated that of the adults, which was preferential autogrooming of the trunk of the body with less attention given to the hands or feet. When the adults groomed an infant, grooming was centered on the face, head, and body, and allogrooming of the infant's hands or feet was rare. The observation of infant hand-mouth behavior in these prosimians suggests that their ontogenetic development is not essentially different from that of anthropoid infants in this respect. A special relationship between mouth and hand as a right hand bias in thumb-sucking by infant chimpanzees has been reported by Bard, Hopkins, and Fort (1990) and is further discussed in Chapter 11 of this volume.

Variations in Species Use of Mouth and Strength of Lateral Hand Bias

Species differences in the propensity to pick up food with the mouth rather than the hand have appeared consistently in the results of prosimian studies of hand laterality in feeding. Some species variation is illustrated in Table 2.1. Dodson et al. (1992) found galagos (*Galago moholi*) to have little inclination to pick up mealworms with the mouth from a shallow open cup: the mean use of mouth for 16 galagos was 7.7%, with a range of 0% to 29.2%. All galagos had lateral hand biases and in 14 the biases were greater than 80% use of preferred hand. In the same study, eight mouse lemurs (*Microcebus murinus*) had such a strong propensity to pick up quartered raisins with the mouth that the hands were never used to reach for food from an open cup. With a special apparatus that required it, all the mouse lemurs did use their hands to retrieve raisin quarters; however, their lateral hand bias was less strong than the galagos. Although seven of eight mouse lemurs had a lateral hand bias, the bias was greater than 80% in only two animals. The species difference in strength of lateral hand bias was

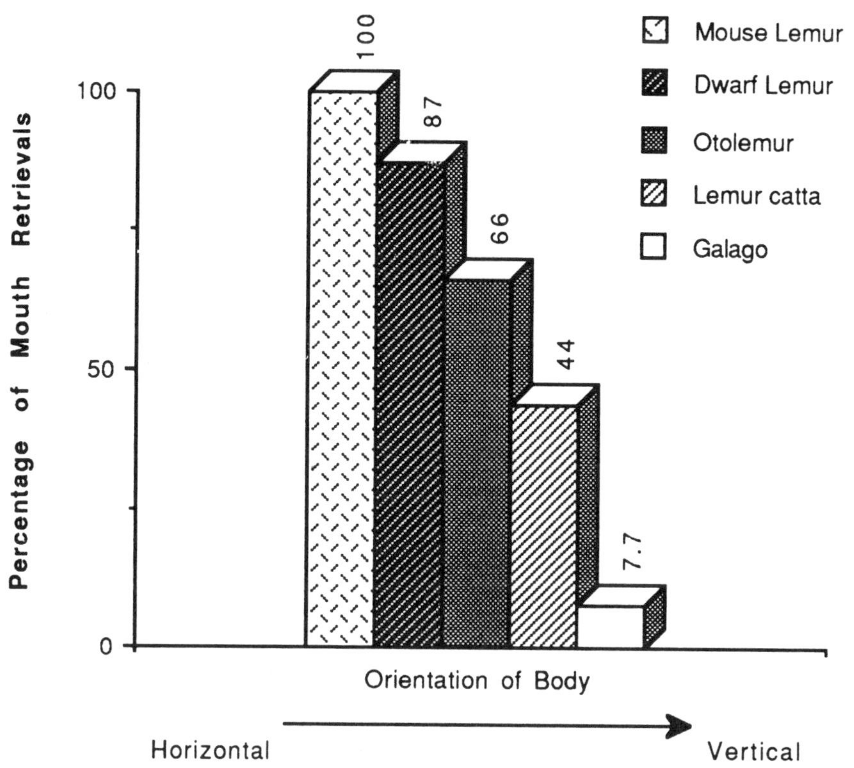

Figure 2.2. Tendencies of five prosimian species to retrieve food with the mouth as a function of typical orientation of the long axis of the body.

interpreted in terms of postural differences between the galago with its characteristic bipedal posture and the more quadrupedal mouse lemur.

The dwarf lemur (*Cheirogaleus medius*) has a quadrupedal habit with a horizontal orientation of the long body axis. This species also has a strong inclination to retrieve food with the mouth and it was again necessary to require reaching through an aperture in order to force the use of hands in reaching for fruit. In this study (authors' unpublished research) lateralization of hand use in feeding was evaluated in three measures: food reaching, duration of food holding, and manipulation of food. The eight dwarf lemurs were biased to the left on the three measures; but again the lateral biases, though statistically significant, were not strong in any measure.

In order to make a direct species comparison of the propensity to use the mouth in food retrieval, we compared the results from three separate studies for this review chapter. Percentage use of mouth in preference to hand was compared in dwarf lemur (authors' unpublished research), bushbaby (Milliken et al., 1991), and the ring-tailed lemur, *Lemur catta*, (Milliken, Forsythe, & Ward, 1989). The mouth use data were not treated in the

Milliken et al. (1989) report of multiple measures of hand use in the feeding behavior of a matriline of 13 ring-tailed lemurs but had been scored and were available for analysis here. Comparison found the dwarf lemur (M = 87%) to have a greater propensity to use the mouth than the bushbabies (M = 66%), $t(29) = 2.09, p < 0.045$, and the mouth bias of the bushbabies to be greater than that of the ring-tailed lemurs (M = 44%), $t(34) = 2.74, p < 0.01$). This species ordering by use of mouth is the inverse of that for strength of lateralization in hand use and for the degree to which the long axis of the body has a characteristic vertical orientation. (A summary of all species evaluated is given in Figure 2.2). More specifically, we would predict that quadrupedal primate species with horizontal orientation of the trunk and head are more likely to prefer mouth retrieval of food and also to be less lateralized in all food-related measures of hand use. To the extent that a species typically consumes food with an upright, and less stable, posture, the hand will be favored over the mouth in reaching and will have a stronger lateral bias. Thus the relationship between the hand and mouth and the development of lateral bias is seen both in ontogeny and phylogeny.

Variations in Strength of Laterality as a Function of the Trait of Species Membership

As comparative psychologists, the search for species differences in lateralized behavior, perhaps related to body structure or to adaptation to specific habitat, has been a matter of continuing interest to us. Species differences in strength of hand preferences were introduced above in relation to age effects and use of mouth and are documented in Table 2.1. The reader should note when inspecting this table that in every case the mean lateral hand bias of the young of each species is less than that of its adult counterpart. However, the range of strengths among these species is such that in many cases the average strength for the young of a more strongly lateralized species is greater than that of the adults of a less lateralized species. This constitutes powerful evidence that behavioral lateralization varies with species. We have not as yet investigated what characteristics of these lemur species and subspecies account for their variable strength of laterality.

A direct comparison of the effect of species typical posture as bipedal or quadrupedal on strength of lateral bias was made in a study of food reaching and whole-body turning in the galago (*Galago moholi*) and the mouse lemur (*Microcebus murinus*) by Dodson et al. (1992). In this study whole-body turning, defined as turning 180° to the right or left about the long axis of the body, was incorporated as a measure of laterality because it had previously been reported in *Galago senegalensis* by Larson et al. (1989). The same test procedure was used, and the galagos reached for food and executed the whole-body turn from a bipedal posture with trunk and head in vertical orientation. Mouse lemurs are quadrupedal in habit and these subjects were tested for laterality in reaching and turning in an apparatus that permitted them to respond from the species typical posture with the long axis in horizontal orientation. All sixteen galagos were lateralized in reaching and turning, whereas seven of eight mouse lemurs had a lateral hand bias in reaching for food and six of eight were lateralized in whole-body turning. More remarkably, the species comparison for strength of lateralization found galagos to be more strongly lateralized in both behavioral measures than the mouse lemurs: *Reaching*; galagos (M = 91.9%), mouse lemurs (M = 69.2%), $t(22) = 4.15, p<0.001$; *Whole-body turning*; galagos (M = 83.1%), mouse lemurs (M = 58.7%), $t(22) = 4.25, p<0.001$. The results of this study support our prediction that the extent to which a species is naturally given to a postural habit which is unstable, such as bipedal posture, behaviors are more strongly lateralized as compared to those of quadrupedal species that are more stable in posture. Olson, et al. (1990) also proposed that difference in species typical posture was the source of the variations in laterality that they found among gorillas, gibbons and orang-utans. The effect of species typical posture on strength of lateralized behavior gains greater significance when considered in the context of the "Postural Origins Theory" (MacNeilage, 1991) which proposes that the functional specializations of the human cerebral hemispheres for spatial abilities and language have their phylogenetic origins in the development of lateral bias from postural instability.

Further insight into the manner by which species typical behaviors may contribute to the strength and consistency of lateral bias was gained by study of a prosimian species specialized for bamboo foraging. The gentle lemur (*Hapalemur griseus griseus*) is indigenous to Madagascar and its diet there is composed of the shoots, stalks, and leaves of the bamboo plant. During the seasons when shoots are available, they are the preferred part of the bamboo but are low in abundance relative to the entire plant mass. The gentle lemur has a rather elaborate and species typical pattern of foraging for bamboo shoots (Milton, 1978); its arm and hand movements are swift and the pattern appears stereotypical in execution. The basic elements of this pattern are: 1) the bamboo stalk is grasped with one hand and pulled in toward the lemur; 2) the terminal shoot is then grasped in the mouth, gripped between the teeth, and pulled free from the stalk with an extensor arm movement; 3) the shoot, now held in the teeth, is grasped in one hand and rotated lengthwise (which has the effect of directing the tender basal portion of the shoot toward the mouth); and 4) the shoot is fed into the mouth by the hand. Initial observation of the simple food reaching behavior of 12 gentle lemurs housed in small family groups at the Duke Primate Center indicated that this species had the strongest lateral bias in reaching we had encountered.

A study was conducted which permitted a within subject comparison of laterality in simple reaching into a bowl for small pieces of fruit, vegetables, and primate chow, to the degree of lateral hand bias expressed in the performance of bamboo shoot eating (Stafford, Milliken & Ward, in press). Eight of twelve gentle lemurs were 100% lateralized in use of a dominant hand for simple food reaching; four others had preference scores of 95%, 90%, 88%, and 77%. A pattern of lateralized hand use was also found in the four components of bamboo shoot eating described above, with strength of lateral bias increasing from the movements that occurred earlier in the pattern to the the final feed-in of the shoot. Most important was the fact that every gentle lemur preferentially used the same hand for the turn and feed-in of the bamboo shoot as it used for simple reaching: eight had a dominant left hand and four a dominant right. The pattern of laterality that emerged from our study of bamboo feeding was as follows: Although three animals did have a hand bias in the initial stalk pull-in, this behavior was more variable and influenced by the lemur's physical position in relation to the stalks of bamboo. The counterforce movement that freed the shoot from the stalk was lateralized in six of the 12 lemurs and was negatively correlated with both the shoot turn (r = -0.70, p< 0.05) and the feed-in (r = -0.77, p< 0.01). The turn and feed-in movements were strongly related (r = 0.99, p< 0.01). Finally, simple food reaching was positively correlated with both shoot turning (r = 0.75, p< 0.01) and feed-in (r = 0.67; p< 0.05). We propose that the exceptionally strong lateral hand bias of the gentle lemur arises from or is promoted by the specialized and stereotypical pattern of bamboo feeding characteristic of this species.

We have observed a gentle lemur infant who had been removed from his mother at birth and we videotaped the infant's first encounters with bamboo mixed with the leaves of several species of deciduous trees (authors' unpublished research). From this young infant's preferential attention to the bamboo, as well as from additional study of the movement patterns of other infants in bamboo eating, we are persuaded that this is a species typical behavior and that the gentle lemurs are biologically prepared to engage in this strongly lateralized behavior. Many studies of laterality in nonhuman primates have used the simple food reaching behavior of captive primates as the basis for estimating lateral hand bias. This practice is occasionally called into question as not representing a "natural" behavior of the species. The gentle lemur study demonstrates that hand lateralization can be based in species typical behavior and that this lateral bias may be expressed in other less naturalistic test situations. It should also be noted that although the gentle lemur is quadrupedal in locomotion, its typical posture in feeding on bamboo shoots or reaching into a bowl of food is seated on the hindquarters with the long axis of the body in vertical orientation.

Evidence of "Mouthedness"

The study of gentle lemur bamboo shoot feeding (Stafford, et al., 1992) permitted the first quantification of a lateral mouth bias in feeding. We refer to this phenomenon as "mouthedness" in the interest of maintaining symmetry with the term handedness. The videotaped records of bamboo shoot eating permitted the careful analysis of the side of the mouth each gentle lemur fed-in each bamboo shoot. The correlation between side of mouth, left or right, and the preferred hand for feeding the shoot into the mouth was $r = 0.93$, $p < 0.001$. The fact that this was not an inevitable consequence of the structure of the hand-mouth interaction was shown by the small percentage of cross hand-mouth feeding and alternate feeding of the same shoot to both sides of the mouth with the same hand that was recorded for every gentle lemur. At this point in our studies it is, of course, not possible to know whether the lateral bias of the mouth arises from the hand preference, develops in concert with it, or whether a favored side of the mouth leads the development of directional bias in hand use. The appropriate ontogenetic studies are needed to clarify this question.

The lateral mouth bias in feeding is not limited to the *Hapalemur* species, however. A detailed analysis of *Lemur catta* feeding behavior from videotaped records has discovered a similar mouthedness in this species; that is, individual ring-tailed lemurs have a favored side of mouth for biting off pieces of primate chow (authors' unpublished research). We think that these findings are especially important because confirmation of a lateral mouth bias, in biting or chewing, that results in unequal tooth wear for extant prosimian species, opens the way for the study of asymmetric tooth wear in fossil specimens and, ultimately, for estimation of the antiquity of the primate laterality phenomenon. The use of the scanning electron microscope to study microwear of the teeth is a methodology that has been used to explore wear patterns as a function of food type in extant species (Kay, 1984; Kay & Hylander, 1978; Ryan, 1979) and is suitable for assessment of tooth wear patterns in extinct species (Covert & Kay, 1981).

Direction of Lateral Bias as Trait and State Factors

The Left Hand Bias of Prosimians

In the early years of our prosimian research we were impressed by the unusual amount of left hand preference recorded but had no ready explanation for the left directional bias. We subsequently became aware of MacNeilage's theory of the evolution of primate laterality that proposes a prosimian left hand preference for visually-guided prey capture in relation to right hemisphere specialization for visuospatial processing (MacNeilage, Studdert-Kennedy & Lindblom, 1987, 1988, 1991). It was, in fact, our interest in this intriguing theory that has inspired our ongoing study of prosimian laterality.

Table 2.2 summarizes our published research on the topic of prosimian lateral hand bias. For each study population, the individuals that were evaluated by use of z-score as left or right preferent or ambilateral are

Table 2.2. Lateral Hand Bias of the Individuals of Various Study Populations of Prosimians as Left (L), Right (R), or Ambilateral (A) and as a Function of Gender

SPECIES	MALE			FEMALE		
(Literature source)	L	A	R	L	A	R
Varecia v. variegata (Forsythe et al., 1988)	5	0	0	0	0	0
Lemur macaco (Forsythe & Ward, 1988)	13	0	7	7	1	5
Lemur catta (Milliken et al., 1989)	6	1	0	1	2	3
Galago senegalensis (Larson et al., 1989)	5	0	2	2	0	1
Lemur species (Ward et al., 1990)	59	21	36	32	17	29
Otolemur garnettii (Milliken et al., 1991)	4	7	3	1	2	6
Galago moholi (Dodson et al., 1992)	4	0	4	1	0	7
Microcebus murinus (Dodson et al., 1992)	1	1	3	2	0	1
TOTALS	97	30	55	46	22	52

separated by sex to facilitate discussion in the following paragraphs. Results of our initial study (Sanford et al., 1984) were not included in the table because individual handedness as a function of sex was not available. The performance of each individual in that study has been analyzed by z-scores ($p < 0.01$). The analysis found 14 to be left preferent, five right preferent, and six had no bias; however, it was not possible to identify the sex of each individual with certainty. In the subsequent studies, lateral hand bias has been reported for a total of 302 prosimians. When the data were collapsed over sex, there were 143 that preferred use of the left hand, 107 the right, and 52 had no hand bias. The fact that this distribution is different from chance (*chi square* = 41.73, $df = 2$, $p < 0.01$) confirms that lateral bias is a factor in prosimian hand use. A total of 250 of 302 prosimians, that is 82.8% of the population evaluated, had a lateral hand preference. Further, there were more left preferent (143) than right preferent prosimians (107), $z = 2.28$, $p < 0.05$.

These results do confirm MacNeilage's proposal of a population level *left* hand reaching bias in prosimians, as do two additional reports. In what may be the first report of a prosimian hand bias, Subramoniam (1957) observed a left hand preference in insect capture for eight slow lorises. More recently, Masataka (1989) has reported a population level left hand preference in ring-tailed lemurs (*Lemur catta*): 20 of 22 had a strong left hand bias in each of four behavioral measures. In addition, a within species comparison of strength of lateralization as percentage use of dominant hand in left and right preferent lemurs found left preferent lemurs ($M = 92\%$) to be more consistent in use of dominant hand than right biased lemurs ($M = 85\%$) of the same species, $t(9) = 2.33$, $p < 0.045$ (Ward et al., 1990). The greater strength of lateralization in left hand preferent lemurs constitutes additional evidence that use of the left hand is strongly developed in lemurs and perhaps somehow basic to their functional organization.

The Effect of Gender on Direction of Lateral Hand Bias

Analysis of our data in terms of gender has added a qualification to the straightforward interpretation of a left hand preference for prosimians as a group. When the male and female groups were considered separately, it was found that the distribution into left, right, and ambilateral was different from chance for both males (*chi square* = 37.79, $df = 2$, $p < 0.01$) and females (*chi square* = 19.13, $df = 2$, $p < 0.01$), with 83.5% of males and 81.7% of females laterally biased in hand use. Thus the incidence of lateral bias was not more characteristic of males than of females. However, when *direction* of lateral bias was considered for those male and female prosimians that had a lateral bias, it was found that there were more left (97) than right (55) hand preferent males ($z = 3.4$, $p < 0.001$) but females did not differ in left (46) and right (52) direction of preference ($z = 0.61$, $p > 0.05$). Rather than take these numbers at face value, which would imply that male prosimians have a population lateral bias to the left whereas females have no population bias, we propose, in light of the discussion that follows, the following theory: that the basic direction of hand lateralization of prosimians is to the left, but that there is an additional factor, perhaps related to age and sex, that tends to increase the incidence of right hand use, especially in the females of the population. The composition of our subject populations has not permitted an evaluation of sex and hand laterality in every study, but whenever a relationship between direction of lateralization and sex has occurred in a given study, the result has always been a left bias for males and/or a right bias for females.

Conclusions based on the seven studies cited in Table 2.2 should be considered to be somewhat tentative because our subject sample is biased. All major prosimian groups are not represented; some are represented by more individuals than others; and the individuals of species are in many cases members of the same family group or closely related. Despite these facts, which are a consequence of the characteristics of the prosimian populations available in the United States, we have been persuaded by the pattern of results that recur from one study to the next across several major prosimian groups that the proposed male left and female right tendencies are real.

Direct Comparison by Sex and Species

The small-eared bushbaby (*Otolemur garnettii*) of Africa and the ring-tailed lemur (*Lemur catta*) of Madagascar have long been separated geographically and have membership in different taxonomic families — *Lorisidae* and *Lemuridae*, respectively. They are similar in that they are among the midsize prosimians and are both principally frugivorous. They differ in that the lemur is more diurnal and terrestrial in habit and congregates in larger social groups (Bearder & Doyle, 1974; Jolly, 1966). Hand preference studies have been reported for the ring-tailed lemur (Milliken et al., 1989) and the small-eared bushbaby (Milliken et al., 1991) that are directly comparable in terms of the multiple measures of hand use in feeding that were scored. Although the matriline of lemurs was tested at a zoological park and the bushbabies were tested in our laboratory, both species were videotaped while feeding in social groups and hand use was scored for lateral bias in the measures of reaching for food, duration of food holding, and manipulation of food between the hand and the mouth. A z-score ($z = 2.54, p < 0.01$) was calculated for each animal in each of the three measures, with a negative score indicating a left hand bias and a positive score a right hand bias. In addition, a composite z-score was constructed by taking the mean of the three scores for each subject. Because the mean was calculated with respect to signed scores, the composite score represented consistency in directional bias across the three measures of hand use.

Species by sex two-way ANOVAs were performed on z-scores for each of the component measures and on the composite index. These comparisons have been reported in Milliken et al. (1991). Sex was a statistically significant main effect for reach, $F(1,23) = 7.08, p < 0.01$; duration, $F(1,23) = 8.92, p < 0.01$; manipulation, $F(1,23) = 10.28, p < 0.01$; and in the composite index comparison, $F(1,23) = 9.94, p < 0.01$. There were no significant main effects for species nor any species by sex interactions. The hand bias for the male group, composed of six bushbabies and six lemurs, was to the left in all four measures, whereas that for the female group of nine bushbabies and six lemurs was to the right. In this comparison, direction of lateral hand bias varied across species as a function of gender but was not different as a function of species membership. This result is all the more amazing given the considerable distance of the two species from each other in geological time, geographical space, and taxonomic classification, and implies that the ancestral condition was for left hand preference in males and right preference in females.

Evaluation of Direction of Lateral Hand Bias as a Function of Sex and Age in a Large Population of Lemurs

The bushbabies and lemurs directly compared in the Milliken et al. (1991) study were all fully adult. In a large *Lemur* group, Ward, et al. (1990) found a relationship between gender and direction of hand preference in food reaching that was somewhat modified in frequency of occurence in the population with increasing age as shown in Figure 2.3. This figure shows that approximately 50% of the males were left hand preferent regardless of age, whereas the proportion of right biased males increased with age. A similar, but less clear, profile may be seen for the females, wherein the level of left hand bias was lower than that of males and there was a dramatic increase in the proportion of right handed females in the group of females that were full adults as compared to the young adult female group. A food reaching study of black lemurs (*Lemur macaco*) found a decreased incidence of left hand preference in older animals, $r = -0.538, p < 0.001$ (Forsythe & Ward, 1988). In this report the results were not analyzed for sex effects. An analysis of the data performed for this review found the correlation of increased use of the right hand in older lemurs characteristic of the 12 females one year of age or older, $r = 0.69, p < 0.013$, but only a trend in the 20 male lemurs, $r = 0.43, p < 0.06$. There is also evidence for an association between sex and direction of lateralization in anthropoids. In a mixed population of hylobatids (siamangs, concolors, and lars) there was a relationship between between increasing age and propensity to retrieve food with the right hand in the group of nine females ($r = 0.90, p < 0.001$), but not in the group of ten males (Stafford et al., 1990). In manual skill tasks human females are more consistently lateralized in hand use to the right (Annett, 1972, 1978) and are more congruent in direction of lateralization in hand, eye, and foot dominance (Porac & Coren, 1981).

Two Views of the Effect of Sex on Direction of Lateral Bias

As yet we have no favored hypothesis for the greater incidence and persistence of a left hand preference in food reaching in male prosimians as compared to females, nor is the direction of theory development any more firmly set regarding the phenomenon of the increase of right hand use with age that may be more characteristic of females. It is possible that the leading ontogenetic development of the right cerebral hemisphere of the brain that has been reported for humans (Taylor, 1969; also, for review see Geschwind & Galaburda, 1985, 1987) is more pronounced in prosimians and is the source of the youthful bias toward use of the left hand which persists into adulthood for some animals. It is also possible that the greater incidence of left hand bias in males is related to an early slowing of left hemisphere development, perhaps implemented by hormonal effects as suggested by Geschwind and Galaburda (1985, 1987). In this same framework, it could be

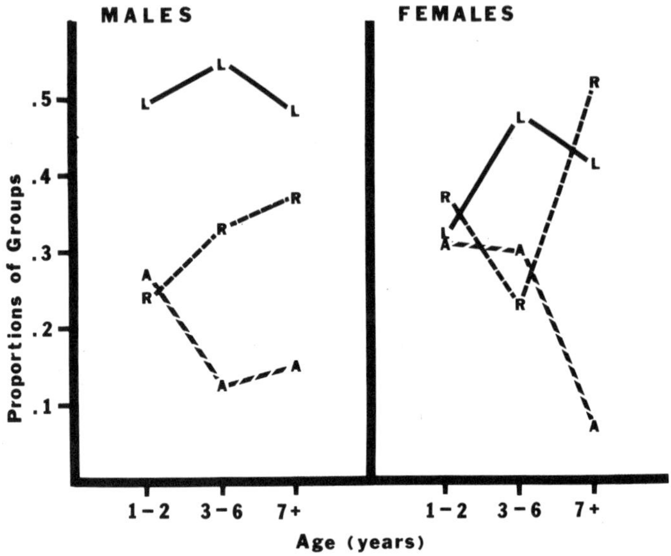

Figure 2.3. Proportion of male and female lemurs in each age group that were left-hand preferent (L), right-hand preferent (R), or ambipreferent (A). (Sample sizes for male age groups were 1-2 years, $n = 41$; 3-6 years, $n = 48$; and 7 years or older, $n = 27$. Sample sizes for female age groups were 1-2 years, $n = 19$; 3-6 years, $n = 30$; and 7 years or older, $n = 29$.)

postulated that the continued increase in incidence of right hand use with age reflects the continuing maturation of function in the left hemisphere. One difficulty with this scenario is that the increase in incidence of right hand bias in the lemur population as assessed by our cross-sectional study (Ward et al., 1990) has such a long time course (see Figure 2.3). It is possible that anatomical development and/or functional maturation of the cerebral hemispheres could extend over years and into full adulthood, but we know of no precedent for such an assumption.

We have also considered the possibility that differential arousal of the left or right hemisphere as a function of social precedence in feeding could influence direction of lateralization in hand use. Some lemur species are matriarchal in social structure (Jolly, 1966). Food is a limiting resource for adult females (Wrangham, 1980) and the precedence of adult females is nowhere more obvious than at a feeding site. Whether food is placed in a bowl or spread around on the ground, we have observed the "matriarch" and her adult female relatives, often daughters, with their infants take physical possession of the feeding area and threaten or energetically repel males or, in some cases, younger females who may approach their area of feeding too closely before they are satiated. This social dynamic varies considerably with species and with individual groups. In our laboratory, we have observed similar maternal precedence in the feeding sessions of family groups of small-eared bushbabies.

An observation of the feeding behavior of a pair of gibbons (*Hylobates syndactylus*) housed at the Memphis Zoological Park suggested to us the possibility that arousal state consequent to social precedence in feeding could influence the directional bias of hand use. The male and female had been co-residents at the zoo for 20 years and were fully mature animals. At feeding time, the food was placed in a pile on the floor and the female calmly took her place beside it and began to eat. The male walked around the outer edge of the pile, occasionally moving closer to grab a piece of food from the edge. This often brought a threatening look or gesture from the female and she would sometimes slap his hand as he reached for food. When she had finished and left the area, the male moved over to the food and ate without interruption. A total of 222 food reaches were scored for the female and she had a right hand bias of 93%. A total of 201 reaches were scored for the male and he was evaluated over all reaches as ambilateral. However, in scoring his reaches the presence or absence of the female was also recorded. Analysis of the lateral bias of his reaches in terms of her presence or absence showed him to have a left hand bias for food reaching when she was present (77.6%, $z = -4.8, p < 0.01$) but a tendency to favor the right hand in her absence (56.5%, $z = 1.4$, n.s.). Could his state of arousal as a function of her presence or absence at the feeding site have influenced his use of left or right hand by selective activation of the right or left cerebral hemispheres? Unfortunately, other zoological parks where we studied hylobatids do not feed their males and females together so it was not possible to make the same type of observation on other pairs and the result was not included in our report (Stafford et al., 1990).

Ethologically valid manipulation of arousal state in animals is difficult, but we have considered that the greater left hand bias in male lemurs, younger lemurs, and less assertive females could be influenced by right hemiphere activation in a situation of uncertainty, such as feeding under threat. By contrast, matriarchal females and other fully mature lemurs that may be more socially assertive in feeding and that have a greater incidence of right hand use could be responding under greater left hemisphere activation. This speculation tenders the hypothesis that a directional effect on lateral hand bias may be brought about by arousal or activation of the left or right cerebral hemisphere.

In summary, whereas the present evidence indicates an association between age, sex, and direction of lateral hand bias in feeding, more studies of prosimian and anthropoid primates are required to clarify these associations.

Population Level Directional Biases in Whole-body Turning

A left whole-body turning bias is the most directionally consistent laterality effect that has been identified in prosimian populations. Three species populations have been found to have a large proportion of members that execute a 180^0 turn about the long axis of their own body preferentially in a leftward direction: 8 of 10 galagos, *G. senegalensis* (Larson et al., 1989), 15 of 16 galagos, *G. moholi* (Dodson et al., in press), and 6 of 8 mouse lemurs, *M. murinus* (Dodson et al., in press). Although other prosimian species must be evaluated to establish the generality of the leftward turning bias, the magnitude of the result in these three studies suggests that the phenomenon may be characteristic of primates. Of the 34 prosimians tested, a total of 29 (85.3%) turned consistently to the left, two (5.9%) turned consistently to the right, and three (8.8%) had no preferred direction of turning.

It may be that these biases in turning are related to the relative dopamine activity in the basal ganglia of the two cerebral hemispheres. Glick and his colleagues have clearly demonstrated the relationship between direction of rotation of rats and the level of dopamine activity in the basal ganglia of the forebrain (Glick, 1983; Glick & Shapiro, 1985). In the rat there is a small, but statistically significant, population level rotational bias (54.8%) toward the right (Glick & Ross, 1981). However, children between the ages of 7 and 9 were found to have a left rotational bias whereas adults had turning biases related to their cerebral dominance (see Chapter 14, this volume).

The identification of a reliable left whole-body turning bias in prosimians will be useful in future studies of neural mechanisms of laterality. At present, it is not possible to guess why the turning bias in prosimians is predominantly to the left. It may be that the lateral turning bias is a component of postural lateralization. In the prosimian studies discussed above, a correlation between hand use and whole-body turning was only found for one species, *G. moholi*, and in this case strength of lateral hand bias was correlated with strength of turning but there was no correlation between direction of turning and direction of hand bias (Dodson, et al., 1992). A similar dissociation of direction of lateral bias in manual, locomotor, and postural measures has been reported for human subjects (Seltzer, Forsythe & Ward, 1990).

It is certain that laterality is not a unitary phenomenon; it is an oversimplification to characterize an individual as lateralized to the left or right. Statements about lateral bias must always be qualified in terms of specific response systems. Geschwind (1985) suggested two levels of lateralization in the nervous system, one involving the cerebral hemispheres and one organized at the neuraxial level. The evidence from studies of multiple response systems confirms that lateralization exists at different levels of neural organization and that these levels are independent with respect to direction of bias.

Performance Research with Prosimians

There is broadly held an implicit assumption that the right population level bias in human handedness arises from the greater inherent capacity of the right hand for fine manipulation, particularly of tools and weapons. In fact, it has been advanced that the use of tools was the driving force in human evolution for the development of handedness (Dart, 1949; Frost, 1980; Toth, 1985). This emphasis on the superior performance capacity of the dominant hand in humans has colored assumptions about nonhuman primate laterality, so much so that it is often expressed that laterality can not be "the same thing" in nonhuman primates unless greater performance efficiency can be demonstrated for the preferred hand. The search for differences between preferred and nonpreferred hand in performance efficiency in prosimians is in an early stage of development but some differences have been identified.

Efficiency of Food Reaching in Galagos

One of the greatest difficulties in obtaining a comparison of the efficiency of the preferred and nonpreferred hands in a strongly lateralized prosimian is to induce the animal to use the nonpreferred hand on enough occasions to furnish a response sample comparable to that easily obtainable from the preferred hand. We require that our subjects respond freely and naturally; these species would not respond favorably to physical constraint by force such as binding an arm. Carl F. Larson, one of our collaborators, had the insight that manipulation of the structure of the environment could influence the animals to use the nonpreferred hand. He designed an apparatus that was successful in accomplishing the desired result. The apparatus was a rectangular, clear Plexiglas box (18 in. long x 10 in. high x 7 in. wide) with three food cups inset and mounted vertically at one end of the box and the start box located at the other end. The galagos were tested in discrete trials. At the beginning of a trial the galago left the start box and proceeded to the food cup area. In this area the three food cups were equidistant from the galago with one directly in front of the subject, one 90^0 to the right, and one 90^0 to the left. All cups were mounted in the wall 8 in. above the floor of the apparatus, a height that required a bipedal posture for good visibility of the mealworm in the cup and for ease of reaching. The structure of this testing situation was such that the galagos used the preferred hand for reaching directly forward into the center cup but reached across the body toward the cups placed laterally: the right hand to the left cup and the left hand to the right cup. The use of this test apparatus made possible the scoring of a large number of reaching responses with both the preferred and nonpreferred hands without constraining or distressing the animal.

Dodson, LeDuc, Larson, and Ward (1992) tested eight galagos in this apparatus and evaluated efficiency in reaching. All subjects had been tested previously in simple food reaching from a single cup and preferred hand determined for each. Four subjects were left hand preferent and four preferred the right hand. Behavioral testing was recorded on videotape and both time required to execute the reach and success in grasping the mealworm on the first attempt were used to assess efficiency. A total of 240 reaching responses, 80 responses to each of the three food cups, was recorded for each galago. All subjects maintained use of the preferred hand for reaching forward into the center cup (M = 83.1%), but employed the right hand to grab mealworms from the left cup (M = 96.7%) and the left hand to reach toward the right cup (M = 94.2%).

Results of the two response measures were analyzed within subjects, both in terms of preferred hand versus nonpreferred hand and right hand versus the left. In the success measure, neither the comparison of left versus right nor that of preferred versus nonpreferred was statistically significant. Evaluation of reach duration found a difference between the time required to execute reaches with the left and right hands. An average time of .1586 seconds (*standard deviation* = .0357 sec) was required by the left hand as compared to .1694 seconds (*standard deviation* = .0343 sec) for the reach of the right hand, $t(7) = 2.74, p < 0.029$. Although the absolute difference in time required for left and right reaches was small, the variability of scores for each animal was also very slight and seven of eight galagos required less time to reach with the left hand. This result supports the hypothesis of MacNeilage et al. (1987; 1988; 1991) that the evolution of primate lateralization includes a phylogenetically early prosimian specialization for visually guided left hand prey capture.

Hand Preference and Efficiency in the Aye-Aye

Studies of the patterns of hand use in the aye-aye (*Daubentonia madagascariensis*) have posed the most interesting research challenge of all the prosimian research. The aye-aye is a most specialized prosimian, as evidenced by the bizarre anatomical characters that are used to forage for insect larvae in wooden galleries. These features include a slender third finger that is used to lightly tap and, with the large ears, localize the galleries of larvae. Continuously growing incisors are used to chew into the wood. The construction of the gracile third finger also facilitates probing into the excavated cavity, and removing the larvae. The aye-aye has a unique ability in comparison to other prosimians in that the third finger exhibits movement that is independent of the other fingers. Independent movement was observed when the aye-aye was tapping and when extracting grubs (Milliken, Ward & Erickson, 1991).

The two apparatuses used to evaluate digit function when inside wood cavities were also used to study preference and performance measures of lateralization (Milliken, 1991). The apparatuses differed from each other in the orientation of the cavities relative to the entry of the finger. The vertical cavity apparatus (VCA) consisted of vertical cavities that were 6, 7.5, 9, and 10 cm deep, and the horizontal cavity apparatus (HCA) had cavities that were located off a runway and oriented in straight, obtuse, right, and acute angles. The two apparatuses also differed in the complexity of task requirements as evidenced by the success rates in each condition. Performance, as indicated by success in retrieving larvae, was better in the VCA than the HCA, indicating that the VCA was a simpler task than the HCA. Measurement of lateral preference in the VCA revealed that the aye-ayes executed more probes into the shallow cavities (6, 7.5, and 9 cm) than the deepest cavity, and that there was a right preference in the shallow cavities (6 and 7.5 cm), but not the deep cavities (9 and 10 cm). There was, however, no evidence of a performance advantage, i.e., superior efficiency, in the less demanding VCA task. By contrast, there was evidence of a right hand efficiency advantage in the more difficult HCA task but no overall lateral hand preference. Thus, hand preference and hand efficiency interacted with task complexity: a right preference was measured by the less complex VCA task whereas right finger efficiency was detected by the more complex HCA task. The fact that the lateral biases of the aye-aye favor the right finger, rather than the left, is likely related to the development of independent digit control in this species and the manipulative nature of the foraging task. The right bias in this case tends to support the evolutionary scenario of MacNeilage, et al. (1987, 1988, 1991). It may be that the essential left preference of prosimians was modified as species altered their feeding niche to include foods that required fine motor control to remove edible portions from protective coverings. Manual division of labor may have progressed from left hand bias for insect capture with right hand postural support to left hand prehension with right hand manipulation.

Preference and Performance in the Evolution of Handedness

We are persuaded that lateral hand use preference, rather than any inherently greater capability of one hand, was the driving force in the evolution of handedness. It is also likely that the ontogenetic development of lateralized hand use in primates, human and nonhuman, is led by preference rather than a potential difference in level of skill between the two hands. Primates use one hand in lieu of the other as a pre-existing lateral bias or preference, perhaps arising in part from the postural biases in the global neuromuscular organization of the individual. To the extent that the preferred hand is repeatedly employed in a task as a consequence of strength of lateral preference, and to the extent that the task is a demanding one requiring a high level of neuromuscular coordination and extensive practice toward its perfection, differences between the two hands in performance skill may develop as the preferred hand is repeatedly practiced in that task and the nonpreferred hand is not. Briefly stated, hand preference determines which hand is used and the practiced hand may develop superior performance ability. When the literature of skill differences between dominant and nondominant hands is examined across primate species, including humans, surprisingly few differences are found and these are of relatively small magnitude.

The galagos were found to complete a reach for food in less time with the left hand. This performance effect was, although statistically significant, of such small magnitude that it is difficult to imagine that the level of these differences in hand skill would have adaptive significance in the survival of the galago in its natural habitat. It should also be mentioned that the one galago that was faster with the right hand than the left was a right preferent animal that had been trained in a previous experiment for hundreds of trials on a much more demanding hand-eye coordination task which required the galago to grab a bit of food from a turntable with the speed gradually increased over sessions. Most of this animal's ballistic grabs toward the turntable were executed with the right hand and it is altogether possible that the greater reaching speed of the right hand in the subsequent study was a consequence of the extensive practice with the turntable. The other seven galagos had no such previous experience and all had shorter reach durations with the left hand. Although there is a slight advantage for the unpracticed left hand of the galago in time required to capture prey, the right hand apparently may exceed that advantage with extended practice.

In studies of the aye-aye both right hand preference and right hand efficiency were measured but again the magnitude of the efficiency effects were not great. Chapters 3 and 5 of this volume present some of the most thoughtfully conceived and carefully executed studies of preference and performance in nonhuman anthropoid primates that we are aware of. The study of crab-eating macaques by Fragaszy and Adams-Curtis and the research of Preilowski with rhesus macaques also identified only small magnitude performance differences. In the human literature there is a need for more studies that compare the actual performance of dominant and nondominant hands. Although there are some studies of relatively simple skills, we are impressed that the absolute magnitude of the differences obtained between right and left, or dominant and nondominant, hands are not great. It would be interesting to know how much the dominant and nondominant hands differ in the asymtotic performance of complex tasks such as writing or tool use both before and after the nondominant hand was extensively practiced.

The lateral hand specialization of humans for skilled acts has been regarded as the yardstick by which the performance differences of the hands of nonhuman primates should be judged. And yet, the magnitude of performance differences between the dominant and nondominant hands of humans may not be as great as imagined, and could easily be attributed to differential practice as a consequence of hand preference. In fact, the concept of a markedly greater manual dexterity of the dominant hand in humans seems shaped more by a language mediated subjective evaluation based in hand preference than in any dramatic differences in the motoric capacities of the two hands. For example, in a study conducted by Seltzer, Forsythe, and Ward (1990), a population of human subjects asked to evaluate their hand preference in motoric activities using a 60-item questionnaire was strongly skewed toward right handedness: 95 of 100 subjects were categorized as right handed, five as ambilateral, and none were left handed. This same population of subjects evaluated in reaching performance showed a large shift toward left hand preference: 68 were right handed, 27 ambilateral, and five were left handed. Could it be that humans amplify the importance of the dominant hand and exaggerate its skill level because they have verbally categorized themselves as right or left handed? Or could it be that the activities of the dominant hand are more closely attended and more clearly recorded in memory because of that hand's association with the dominant cerebral hemisphere?

Postscript

The incongruence of the occurrence of left-right behavioral asymmetry in organisms that are anatomically symmetrical across their bilateral plane of organization is the essence of the laterality enigma. We expect that form and function will be congruent, and it is this assumption that makes the laterality puzzle so endlessly fascinating. Is there more, and more meaningful, anatomical asymmetry than we have been able to find or is the behavioral asymmetry less, and less important, than we have believed?

All living things are constructed and move about not in this one lateral plane but in three planes of organization. The laterality mystery is heightened by consideration of structure-function relationships in the other two planes of organization: dorsal-ventral and anterior-posterior. For most species, and certainly all vertebrates, there is a clear congruence of asymmetries in structure and function within the dorsal-ventral and anterior-posterior planes. In both structure and function, our backs are not like our fronts and our tops are not like our bottoms. Neither do they engage in the same activities. This makes the distinction easy to perceive and to remember, unlike our right and left sides which are frequently confused.

It is likely that both the structural differences and the behavioral specializations in the dorsal-ventral plane and the anterior-posterior plane were shaped together by adaptation to the conditions of existence. For most species the arrangement of the dorsal-ventral plane suggests that its structure and function was modified in the interest of maintaining a stable directional orientation to the world in relation to the force of gravity. Many adaptive advantages could be cited for the development of an asymmetrical anterior-posterior plane: directional course of food processing, and directional locomotion with the anterior pole leading, and with preferential anterior placement of the sense receptors used to locate distal stimuli. For most species the anterior-posterior asymmetry directionally organizes and orients an organism, particularly the moving organism, within the environment. Given the adaptive significance of appropriate and reliable orientation to the horizontal and vertical dimensions of the spatial environment, it is reasonable that both the structures and behaviors attached to the dorsal-ventral and anterior-posterior dimensions of an organism would develop congruent asymmetries in accomodation of these important functions.

Now, as we return to consideration of the bilateral dimension, it seems even more difficult to reconcile the basic symmetry of external structure with the asymmetry of lateral bias in behavior. If adaptive advantage was gained along the other two dimensions of organization by structural and functional congruence in asymmetry, why was there no selection pressure for a similar development across the bilateral plane? Why is there obvious behavioral asymmetry but no obvious structural asymmetry? We speculate that there may have been two selection pressures working at cross purposes: one served by external anatomical symmetry in the bilateral plane and a second served by behavioral asymmetry. The symmetrical structure of the bilateral organism may have been developed and maintained by the advantages gained from broad and balanced sensory monitoring of the surrounding environment, whereas the adaptive advantages of asymmetry are realized in directional preference, a lateral bias in action. Lateral behavioral bias facilitates the orientation and execution of responses, promotes a division of labor between the two sides, and thereby enhances the efficiency of an organism's interactions with the environment. Such behavioral asymmetries must depend upon neural asymmetries at some level of neural organization, and yet here too we find a basic structural plan of symmetry with some, as yet to be fully determined, asymmetrical modifications. We suggest that interactive tensions between the conflicting selection pressures for symmetry and asymmetry in the lateral plane may have contributed to the evolutionary development of the forebrain and the emergence and elaboration of cognitive abilities.

Acknowledgements

The research reviewed herein was supported in part by National Science Foundation Grant BNS-8707754 to Jeannette P. Ward. Support for the research and for the preparation of this report was received from a Centers of Excellence grant awarded to the Department of Psychology, Memphis State University, by the State of Tennessee. The authors would like to express gratitude to the many graduate and undergraduate student collaborators who have contributed to the collection and analysis of the research data reported here.

References

Annett, M. (1972). The distribution of manual asymmetry. *British Journal of Psychology*, *63*, 343-358.

Annett, M. (1978). Genetic and nongenetic influences on handedness. *Behavior Genetics*, *8*, 227-249.

Bard, K.A., Hopkins, W.D., & Fort, C.L. (1990). Lateral bias in infant chimpanzees (*Pan troglodytes*). *Journal of Comparative Psychology*, *104*, 309-321.

Bearder, S.K., & Doyle, G.A. (1974). Ecology of bushbabies, *G. senegalensis* and *G. crassicaudatus*, with some notes about their behaviour in the field. In R.D. Martin, G.A. Doyle, & A. C. Walker (Eds.), *Prosimian behaviour* (pp.109-130). London: Duckworth Press.

Charles-Dominique, P. (1977). *Ecology and behaviour of nocturnal primates*. New York: Columbia University Press.

Charles-Dominique, P., Cooper, H.M., Hladik, A., Hladik, C.M., Pages, E., Pariente, G.F., Petter-Rousseaux, A., Petter, J.J., & Schilling, A. (Eds.) (1980). *Nocturnal malagasy primates*. New York: Academic Press.

Collins, R.L. (1969). On the inheritance of handedness. II. Selection for sinistrality in mice. *Journal of Heredity*, *60*, 117-119.

Collins, R.L. (1975). When left-handed mice live in right-handed worlds. *Science*, *187*, 181-184.

Covert, H.H., & Kay, R.F. (1981). Dental microwear and diet: Implications for determining the feeding behaviors of extinct primates, with a comment on the dietary pattern of *Sivapithecus*. *American Journal of Physical Anthropology*, *55*, 331-336.

Dart, R.A. (1949). The predatory implemental technique of Australopithecus. *American Journal of Physical Anthropology*, *7*, 1-38.

Dodson, D.L., Leduc, P., Larson, C.F., & Ward, J.P. (1992). Left hand advantage in reaching in the galago (*Galago moholi*). Manuscript submitted for publication.

Dodson, D.L., Stafford, D.K., Forsythe, C., Seltzer, C.P. & Ward, J.P. (1992). Laterality in quadrapedal and bipedal prosimians: Reach and whole-body turn in the mouse lemur (*Microcebus murinus*) and the galago (*Galago moholi*). *American Journal of Primatology*, *26*, 191-202.

Doyle, G.A. (1974). Behavior of prosimians. In A.M. Schrier & F. Stollnitz (Eds.), *Behavior of nonhuman primates* (pp.155-353). New York: Academic Press.

Doyle, G.A., & Martin, R.D. (1979). *The study of prosimian behavior*. New York: Academic Press.

Forsythe, C., Milliken, G.W., Stafford, D.K., & Ward, J.P. (1988). Posturally related variations in the hand preferences of the ruffed lemur (*Varecia variegata variegata*). *Journal of Comparative Psychology, 102*, 248-250.

Forsythe, C., & Ward, J.P. (1988). Black lemur (*Lemur macaco*) hand preference in food reaching. *Primates, 29*, 369-374.

Fortuyn, J.D. (1982). On the organization of action: A new synthesis. *Human Neurobiology, 1*, 145-151.

Frost, T.F. (1980). Tool behavior and the origins of laterality. *Journal of Human Evolution, 9*, 447-459.

Geschwind, N. (1985). Implications for evolution, genetics, and clinical syndromes. In S.D.Glick (Ed.), *Cerebral lateralization in nonhuman species* (pp.247-278). Orlando, Florida: Academic Press.

Geschwind, N., & Galaburda, A.M. (1985). Cerebral lateralization. Biological mechanisms, associations, and pathology: I. A hypothesis and a program for research. *Archives of Neurology, 42*, 428-459.

Geschwind, N., & Galaburda, A.M. (1987). *Cerebral lateralization: Biological mechanisms, associations, and pathology*. Cambridge: The MIT Press.

Glick, S.D. (1983). Cerebral lateralization in the rat and tentative extrapolations to man. In M.S.Myslobodsky (Ed.), *Hemisyndromes: Psychobiology, neurology, psychiatry* (pp.7-26). Orlando, Florida: Academic Press.

Glick, S.D., & Ross, D.A. (1981). Lateralized effects of bilateral frontal cortex lesions in rats. *Brain Research, 210*, 379-382.

Glick, S.D., & Shapiro, R.M. (1985). Functional and neurochemical mechanisms of cerebral lateralization in rats. In S.D.Glick (Ed.), *Cerebral lateralization in nonhuman species* (pp.157-183). Orlando, Florida: Academic Press.

Hamilton, C.R., & Vermeire, B.A. (1988). Complementary hemispheric specialization in monkeys. *Science, 242*, 1691-1694.

Hamilton, C.R., & Vermeire, B.A. (1991). Functional lateralization in monkeys. In F.L.Kitterle (Ed.) *Cerebral laterality: Theory and research* (pp.19-34). Hillsdale, NJ: Lawrence Erlbaum.

Heffner, H.E., & Heffner, R.S. (1984). Temporal lobe lesions and perception of species-specific vocalizations by macaques. *Science, 226*, 75-76.

Heffner, H.E., & Heffner, R.S. (1986). Effect of unilateral and bilateral auditory cortex lesions on the discriminations of Japanese macaques. *Journal of Neurophysiology, 56*, 683-701.

Hopkins, W.D., & Morris, R.D. (1989). Laterality for visual-spatial processing in two language-trained chimpanzees (*Pan troglodytes*). *Behavioral Neuroscience, 103*, 227-234.

Hopkins, W.D., Washburn, D.A., & Rumbaugh, D.M. (1990). Processing of form stimuli presented unilaterally in humans, chimpanzees (*Pan troglodytes*) and monkeys (*Macaca mulatta*). *Behavioral Neuroscience, 104*, 577-582.

Jolly, A. (1966). *Lemur behavior*. Chicago: Chicago University Press.

Jouffroy, F.K. (1962). La musculature des membres chez les Lemuriens de Madagascar. Etude descriptive et comparative. *Mammalia, 55*, 1-326.

Jouffroy, F.K. (1975). Osteology and myology of the lemuriform postcranial skeleton. In I. Tattersall & R.W. Sussman (Eds.), *Lemur biology* (pp.149-192). New York: Plenum.

Kay, R.F. (1984). On the use of anatomical features to infer foraging behavior in extinct primates. In P.S. Rodman & J.G.H. Cant (Eds.) *Adaptations for foraging in nonhuman primates* (pp.21-53). New York: Columbia University Press.

Kay, R.F. & Hylander, W.L. (1978). The dental structure of mammalian folivores with special reference to Primates and Phalangeroidea (Marsupialia). In G.G. Montgomery (Ed.), *The ecology of arboreal folivores* (pp.173-196). Washington, D.C.: Smithsonian Institution Press.

Larson, C.F., Dodson, D., & Ward, J.P. (1989). Hand preferences and whole-body turning biases of lesser bushbabies (*Galago senegalensis*). *Brain, Behavior and Evolution, 33*, 261-267.

MacNeilage, P.F. (1991). The postural origins theory of neurobiological asymmetries in primates. In N. Krasnegor, D. Rumbuagh, M.G. Studdert-Kennedy, & R. Schiefelbusch (Eds.), *The biological foundations of language development* (pp. 165-188). Hillsdale, NJ: Lawrence Erlbaum.

MacNeilage, P.F., Studdert-Kennedy, M.G., & Lindblom, B. (1987). Primate handedness reconsidered. *Behavioral and Brain Sciences, 10*, 247-303.

MacNeilage, P.F., Studdert-Kennedy, M.G., & Lindblom, B. (1988). Primate handedness: A foot in the door. *Behavioral and Brain Sciences, 11*, 737-744.

MacNeilage, P.F., Studdert-Kennedy, M.G., & Lindblom, B. (1991). Primate handedness: The other theory, the other hand and the other attitude. *Behavioral and Brain Sciences*, 14, 344-347.

Martin, R.D. (1990). *Primate origins and evolution.* Princeton, NJ: Princeton University Press.

Martin, R.D., Doyle, G.A., & Walker, A.C. (1974). *Prosimian biology.* London: Duckworth.

Masataka, N. (1989). Population-level asymmetry of hand preference in lemurs. *Behaviour, 110*, 244-247.

Milliken, G.W. (1991). Analysis of hand use in foraging by the aye-aye (*Daubentonia madagascariensis*). Unpublished doctoral dissertation, Memphis State University, Memphis, TN.

Milliken, G.W., Forsythe, C., & Ward, J.P. (1989). Multiple measures of hand-use lateralization in the ring-tailed lemur (*Lemur catta*). *Journal of Comparative Psychology, 103*, 262-268.

Milliken, G.W., Stafford, D.K., Dodson, D.L., Pinger, C.D., & Ward, J.P. (1991). Analyses of feeding lateralization in the small-eared bushbaby (*Otolemur garnettii*): A comparison with the ring-tailed lemur (*Lemur catta*). *Journal of Comparative Psychology, 105*, 274-285.

Milliken, G.W., Ward, J.P., & Erickson, C.J. (1991). Independent digit control in foraging by the Aye-Aye (*Daubentonia madagascariensis*). *Folia Primatologica, 56*, 219-224.

Milton, K. (1978). Role of the upper canine and P2 in increasing the harvesting efficiency of *Hapalemur griseus* Link, 1795. *Journal of Mammology, 59*, 188-190.

Napier, J.R., & Walker, A.C. (1967). Vertical clinging and leaping: A newly recognized locomotor behavior among primates. *Folia Primatologica, 6*, 180-203.

Olson, D.A., Ellis, J.E., & Nadler, R.D. (1990). Hand preferences in captive gorillas, orang-utans and gibbons. *American Journal of Primatology, 20*, 83-94.

Pereira, M.E., Seeligson, M. L., & Macedonia, J.M. (1988). The behavioral repertoire of the black-and-white ruffed lemur, *Varecia variegata variegata* (Primates: Lemuridae). *Folia Primatologica, 51*, 1-32.

Petersen, M.R., Beecher, M.D., Zoloth, S.R., Moody, D.B., & Stebbins, W.C. (1978). Neural lateralization of species-specific vocalizations by Japanese macaques (*Macaca fuscata*). *Science, 202*, 324-327.

Porac, C., & Coren, S. (1981). *Lateral preferences and human behavior*. New York: Springer-Verlag.

Ryan, A.S. (1979). Wear striation direction on primate teeth: A scanning electron microscope examination. *American Journal of Physical Anthropology, 50*, 155-168.

Sanford, C., Guin, K., & Ward, J.P. (1984). Posture and laterality in the bushbaby (*Galago senegalensis*). *Brain, Behavior and Evolution, 25*, 217-224.

Seltzer, C., Forsythe, C., & Ward, J.P. (1990). Multiple measures of motor lateralization in human primates (*Homo sapiens*). *Journal of Comparative Psychology, 104*, 159-166.

Sherrington, C.S. (1906). *The integrative action of the nervous system*. New Haven: Yale University Press.

Stafford, D.K., Milliken, G.W., & Ward, J.P. (1990). Lateral bias in feeding and brachiation in Hylobates. *Primates, 31*, 407-414.

Stafford, D.K., Milliken, G.W., & Ward, J.P. (in press). Patterns of hand and mouth lateral biases in bamboo leaf shoot feeding and simple food reaching in the gentle lemur (*Hapalemur griseus*). *American Journal of Primatology.*

Subramoniam, S. (1957). Some observations on the habit of the slender loris, *Loris tardigradus* (Linnaeus). *Journal of Bombay Natural History Society, 54*, 386-398.

Talbott, R.E. (1979). Ferrier, the synergy concept, and the study of posture and movement. In R.E. Talbott & D.R. Humphrey (Eds.), *Posture and movement* (pp.1-12). New York: Raven Press.

Tattersall, I. (1982). *The primates of madagascar*. New York: Columbia University Press.

Taylor, D.C. (1969). Differential rates of cerebral maturation between sexes and between hemispheres. *Lancet, 2*, 140-142.

Toth, N. (1985). Archaeological evidence for preferential right-handedness in the lower and middle Pleistocene, and its possible implications. *Journal of Human Evolution, 14*, 607-614.

Walker, A.C. (1974). A review of the Miocene Lorisidae of East Africa. In R.D. Martin, G.A. Doyle & A.C. Walker (Eds.), *Prosimian behaviour* (pp. 435-448). London: Duckworth.

Ward, J.P. (1991). Prosimians as animal models in the study of neural lateralization. In F.L. Kitterle (Ed.), *Cerebral laterality: Theory and research* (pp.1-17). Hillsdale, NJ: Erlbaum.

Ward, J.P., Milliken, G.W., Dodson, D.L., Stafford, D.K., & Wallace, M. (1990). Handedness as a function of sex and age in a large population of Lemur. *Journal of Comparative Psychology, 104*, 167-173.

Wrangham, R.W. (1980). An ecological model of female-bonded primate groups. *Behaviour, 75*, 262-300.

3
An Exploration of Manual Preference and Performance in Crabeating Macaques

Dorothy M. Fragaszy and Leah E. Adams-Curtis

An evolutionary perspective suggests that precursors of human lateral asymmetry should be present in nonhuman species. Indeed, many forms of asymmetry, including behavioral and anatomical, are found in nonhumans (e.g, Geschwind & Galaburda, 1984; Glick, 1985). Development of nonhuman models of asymmetric function are deemed critical to advances in our understanding of human asymmetry (Geschwind, 1985). Nonhuman primates have been a natural target of comparative studies of manual asymmetry (e.g., see references in Williams, 1985) because they share with humans a reliance on eye-hand coordination to obtain food, a reliance unique to the primate order (Gibson, 1986). This shared fundamental feature of motor organization makes it reasonable to expect that nonhuman primates might share other features of manual control evident in humans, including manual asymmetry.

What questions about manual asymmetry should studies of nonhuman primates address? We can be assisted in developing a productive approach to laterality in nonhuman primates by looking to current developments in the study of laterality in humans. Handedness in humans is no longer considered a unitary trait which can be assessed with a quick observation (Healey, Liederman & Geschwind, 1986; Salmoso & Longoni, 1985), even though we immediately and confidently identify ourselves as "right-handed" or "left-handed" when asked. Consequently, although one still encounters studies which are nominal in nature (i.e., in which the data are collected via questionnaires or observation of preference), a great deal of research on motor asymmetry in humans concerns performance. In these studies, the hand used for a particular act, usually a closely defined act, is specified by the experimenter, and the variables of interest are reaction time, movement time, accuracy, the trajectory and kinetics of movement, and so forth. Performance of both hands in the experimental task can be assessed. This approach allows one to study the relationship between motor and cognitive skills, the extent of differential skill in the two hands, patterns of interlimb coordination, and so forth - a rich array of topics not touched by studies of nominal preference (Kelso, Putnam, & Goodman, 1982; Prinz & Sanders, 1984; see also any recent issue of *Journal of Motor Behavior*).

The findings of studies looking for evidence of unitary handedness in nonhuman primates have, by and large, been disappointing. Some believe that the whole enterprise is fruitless, and that humans possess a unique form of manual asymmetry associated with unique structural asymmetries in the central nervous system (CNS) (e.g., Peters, 1988; Warren, 1980). But recent developments have breathed new life into the comparative program, as the chapters in this book attest. We now appreciate that lateralization of motor function may exist in nonhuman primates, as it does in humans, in more subtle forms than a unitary bias for one hand. That is, asymmetry may be more context-dependent than we commonly appreciate. For example, bias for the use of one hand to reach may depend critically upon posture at the time of initiation of the movement, as nonhuman primates (particularly arboreal species) use the hands for postural support as well as for prehension (Forsythe & Ward, 1988; Forsythe, Milliken, Stafford & Ward, 1988; Sanford, Guin & Ward, 1984). Fagot and Vauclair (1988a, b, 1991) argue persuasively that bias for the use of one hand in nonhuman primates depends upon the details of sensorimotor control demanded by the task. A task requiring visually-controlled manual alignment of two objects elicited strong lateral bias in baboons in Fagot and Vauclair's study (1988b).

We think that studies of performance are a necessary complement to studies of postural and contextual influences on manual asymmetry in humans and nonhumans alike. Studies of performance open up rich questions about laterality which studies of choice can not address, and they provide a direct bridge to the human literature.

We present here observations on manual activity in crabeating macaques which combine studies of both choice and performance. We first evaluated preference during spontaneous manipulations, constructing a detailed ethogram in which we attempted to take the degree of fine digital control involved in movement into account. The ethogram was followed by analysis of performance of each hand on pointing tasks where each hand was used equally often. The design of the study minimized the contribution of differential practice to performance of the two hands on the experimental tasks. The purpose of the current chapter is to explore different ways of looking at manual asymmetry and the relations among different aspects of asymmetry.

Observations of Preference

Design

The purposes of this study were to develop a profile of manipulative activity (a manipulation ethogram) and to evaluate preference during spontaneous manipulations. Behaviors involving the use of the hands in singly housed monkeys in their home cage were noted in three different conditions including: a) baseline condition, b) during the presentation of a mechanical food dispenser, and c) during the presentation of an unbaited mechanical puzzle. We also evaluated the effects of expanded manipulative opportunity in the latter two conditions on activity budgets, including self-destructive behaviors and stereotypic motions. We take up the time-budget aspect of the study elsewhere (Adams-Curtis & Fragaszy, in preparation).

Subjects

Four male crabeating macaqes (*Macaca fasicularis*), approximately four years of age, served as subjects for this study (Malcolm, Jim, John, and Ted). Each subject was housed singly in a metal and wire mesh cage measuring 71 cm x 61 cm x 76 cm. The front of each cage contained a door, 13 cm x 18 cm, consisting of wire bars 2.6 cm apart. Subjects could reach through these bars and manipulate items presented outside the front of the cage.

The subjects were housed in one room on two cage racks with two or three animals per rack. The two racks were placed at a 90 degree angle from each other so that the animals could see the animals on the other rack. Animals housed next to each other could make limited tactile contact with their hands, but not visual contact. The subjects were fed Purina monkey chow once daily; water was available ad libitum. Each subject had a rawhide chewbone of the type provided for large dogs in its home cage. Aside from food and the chewbone, the housing provided extremely limited manipulative opportunities.

Materials

The test materials for the two experimental conditions were a mechanical food dispenser and a three-step mechanical puzzle. Depicted in Figure 3.1a is an illustration of the dispenser made of clear Plexiglas (38 cm x 38 cm x 8 cm), and attached to the front of the cage using rubber cords. To activate the dispenser, a short spring-operated lever (3 cm long) had to move 11 cm from left to right. Movement of the lever dispensed a small amount of cereal or sunflower seeds into a drawer (5 cm x 5.5 cm x 1.5 cm). The drawer opened by a hinged lid (2.8 cm by 6.5 cm) and was placed directly in front

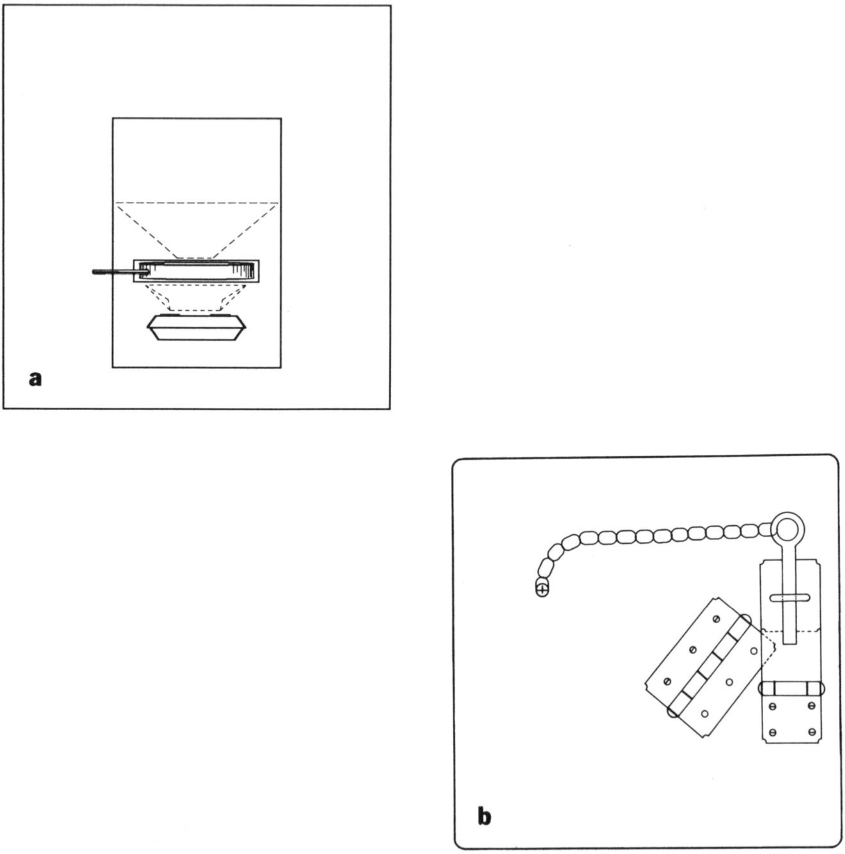

Figure 3.1. (a) The mechanical food dispenser; (b) the mechanical puzzle board

of the largest opening in the cage door. Both hands could be used simultaneously to open the drawer and retrieve food.

This task requires at least 3 different manipulations: moving the lever, lifting the lid, and retrieving the small pieces of food. A precision grip was required for the last action. Bimanual actions were not required, although they could be used in order to retrieve the food efficiently.

The mechanical puzzle consisted of a board, 23 cm x 23 cm x 1 cm, painted either black or white (Figure 3.1b). Attached to the board was an eye bolt, a hasp, and a hinge. To open the entire puzzle, the eyebolt had to be removed, the hasp opened, and the hinge lifted. There was no extrinsic reward associated with the solution of the puzzle. Any opened parts were reset prior to the start of each 2.5 minute observation period. The puzzles were presented to the subjects by attaching them to the end of a cinder block (39 cm X 19 cm X 19 cm) using two C-clamps and placing the cinder block in front of the cage. In order to manipulate the puzzles, subjects reached through the bars forming the door of the cage. Like the previous task, this apparatus was designed to elicit a variety of manipulative actions, although unlike the first, it did not require precision or encourage bimanual manipulations.

Methods

Each animal was observed for ten 10-minute sessions in the baseline condition and five 10-minute sessions in each of the two experimental conditions. Data for each animal were collected in 2.5 minute blocks within sessions (4 per session), with 2.5, 5, or 7.5 minutes separating each block, depending on the number of subjects observed that session. At least two animals were observed per session. The order in which the animals were observed was constant within sessions and quasi-random across sessions. The operational definitions of the fourteen acts scored can be found in Table 3.1. These acts were selected based on prior observations of manual activity in these monkeys. For each of the experimental conditions, four additional behaviors, defined according to the part of the apparatus manipulated, were also scored. These additional behaviors are described in Table 3.2.

The data were collected using a one-zero sampling method (Altmann, 1974) at 5-second intervals. At the end of each interval, the observer recorded all acts which occurred in the previous 5 seconds and hand(s) used (left, right or both). Using a foot for a behavior was also recorded but left or right foot use was not noted. An audible electronic timer signaled the 5-second intervals. The sampling rate produced 1,200 intervals per subject in the baseline condition and 600 intervals per subject in each experimental condition.

The data for the baseline condition were collected by two observers. Interobserver reliability was assessed prior to the start of data collection. Reliability was calculated as the percentage of lines in which the two observers agreed on all acts (using only lines in which either observer scored any act; called "active lines"), and the percentage of active lines in which the observers agreed on at least one act. The observers were considered reliable when 80% of active lines were in total agreement and 90% of active lines agreed for at least one act.

Table 3.1. Behaviors Scored for the Baseline and Two Experimental Conditions.

Behavior	Operational Definition
GROOM SELF	Use of the hands or feet to groom coat, scratch.
MASTURBATION	Manipulation of the genitals.
BITE/SUCK FINGERS/TOES	Placing either fingers or toes in the mouth & nibbling or sucking on digits.
SHAM BITE	Vigorous biting of the hands, arms, legs, or feet, usually with rapid locomotion.
HAND PLAY WITH NEIGHBOR	Placing hand through cage bars or mesh and hitting, pulling, or otherwise manipulating the hands of neighbor.
SOLICIT HAND PLAY	Placing hand through cage bars or mesh and moving the hand. Often observed prior to hand play.
MANIPULATE CAGE LOCKS/LATCHES	Contacts with cage padlocks or latches with either hands or feet. Subjects often pulled on padlocks and opened latches. Contact with mouth not scored.
TOUCH,EXPLORE CAGE	Contact to parts of cage other than latches and locks.
REACH OUT	Reaching out of cage, distinguished from SOLICIT HAND PLAY by no hand movement.
SQUEEZE CAGE BARS	Grasping two adjacent cage bars and exerting force.
PUSH CHUTE	Moving the squeeze chute back and forth.
STEREOTYPE LOCOMOTE	Moving about the cage. Also included whole body movements, such as rocking and jumping in place.
HANDLE/HOLD INTRODUCED OBJECTS	Holding or manipulating introduced objects including chow, rawhide chewbones, rubber cords of food dispensers.
PICK UP OBJECTS	Picking up objects with hands or feet, does not include pick up with mouth.

Table 3.2. Behaviors Directed Toward the Experimental Objects.

Behavior	Operational Definition
	FOOD DISPENSER
TOUCH/NONSPECIFIC	Contact with any part of the dispenser other than the lever or the drawer.
FIDDLE WITH LEVER	Manipulation of the lever that activated the food dispenser.
FIDDLE WITH DRAWER	Manipulate any part of the drawer into which food was dispensed.
PICK UP FOOD	Taking food either from the drawer or from areas where food had fallen.
	MECHANICAL PUZZLE
TOUCH PUZZLE NONSPECIFIC	Contact to the board, the cinder blocks or the C-Clamps.
FIDDLE WITH EYEBOLT	Contact to the eyebolt.
FIDDLE WITH HASP	Contact to the hasp.
FIDDLE WITH HINGE	Contact to the hinge.

Table 3.3. Behavioral Categories for the Ethogram.

Behavioral categories	Acts included in each
FINE MANIPULATION	Touch, explore cage; Manipulate cage locks and latches; Pick up objects.
GROSS MANIPULATION	Push squeeze chute; Squeeze cage bars; Handle/hold introduced objects.
SELF DIRECTED	Groom self; Masturbate.
SELF DESTRUCTIVE	Sham bite; Bite/suck fingers/toes.
MANIPULATE NOVEL OBJECTS	The four behaviors specific to each experimental object.

Data Analysis

We tabulated intervals in which the right, left, and both hands were used in individual acts, and the intervals in which the animals were not active. The individual acts were subsequently combined into five behavioral categories. These categorizations can be seen in Table 3.3 and included fine manipulation, gross manipulation, self-directed, self-destructive, and novel object manipulation.

Hand preference was examined within subjects using *Chi-square* tests of observed frequency counts (Siegal, 1954) of left and right usage (ignoring bimanual acts). First, the frequency of left and right hand use was compared for each condition and across all three conditions. Second, preference was evaluated for fine manipulations in the baseline condition. Finally, we compared the frequency of left and right hand use for acts directed toward the introduced objects.

Results

General Activity

Despite the sterility of their surroundings in the baseline condition, the animals were active (including locomotor activity) during at least 60% of the observed intervals, and a variety of hand use acts occurred. These data can be seen in Figure 3.2. Although each animal had its own pattern of manipulations, the most common acts were handle/hold introduced objects, groom self, and explore cage. Many of the acts occurred for at least one percent of the intervals.

The effect of providing objects to manipulate, as expected, was to increase the overall frequency of hand use. The animals were much more active during the two experimental conditions (*Mean* = 98%, 94% vs 83% of intervals, baseline). For the most part, self-directed activities - groom self, masturbate, bite fingers/toes and sham bite - decreased from baseline levels during the experimental conditions and handling and exploring increased. Figure 3.3 shows that both the experimental objects elicited a great deal of manipulation. Of the eight behaviors scored that were directed toward the experimental objects, seven of these occurred for at least 10% of the intervals (at least 60 times) for all animals. Only the hinge of the puzzle board failed to elicit this high level of manipulative activity.

Preference

We turn now to hand preferences. First, the overall proportion of left, right, and bimanual acts varied across subjects, as seen in Figure 3.4, and no single

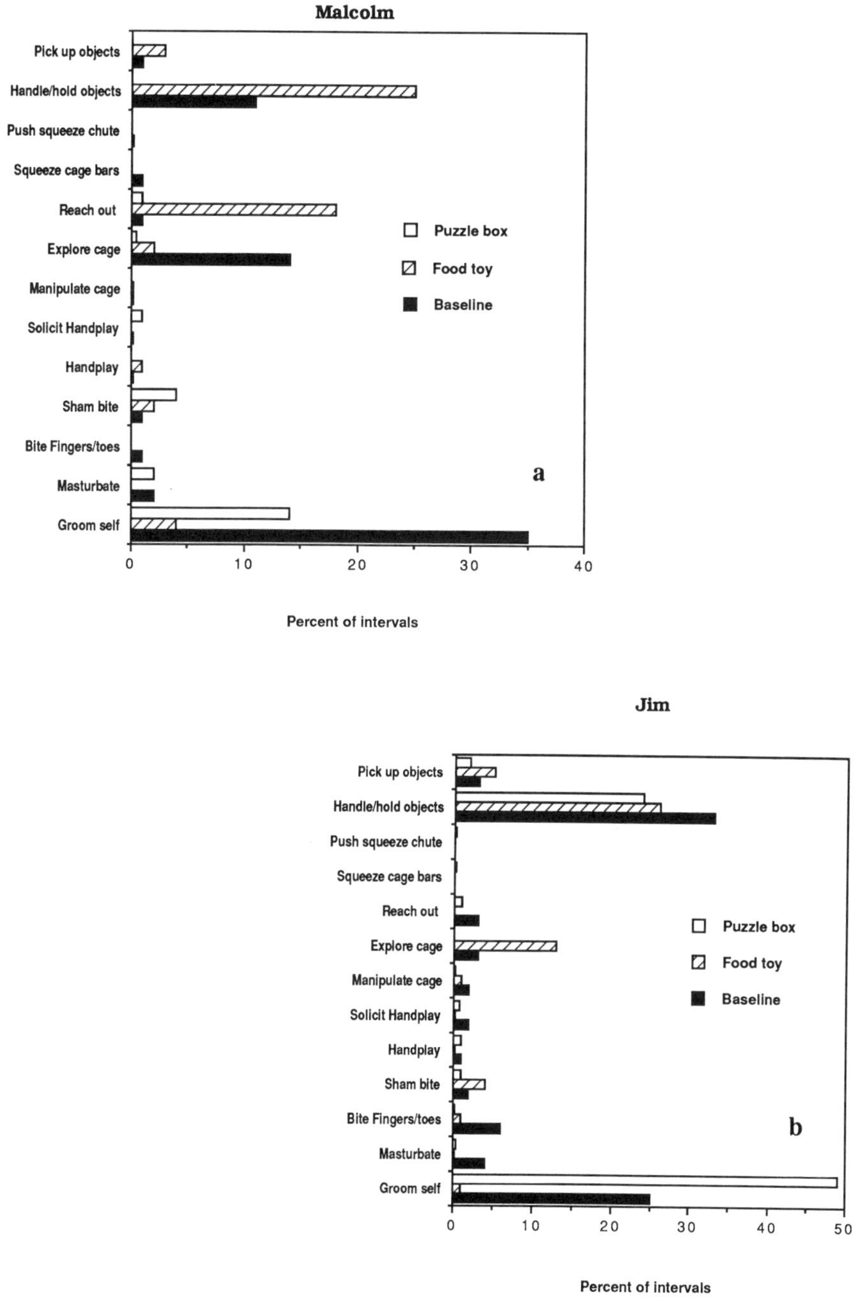

Figure 3.2(a and b). Percent of intervals subjects Jim (a) and Malcolm (b) engaged in various actions common to all conditions.

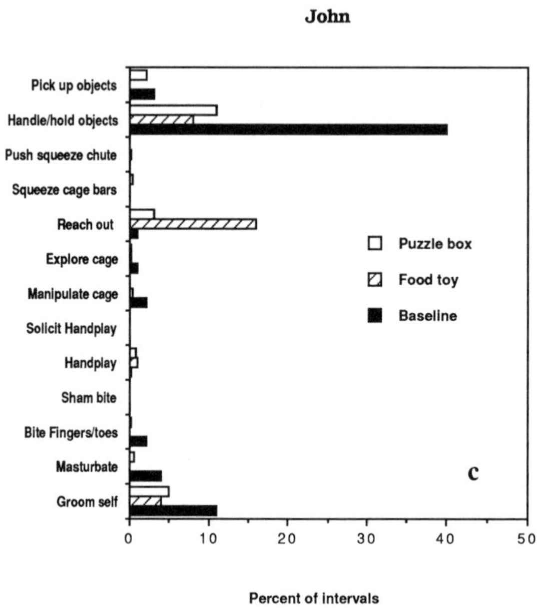

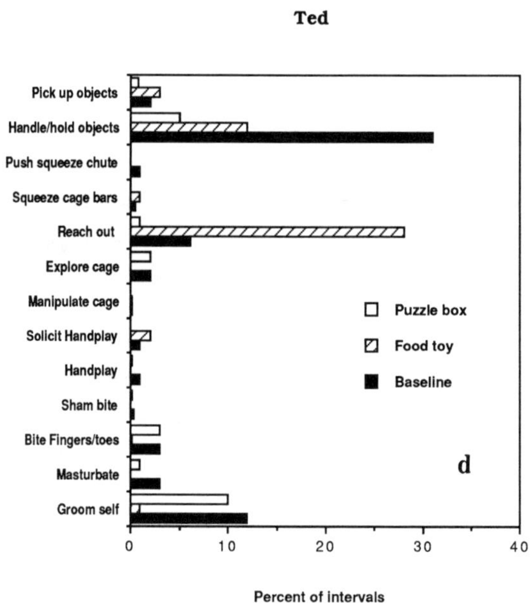

Figure 3.2 (c and d). Percent of intervals subjects John (c) and Ted (d) engaged in various actions common to all conditions.

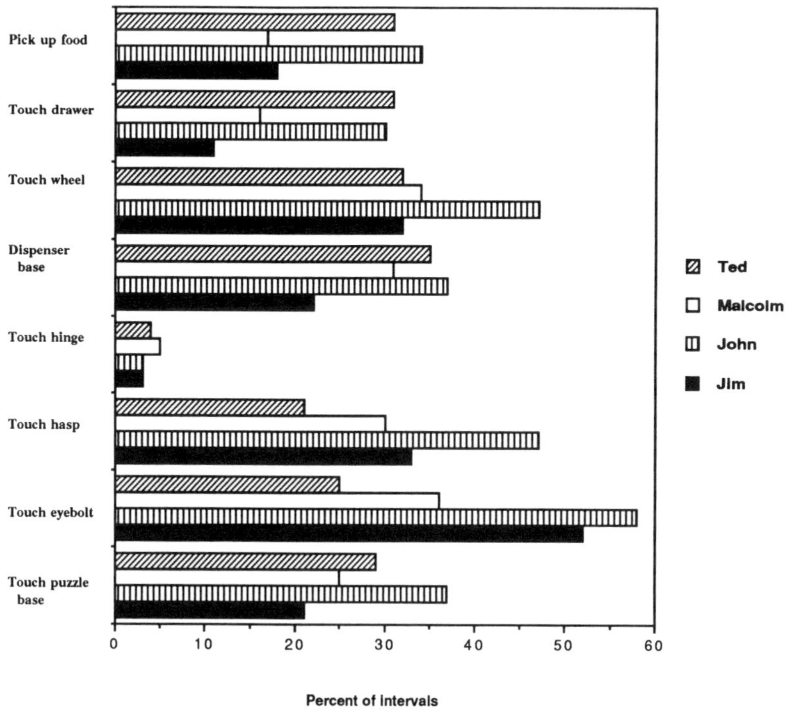

Figure 3.3. Percent of intervals each subject engaged in various manipulative actions in the experimental conditions when a food dispenser or puzzle were presented.

act elicited consistent bias either within or across subjects. Only the subject John exhibited a significant preference for one hand (his left) over all conditions and acts. The overall distribution of preferences for each subject are depicted in Table 3.4.

When hand preferences within conditions are examined, no individual displayed consistently significant preferences. John's significant left hand preference was evident only in the two experimental conditions. In the baseline condition, John showed a significant right hand preference. The frequency of manipulations in the two experimental conditions was great enough to mask the switch of preference in the baseline condition as compared to the experimental condition. In the other individuals, a similar picture of contrasting preferences emerges. One of the four animals (Malcolm), did not show a significant hand preference in any of the three conditions; one (Jim) showed a significant left hand preference in the dispenser condition. The fourth subject (Ted) exhibited three significant hand preferences. In the baseline and the puzzle board conditions, Ted used his left hand significantly more often than his right. However, in the dispenser condition Ted preferred his right hand. This decidedly mixed outcome indicates that frequencies of spontaneous use do not give an interpretable picture of lateral preference in nonhuman primates.

Preference in Fine Manipulation

It has been suggested that performance in tasks requiring certain kinds of manipulations may be the most useful measures of lateral bias in primates (Fagot & Vauclair, 1991; MacNeilage, Studdert-Kennedy, & Lindblom, 1987). MacNeilage et al. suggested "that monkeys have a left hand preference for visually-guided movement and a right hand preference for manipulation, broadly defined." (p. 258). To examine this idea, we evaluated hand preference for the behavioral category "fine manipulation" for the baseline condition, and compared it to preferences in three manipulations directed toward the dispenser that all involved fine manipulation (Fiddle with lever, Fiddle with drawer, and Remove food from drawer). The overall preferences for each of these behaviors can be seen in Table 3.5. Initially these findings seemed promising. All four animals expressed a significant hand preference for fine manipulation in the baseline condition (three for the right hand and one for the left) and three of the four animals exhibited a significant preference in all three acts equated with fine manipulations of the dispenser. However, for the three acts toward the dispenser, only John expressed a preference that corresponded to his preference for fine manipulation shown in the baseline condition, and that for only one act (Fiddle with the wheel). For the remaining two actions, John preferred his left hand. The other three subjects all displayed opposite preferences for at least one of the fine manipulations of the dispenser compared to fine manipulation in baseline. The most extreme example is Ted, who preferred the left hand for fine manipulation in the baseline condition but used his right hand significantly more often than his left for all three manipulations directed toward the dispenser.

One could argue that these differences could be due to the novelty of the manipulations of the dispenser, and that given more opportunities to manipulate it, the subjects' preferences would converge with their baseline preferences for fine manipulations. If novelty per se facilitated alternate patterns of preferences in the dispenser condition than in the baseline condition, as seen in some degree in all subjects, then we would expect to see a similar pattern of hand preferences within subjects for the food toy and the puzzle board, regardless of the type of manipulation. This was not the case. While Malcolm showed a significant left hand preference for manipulation of the food toy, he had both a significant left hand preference (for Fiddle with the eyebolt) and a significant right hand preference (for Fiddle with the hinge) for manipulation of the puzzle board. Jim showed no significant preferences for any act with the puzzle board. John preferred his left hand for three of the acts (Touch non-specific, Fiddle with the eye

a

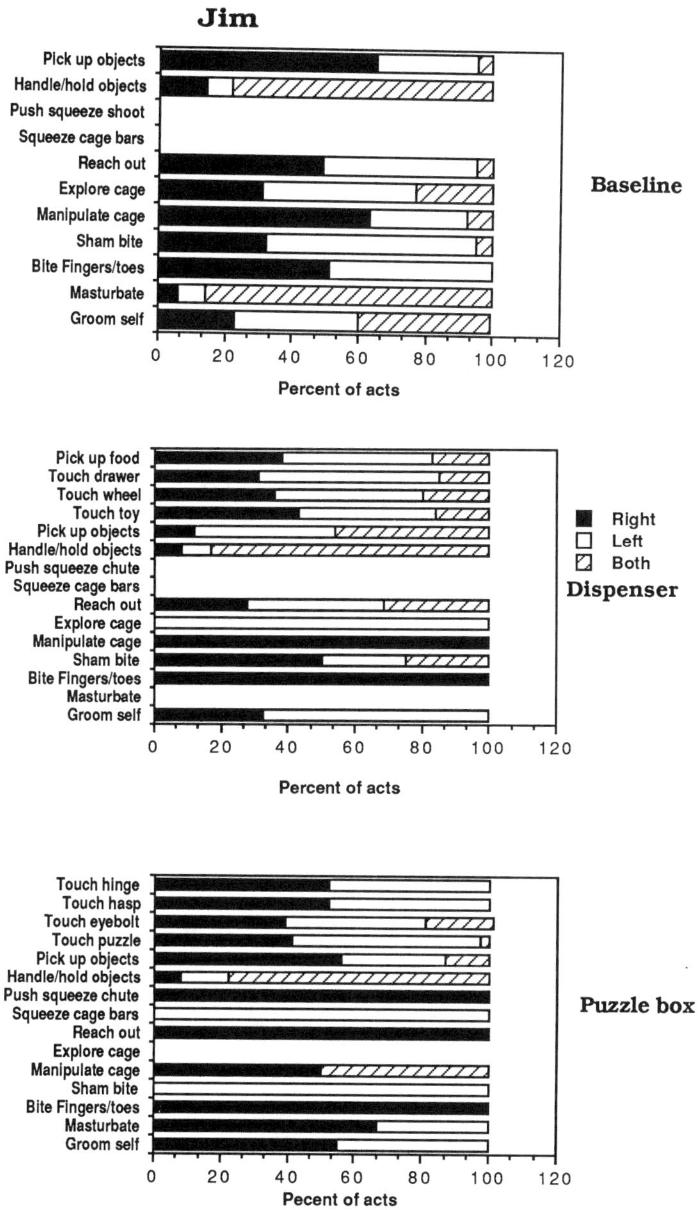

Figure 3.4(a). Distribution of hand choice for actions in common across conditions for subject Jim.

b

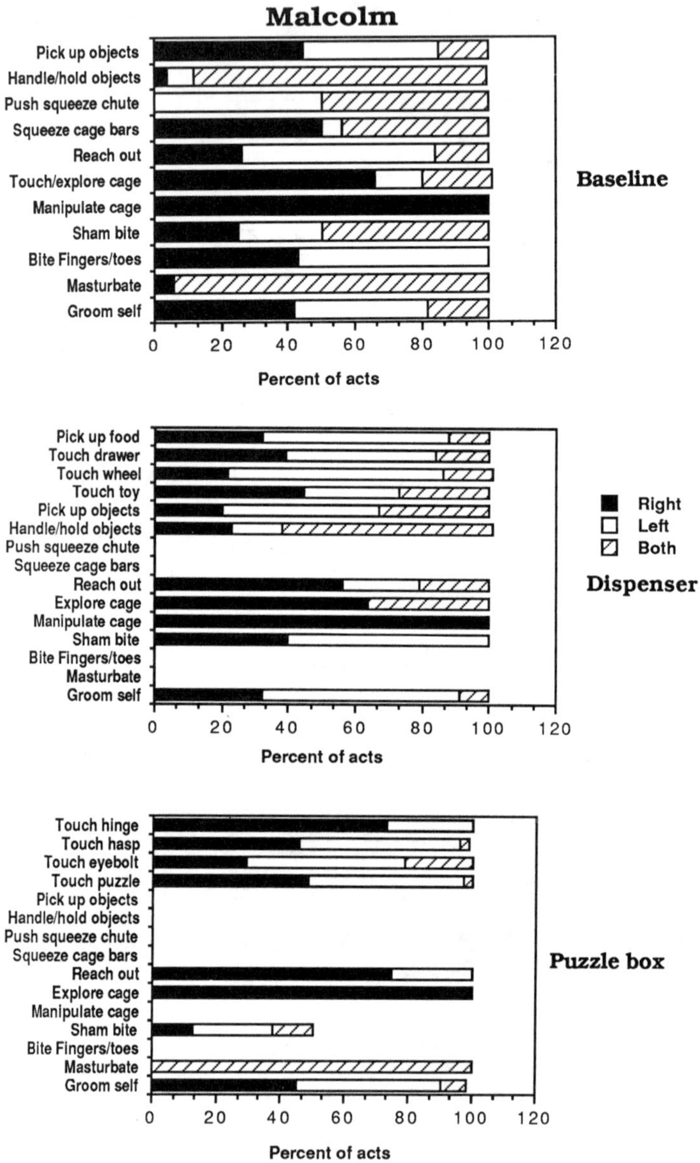

Figure 3.4(b). Distribution of hand choice for actions in common across conditions for subject Malcolm.

c

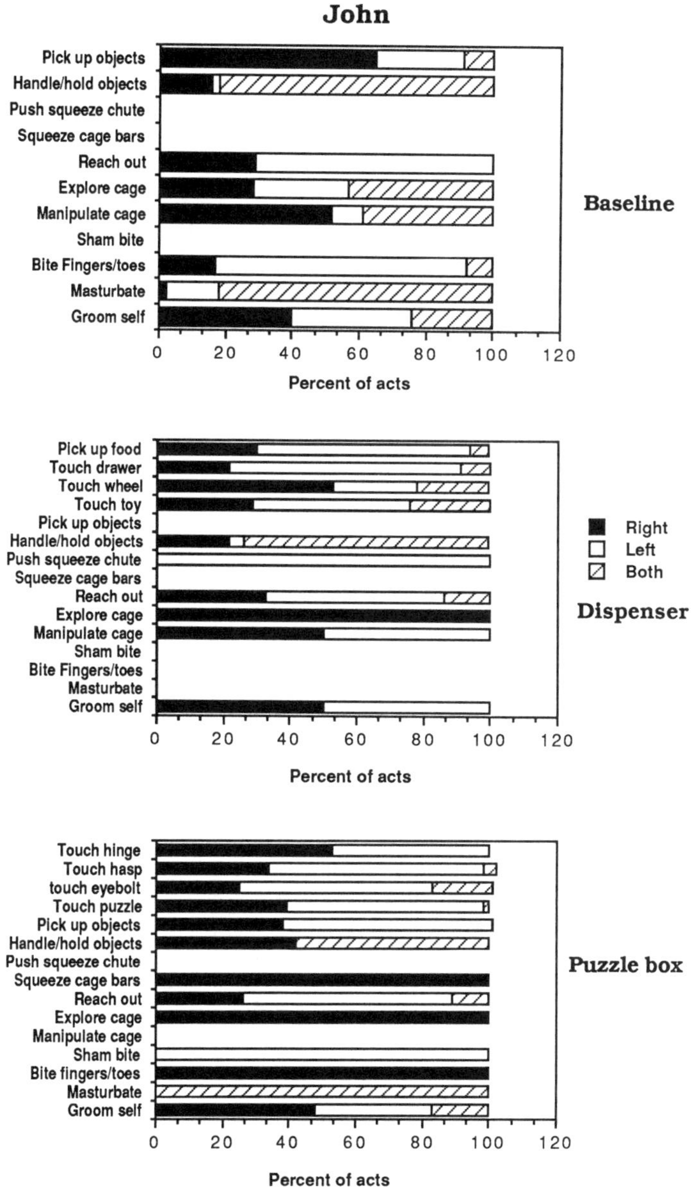

Figure 3.4(c). Distribution of hand choice for actions in common across conditions for subject John.

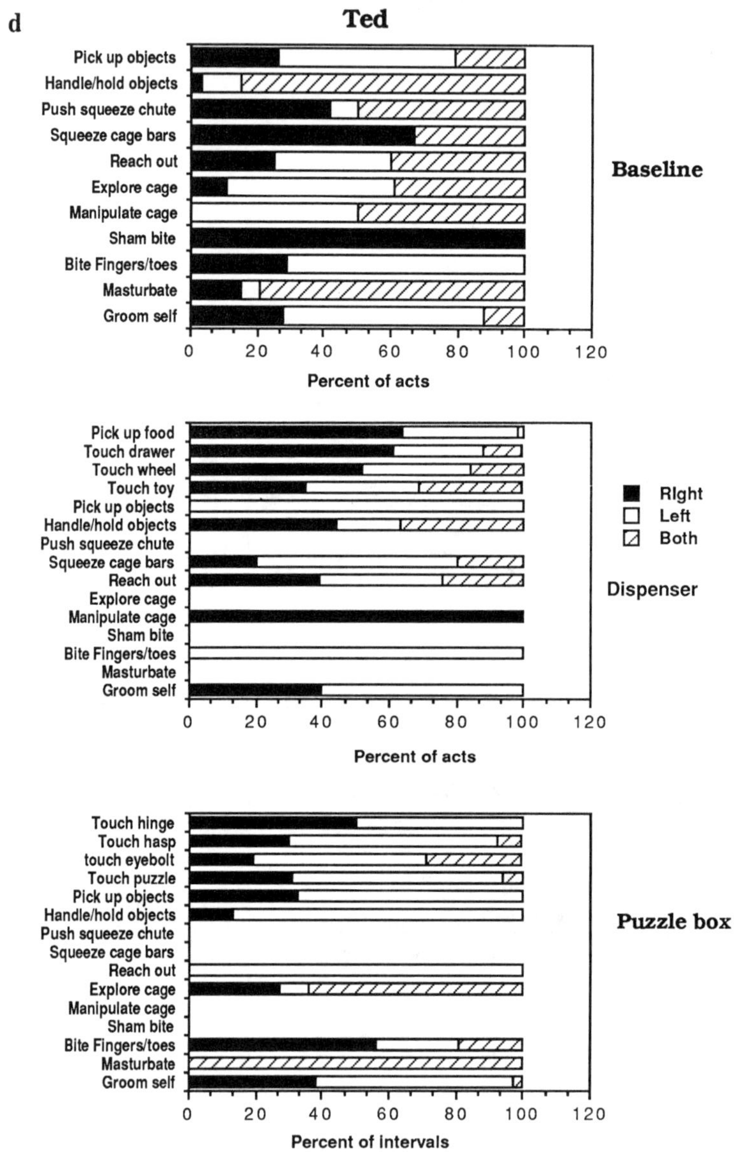

Figure 3.4(d). Distribution of hand choice for actions in common across conditions for subject Ted.

Table 3.4. Frequency of Use for Each Subject Across and Within Conditions. Significant Contrasts Between the Hands Are Indicated.

	Subjects											
	Malcolm			Jim			Ted			John		
Conditions	R	L	X^2	R	L	X^2	R	L	X^2	R	L	X^2
Baseline	265	222	3.6	220	232	0.3	106	163	12.1*	171	100	132.8*
Dispenser	314	348	1.6	235	295	6.8*	507	331	37.0*	364	487	324.1*
Puzzle	265	305	2.8	393	395	0	160	305	45.2*	303	540	334.6*
Overall	844	875	0.6	848	922	3.1	733	799	0.4	838	1127	42.5*

* $p < .05$
Note. R = right hand; L = left hand

Table 3.5. Frequencies of Fine Manipulation in the Baseline Condition and Behaviors Directed Toward the Food Dispenser.

	Behaviors											
	Fine Manipulation in Baseline			Lever Play			Play with Drawer			Remove Food		
Animal	R	L	X^2	R	L	X^2	R	L	X^2	R	L	X^2
Malcolm	**48**	16	16.0*	43	**127**	41.5*	36	42	0.5	32	**56**	6.5
Jim	**41**	30	6.3*	69	84	1.5	21	37	42.3*	40	47	0.6
John	**36**	13	10.8*	**151**	72	28.0*	40	**123**	4.4*	61	**130**	24.9*
Ted	7	**20**	6.3*	**98**	60	9.1*	**113**	50	24.5*	**118**	63	16.7*

* $p < .05$
Note. R = right hand; L = left hand

Table 3.6. Button Tasks Presented to Subjects.

Unimanual tasks	Bimanual tasks
Left near	Both near
Left far	Both far
Right near	Left near and Right far
Right far	Left far and Right near

bolt, and Fiddle with the hasp). Once again Ted represents the most extreme case, with three significant right hand preferences for manipulation of the food dispenser, and three significant left hand preferences (Touch non-specific, Fiddle with the eye bolt, and Fiddle with the hasp) for manipulation of the puzzle board!

These data show that for common manipulative behaviors, there is no systematic variation of hand preference across animals, nor across acts within subjects. Furthermore, regardless of which level (task, action) we chose to examine, we could not predict hand preference for either another act or for a different task. We conclude that spontaneous manipulations do not reveal lateral preferences in these monkeys, even when some effort is made to restrict consideration to actions requiring more coordination than simple reaching and grasping. Performance measures or preferences on very constrained tasks (e.g. Fagot & Vauclair, 1988a, b) are more likely to reveal differences in hand use for these animals.

Analysis of Performance

Methods and Design

Performance in unimanual and bimanual reach and point tasks were examined. The tasks were similar to those used by Kelso and colleagues (Kelso, Southard, & Goodman, 1979; Kelso, Putnam, & Goodman, 1982) with humans. Reaching distance and involvement of one or both hands were varied in a completely crossed design and are detailed in Table 3.6. Reaction time and movement time were assessed in these conditions. We consider here: a) whether lateral bias was evident in either of these variables, and if so, in what conditions, and b) whether variation in the task affected performance of the two hands in a similar fashion.

Subjects

The subjects for this study were three young adult male *Macaca fascicularis*. Two of the three subjects, Jim and Malcolm, were also subjects in the first study. The third, Russell, participated only in this study. Housing and care were the same as those described for the observational study.

Materials

The test apparatus was presented at the front of the test box and placed in front of each subject's cage prior to the start of the test session. The subject entered the test box from the home cage and contacted the test tray by reaching between the bars at the front of the test box. The subject's position during testing was standardized by a Plexiglas panel at the front of the test box that restricted vision and manual access to the test tray. To see the test tray clearly and reach it easily, the subject had to position its head in a face opening and center its torso behind the panel. A sipper tube through which juice was delivered was centered at the bottom lip of the face opening. Openings on both sides of the panel allowed the monkey to use either hand with equal facility. The centering requirement assured that the subject positioned itself uniformly across trials.

The test board (31 cm X 17.25 cm X 8.5 cm) was mounted in front of the test box. There were two important features of the test tray: 1) the ready buttons, which had to be depressed concurrently to activate a ready signal (a tone); and 2) the target button(s), positioned farther from the subject than the ready buttons. The left and right ready buttons (each 2.8 cm in diameter) were 17.5 cm from each other and were located 3 cm from the front and 1 cm from the plane of the test board. The ready buttons had to be depressed 0.5 cm to activate the timer. The target buttons (2.4 cm diameter) were mounted on small boxes (5.8 cm X 12 cm X 5.6 cm) 6.8 cm above the test board and 7.5 cm (Near position) or 14 cm (Far position) behind the ready buttons. For bimanual trials, the left and right target buttons were 19.5 cm from each other when the target buttons were in the same position, and 20 cm from each other when in a Near-Far combination. In order to complete the task, the target buttons had to be depressed 0.7 cm, so as to be flush with the top of the mounting box. As soon as this occurred, a tone signalled completion of the trial. An opaque guillotine door descended between the test box and the test tray at the end of each trial.

Procedure

The task consisted of two parts: The subject first pressed the ready buttons for one second (the ready position), and then reached to and depressed the target button(s) to receive juice. The target button was presented to either the right or the left hand (Unimanual tasks) or two target buttons were presented together (Bimanual tasks). In the bimanual tasks, the subject had to depress the target buttons with both hands, concurrently or sequentially. The position of the target buttons, near or far from the ready buttons, was varied across tasks. There were four unimanual tasks - one for each hand in each position. For bimanual tasks, four combinations of position and hand were possible: Both near, Both far, Left Near-Right Far, and Left Far-Right Near. Reaction time and movement time for each task were recorded electronically to the nearest 1/10,000 of a second. Reaction time was the time elapsed from the completion of the 1 sec. ready position (signalled by a tone) to the release of the ready button. Movement time started at the time of release of the ready button and ended when the target button was depressed. In the bimanual tasks, reaction time and movement time were recorded for each hand independently.

Subjects routinely completed between 15 and 20 trials per day, with about a 15 second intertrial-interval between tasks. On each day of testing, the subjects received a number of different tasks (some not reported here), one trial per task, presented in a fixed order. If the subject did not respond to the task being presented within five minutes, the next task was presented after a normal intertrial-interval. If the subject did not respond on the second task, testing was discontinued for the day. The button tasks were presented for 6 to 8 sessions in sequence in the following order: Right Near, Left Near, Both Near, Right Far, Left Far, Both Far, Left Near-Right Far, and Left Far-Right Near.

Analysis

As in the study of spontaneous manipulation, all analyses were performed within individuals. We examined both reaction time and movement time for each of the tasks, excluding all trials in which either the reaction time or the movement time was greater than 5.9 sec. (This rule excluded five trials for Russell and one trial for Jim). Values were rounded to the nearest .001 of a second. First, times for the left and right hands were compared using t-tests on a number of different comparisons to assess differential performance in a direct manner. Second, we compared the times for the same hand in different tasks to determine the effect of task demands on reaction and movement times within hands. For these comparisons, *t*-tests were used. Lateral bias in adjustment to different tasks would be evident in these analyses as greater differences across tasks in one hand than in the other. Finally, we considered the relation between reaction time and movement time within hands in bimanual and unimanual tasks separately using Pearson Product Moment correlations. This analysis relates to the possibility that the two variables might be more tightly linked on one side of the body than on the other. Reaction time and movement time have been

Table 3.7. Variance Ratio, Bimanual/Unimanual Reaction and Movement Times. Values Greater Than 1 Indicate More Variable Bimanual Scores.

	All bimanual tasks/ Unimanual tasks			Bimanual same position tasks/Unimanual tasks		
	Jim	Malcolm	Russell	Jim	Malcolm	Russell
Right						
Reaction	0.16	0.03	0.74	0.06	0.04	1.22
Movement	0.01	2.77	0.09	0.01	0.004	1.16
Left						
Reaction	0.29	0.59	0.11	0.30	0.26	0.10
Movement	21.91	0.09	0.51	41.73	0.26	0.10

considered to co-vary, in line with the degree of "planning" necessary to achieve the end point of the behavior (Sheridan, 1984).

Results

General Comparisons

The ranges of times for each hand in each condition (bimanual and unimanual) across animals were quite similar, both to each other and for each hand. For bimanual conditions, the reaction times for the the left hand ranged from less than 0.01 sec. to 5.80 sec. (individual means ranged from 0.85 to 3.08), and for the right hand from 0.04 sec. to 4.48 sec. (individual means ranged from 0.46 to 2.08). The movement times in the bimanual condition ranged from 0.18 sec. to 5.02 sec. (individual means from 0.22 to 1.94) for the left hand and from 0.19 to 3.45 sec. (individual means from 0.26 to 1.33) for the right. For unimanual tasks, the reaction times ranged from 0.04 to 4.59 sec. and movement from 0.18 to 5.02 sec. for the left hand. For the right hand, values ranged from less than 0.01 to 1.16 sec. for reaction time and from 0.22 to 4.37 sec. for movement time. Means overlapped completely in the two hands for both reaction time and movement time in the unimanual tasks: from 0.44 to 1.75 sec. for reaction time; from 0.22 to 4.37 sec. for movement time. Thus, no consistent differences in performance of the right or the left hand were apparent.

Overall, mean movement times were faster than mean reaction times (see Tables 3.8 and 3.9). The range of both movement times and reaction times, however, overlapped considerably (mean movement time/task/subject from 0.22 to 1.94 sec.; mean reaction time from 0.44 to 2.08 sec.). The ranges of movement time and reaction time were similar in all three subjects (i.e., no animal was faster overall than the others).

One general statement can be made about the amount of variance that different tasks produced. Performance on bimanual tasks was less variable than performance on unimanual tasks (See Table 3.7). Out of 24 possible comparisons, variance was greater in the bimanual version than in the unimanual version of the task only 5 times, 4 times for movement and once for reaction time. The constraint of using both hands appears to canalize performance across trials, especially, in reaction time, and to do so equally for each hand. We pursue the question of bimanual coordination in more detail elsewhere (Fragaszy and Adams-Curtis, in preparation).

Contrasts Between Hands

We turn now to specific comparisons between hands. First we consider whether any subject showed a consistent difference between the left and the right hand in reaction time for the three different types of tasks. For bimanual tasks we compared performance of the two hands during the same task. For the unimanual tasks we compared the hands for each position (see Table 3.8). Out of six possible contrasts, Jim and Malcolm had a faster reaction time for the right hand for one contrast: the Left Far-Right Near bimanual task. Additionally, there were two other comparisons in which Malcolm's right hand responses were faster than his left: Both far bimanual task, and for the unimanual tasks in which the button was in the far position. A fourth contrast for Malcolm approached significance (Left Near-Right Far $p = .051$), again with the right hand being faster. Reaction time for Russell did not vary significantly between hands on any comparison. In short, only Malcolm displayed any consistent difference between the hands for reaction time, favoring his right hand.

Movement times varied between hands more often than did reaction times (9 vs. 4 significant comparisons, out of 18 possible, compare Tables 3.8 and 3.9; see Table 3.10 for summary). Malcolm's right hand moved faster than his left hand in three of the six comparisons: Both near, Both far, and unimanual near button tasks. However, Malcolm's left hand moved faster than his right hand for one comparison (the Left Near-Right Far tasks). A similar outcome was observed for Jim. On two of the comparisons, Jim's right hand moved faster than his left: Both Near, and Left Near-Right Far bimanual tasks. His left hand moved faster, however, in unimanual far position tasks (this difference approached significance, $p = .09$). Russell's right hand moved faster than his left for three of the movement time comparisons: Left Near-Right Far bimanual tasks, Left Far-Right Near bimanual tasks, and Unimanual Far tasks.

When we compare the outcome of pair-wise (left and right) comparisons for reaction time and movement time, an interesting overall pattern emerges. For twelve of the thirteen significant comparisons the right hand was faster than the left. These data are found in Table 3.10. However, even with this overall trend, no one task produced significant differences in all

Table 3.8. Mean Reaction Time for Comparisons of Left and Right Hand

	Jim					Malcolm					Russell				
	N	Right	N	Left	*p*	N	Right	N	Left	*p*	N	Right	N	Left	*p*
Bimanual Same Task															
Near	4	0.74	4	1.12		5	0.64	5	0.85		3	1.09	3	0.85	
Far	4	0.84	4	3.08		6	**1.18**	6	1.42	**	5	1.16	5	1.59	
Mixed Bimanual Tasks															
LNRF	3	1.58	3	1.41		5	0.70	5	0.90		4	1.67	4	0.51	
LFRN	4	**0.46**	4	2.19	*	5	**0.72**	5	2.51	**	5	2.08	5	1.32	
Unimanual Tasks															
Near	3	0.44	5	0.43		6	0.56	6	0.49		4	0.72	5	0.40	
Far	5	0.47	3	0.56		6	**1.14**	6	1.75	**	3	0.50	5	0.84	

* $p<.05$
** $p<.01$

Table 3.9. Mean Movement Times for Comparisons of Left and Right Hand

	Jim					Malcolm					Russell				
	N	Right	N	Left	*p*	N	Right	N	Left	*p*	N	Right	N	Left	*p*
Bimanual Same Task															
Near	4	**0.41**	4	0.86	*	5	**0.30**	5	0.50	**	4	0.23	4	0.23	
Far	4	0.45	4	0.30		6	**0.45**	6	0.63	**	5	1.33	5	0.78	
Mixed Bimanual Tasks															
LNRF	3	**0.26**	3	0.31	*	5	0.61	5	**0.22**	**	4	**0.29**	4	0.41	**
LFRN	4	0.33	4	0.27		5	0.33	5	0.33		5	**0.27**	5	1.94	**
Unimanual Tasks															
Near	3	0.37	5	0.27		6	**0.26**	6	0.28	**	4	1.35	4	1.44	
Far	5	0.67	3	0.40		6	0.44	6	0.47		3	**0.48**	3	1.35	*

* $p<.05$
** $p<.01$

Table 3.10. Summary of Contrasts between Left (L) and Right (R) Hands in Reaction (R_x) and Movement (M) Times for Bimanual Same Position (BSP), Mixed Bimanual (MB), and Unimanual (U) Responses. The Indicated Hand (R or L) Was Significantly Faster than the Other Hand in That Condition.

	Jim		Malcolm		Russell	
	R_x	M	R_x	M	R_x	M
BSP Near	-	R	-	R	-	-
BSP Far	-	-	R	R	-	-
MB LNRF	-	R	-	L	-	R
MB LFRN	R	-	R	-	-	R
U Near	-	-	-	R	-	-
U Far	-	-	R	-	-	R

animals for all measures. The contrast favoring the left hand in one case for Malcolm and Jim (significant only for Malcolm), in opposition to their overall pattern of favoring the right hand, emphasizes that a general bias for one hand does not rule out a strong bias for the other hand.

The weak pattern of overall right hand bias was investigated further by recasting the data for the bimanual tasks. These data can be seen in Table 3.11. When we compare performance in bimanual tasks for each hand in each position (e.g. all left near bimanual vs. all right near bimanual), ignoring the task requirements for the other hand, there is a significant performance difference for all animals on both movement time and reaction time when the hands are reaching for the near buttons. For five of these six comparisons, the right hand was faster than the left (Russell's reaction time for his left hand was faster than his right). However, when the button was in the far position, only Malcolm's reaction time and movement time differed for the two hands. Furthermore, although Malcolm's reaction time was consistent with his overall bias in this task (the right hand being faster), movement time was not. Even with Malcolm's performance taken into account, these results indicate promise for bimanual tasks, especially when the target is fairly close to the subject, for detecting consistent lateral bias in performance between hands.

Contrast Within Hands

Greater sensitivity of movement time than reaction time to task demands was evident when performance by the same hand was compared across tasks, as seen in Table 3.12. For the left hand, ten of eighteen comparisons produced significant differences in movement times, and eleven differences were observed for the right hand. The same comparisons for reaction time

Table 3.11. Comparisons Between Hands for All Bimanual Tasks.

	Jim					Malcolm					Russell				
	N	Right	N	Left	*p*	N	Right	N	Left	*p*	N	Right	N	Left	*p*
Reaction Time															
Near	8	**0.60**	7	1.25	*	10	**0.68**	10	0.87	*	8	1.71	7	**0.66**	**
Far	7	1.16	8	2.64		11	**0.96**	11	1.92	**	9	**1.39**	10	1.45	
Movement Time															
Near	8	**0.37**	7	0.63	**	10	**0.32**	10	0.36	**	8	**0.26**	7	0.33	*
Far	7	0.37	8	0.29		11	0.52	11	**0.49**	*	5	0.87	10	1.36	

* $p<.05$
** $p<.01$

produced far fewer significant differences, with *alpha* = .06 (five for the left hand and two for the right). The tasks that produced these significant differences within hands are different from the tasks which produced significant differences between the left and right hands (compare Tables 3.10 and 3.12).

Movement time differed more often than reaction time across conditions in the same hand, as it did in comparisons between the left and right hand on the same task. In fact, significant differences in movement time within hands were observed in most comparisons between tasks. This points to the effectiveness of distance manipulations in affecting movement. In relation to lateral bias, however, there is no suggestion that one hand is more affected by changing task demands than the other.

Correlations Between Reaction and Movement Times Within Hands

To examine whether reaction time is related to movement time in the same hand, and whether the relation is stronger in one side than the other, we correlated reaction time with movement time within hands. In bimanual tasks, there were no significant correlations for reaction time with movement time within hands (that is, left reaction times did not correlate with left movement times, and right reaction times did not correlate with right

Table 3.12. Within Hand Comparisons for Reaction Time and Movement Time.

	Jim		Malcolm		Russell	
Reaction Time	Left	Right	Left	Right	Left	Right
Both near vs Mixed	-	-	-	-	-	-
Both far vs Mixed	+Mixed	-	-	-	-	-
Far unimanual vs Near unim.	-	-	**Near	-	-	-
Near all bimanual vs Far all bim.	-	+Near	**Near	-	*Near	*Far
Near unimanual vs Near bimanual	*Uni.	-	-	-	-	-
Far unimanual vs Far bimanual	-	-	-	-	-	-
Movement Time						
Both near vs Mixed	*Mixed	-	**Mixed	-	**B.Near	-
Both far vs Mixed	-	*Mixed	**Mixed	*Mixed	-	**Mixed
Far unimanual vs Near unimanual	-	**Near	-	**Near	-	*Near
Near all bimanual vs Far all bim.	**Far	-	-	**Near	**Near	**Near
Near unimanual vs Near bimanual	**Uni.	-	**Uni.	-	**Bi.	**Bi.
Far unimanual vs Far bimanual	-	*Bi.	**Uni.	-	-	*Uni.

+ $p < .06$, * $p < .05$, ** $p < .01$

movement times). For all unimanual tasks, right hand reaction times correlated significantly with right hand movement times for Jim ($r = .725$, $p<.05$, n = 10) and for Russell ($r=.792, p<.05$, n = 9). For Malcolm, the correlation between left hand reaction times and left hand movement times approached significance ($r=.551$, $p=.051$, n = 13). Thus, a fairly strong relation between movement and reaction time exists for one hand in each subject, but only in unimanual tasks. Unfortunately for simple models of laterality, the relation between the two components of performance appears to be unrelated to the individual's overall bias in performance. That is, movement time is not necessarily more strongly related to reaction time in the generally faster hand than it is in the other hand. For example, Malcolm exhibited more significant right hand biases than either Russell or Jim, but the significant correlation between reaction time and movement time for this subject was for the left hand.

Discussion

Our data on manual preference do not support the notion that monkeys possess a consistent bias for the use of one hand for any particular action. We observed a more or less random pattern of preferences for one hand or the other across conditions and across actions, with only a few comparisons out of the many possible reaching statistical significance for any subject. Individual differences were the most prominent feature of spontaneous hand use, with each animal showing his own pattern of preference.

These data indicate a caveat for the use of preference measures in studies of lateral bias. Because we examined many variables under a variety of conditions, we saw variables in which no preference, shifting preference, and consistent preference were evident. Had we limited our examination to a few actions, the picture would have been quite different. The variety of patterns produced in these four subjects indicate that preference data must be viewed conservatively when the research aim is to detect consistent bias.

Even when activities involving fine manipulation were considered separately, directional preferences varied across conditions within subjects to the same extent as for other actions. Each animal showed at least one shift from a statistically significant preference for one hand in the baseline condition to a statistically significant preference for the other hand in the two test conditions on an action involving fine manipulation. However, given that fine manipulations consistently produced preferences in these animals, tasks involving fine manipulations may be better than the reach and point tasks examined here for revealing lateral bias in performance. Reaction and movement times are logical variables to explore in studies of performance on fine manipulation tasks, in order to provide a bridge to the literature on these variables in simpler tasks. We compared performance (reaction and movement time) on tasks requiring fine manipulation and whole-hand manipulation elsewhere (Fragaszy and Adams-Curtis, in preparation). However, we suggest that other measures of performance will also be appropriate, such as kinematics of hand shaping, control of pressure between the digits (Preilowski, Reger, & Engele, 1986), or smoothness of sequencing of movement (Viviani & Terzuolo, 1980).

Asymmetries between the hands were more evident in performance on pointing tasks than in preference during spontaneous activity. A general bias for more rapid movement to a target by the right hand was observed. Movement time was a more sensitive indicator of differences between the hands than was reaction time. However, movement time was also more sensitive to variations in the position of the target and the involvement of one or two hands than was reaction time. Although our data show a fairly consistent right hand bias for movement time, the sensitivity of this measure to variations in the task indicates that a much larger sample is needed before any inferences about population-wide biases are merited.

Bimanual tasks appear more promising for studies of lateral bias than unimanual tasks. Both reaction times and movement times varied significantly between the hands more often when one target was near to the subject in bimanual tasks than in the equivalent comparisons in the unimanual tasks. Furthermore, measures of both reaction time and movement time for bimanual tasks were less variable overall than in unimanual tasks. Taken together, these results indicate that bimanual tasks are more likely to show lateral differences than unimanual tasks, and at the same time be less vulnerable to spurious results with limited sample sizes.

Each subject exhibited a tendency for reaction and movement times to covary more consistently in one hand than the other. This aspect of asymmetry was independent of asymmetry for movement times or reaction times considered alone. We recognize the dependence of hand preference and performance on tasks and activities. Asymmetry in the covariance within hands of these variables may also vary as a function of task demands or activities. We suggest that this dimension of asymmetry deserves further investigation.

In summation, we recommend a broadened research agenda in the realm of lateral bias in manual activity. We have found performance measures in bimanual tasks and choice measures in fine manipulation to be most promising for revealing lateral bias, although our results do not support any existing view of consistent (structural) bias in these actions within an individual.

Acknowledgments

This work was supported by a Visiting Professorship from the National Science Foundation to D. Fragaszy, and by the Dean, College of Veterinary Medicine, Washington State University. We thank D. Fisher, G. Thompson, J. Baer, and G. van Orden for their assistance in collecting the data, S. Samuels for programming assistance, Alan Wing for commenting on the manuscript during its preparation, and J. Ward for careful editing.

References

Altmann, J. (1974). Observational study of behavior: Sampling methods. *Behaviour, 49*, 227-268.

Corballis, M., & Beale, I. L. (1976). *The psychology of left and right*. Hillsdale, New Jersey: Erlbaum.

Fagot, J., & Vauclair, J. (1988a). Handedness and bimanual coordination in the lowland gorilla. *Brain, Behavior, and Evolution, 32*, 89-95.

Fagot, J., & Vauclair, J. (1988b). Handedness and manual specialization in the baboon. *Neuropsychologia, 32*, 89-95.

Fagot, J., & Vauclair, J. (1991). Manual laterality in nonhuman primates: A distinction between handedness and manual specialization. *Psychological Bulletin, 109*, 76-89.

Forsythe, C., Milliken, G., Stafford, D., & Ward, J. (1988). Posturally related variations in the hand preferences of the ruffed lemur (*Varecia variegata variegata*). *Journal of Comparative Psychology, 102*, 248-250.

Forsythe, C., & Ward, J. P. (1988). Black lemur (*Lemur macaco*) hand preference in food reaching. *Primates, 29*, 369-374.

Geschwind, N. (1985). Implications for evolution, genetics, and clinical syndromes. In S. Glick (Ed.), *Cerebral lateralization in nonhuman species* (pp. 247-278). Orlando, Florida: Academic Press.

Geschwind, N. & Galaburda, A. (1984). *Cerebral dominance: The biological foundations*. Cambridge, MA: Harvard University Press.

Gibson, K. (1986). Cognition, brain size, and the extraction of embedded food resources. In J. Else and P. Lee (Eds.), *Primate ontogeny, cognition, and social behavior* (pp.93-103). Cambridge, MA: Cambridge University Press.

Glick, S. D. (1985). *Cerebral lateralization in nonhuman species*. Orlando, Florida: Academic Press.

Healey, J., Liederman, J., & Geschwind, N. (1986). Handedness is not a unidimensional trait. *Cortex, 22*, 33-53.

Kelso, J. A. S., Putnam, C. A., & Goodman, D. (1982). On the space-time structure of human interlimb coordination. *Quarterly Review of Experimental Psychology, 33A*, 347-375.

Kelso, J. A. S., Southard, D. L., & Goodman, D. (1979). On the nature of human interlimb coordination. *Science, 203,* 1029-1031.

MacNeilage, P., Studdert-Kennedy, M., & Lindblom, B. (1987). Primate handedness reconsidered. *Behavioral and Brain Sciences, 10*, 247-303.

Peters, M. (1988). The primate mouth as an agent of manipulation and its relation to human handedness. *Behavioral and Brain Sciences, 11*, 729.

Preilowski, B., Reger, M., & Engele, H. (1986). Handedness and cerebral asymmetry in nonhuman primates. In D. Taub & F. King (Eds.), *Current perspectives in primate biology* (pp. 270-282). New York: van Nostrand Reinhold.

Prinz, W., & Sanders, A. F. (1984). *Cognition and motor processes*. Berlin: Springer-Verlag.

Salmoso, D., & Longoni, A. (1985). Problems in the assessment of hand preference. *Cortex, 21*, 533-549.

Sanford, C., Guin, K., & Ward, J. P. (1984). Posture and laterality in the bushbaby (*Galago senegalensis*). *Brain, Behavior, and Evolution, 25*, 217-224.

Sheridan, P. (1984). Planning and controlling simple movements. In M. Smyth & A. Wing (Eds.), *The psychology of human movement* (pp. 47-82). London: Academic Press.

Siegal, S. (1956). *Nonparametric statistics for the behavioral sciences*. New York: McGraw-Hill.

Viviani, P., & Terzuolo, V. (1980). Space-time invariance in learned motor skills. In G. E. Stelmach & J. Requin (Eds.), *Tutorials in motor behaviors*. Amsterdam: Elsevier.

Warren, J. (1980). Handedness and laterality in humans and other animals. *Physiological Psychology, 8*, 351-359.

Williams, J. B. (1985). Handedness, cerebral dominance, and brain asymmetry in nonhuman primates: A bibliography, 1970-1985. Seattle, WA: Primate Information Center.

4
Manual Preference in Varieties of Reaching in Squirrel Monkeys

James E. King and Virginia I. Landau

Primates, unlike other mammals, have the ability to grasp objects with one hand. Capability for one-handed prehension evolved as part of the early primate specialization for manual dexterity that eventually included development of pseudo and fully opposable thumbs and the ability for precision gripping (Napier & Napier, 1967). The subsequent intensive and skilled hand use provided an adaptive basis for lateralized specialization of function for the two hands. This manual lateralization is manifested by the right-handedness in about 90% of the human population (Hecaen & de Ajuriaguerra, 1964). Important theoretical questions involve whether motor or cognitive asymmetries were the first to appear during primate evolutionary development and the causal relationship between the two types of asymmetry during evolutionary development. If nonhuman primates display varieties of lateralized handedness, they may be preadaptations not only for highly skilled human right-handedness but also for some language related skills (MacNeilage, 1987). Progress in answering these questions has been impeded by the common assumption that nonhuman primate handedness is nonexistent. When nonhuman handedness has been reported, the finding has often been dismissed as a laboratory artifact of trivial importance. It is further reported to be situation specific and not stable over long periods of time (Warren, 1977, 1980). Individual animals do frequently show reliable hand preferences on particular manual prehension problems. However, the number of animals displaying right hand preferences is often about equal to the number displaying left hand preferences (e.g. Brookshire & Warren, 1962; Cole, 1957; Lehman, 1978). A common generalization is that right and left hand preferences will each be expressed by about 25% of animal subjects with no statistically significant preference expressed by the remaining 50% (Annett, 1985). The presumed distribution of hand use across animals thus resembles a symmetrical U, centered at equal use of each hand with highly lateralized individuals concentrated at the two extremes.

Yet reports of significantly asymmetric hand use across samples of subjects have regularly appeared against a background of influential negative reports. These reports of asymmetric hand use have, perhaps, seemed like the smiling Cheshire Cat that exasperated Alice by appearing and then disappearing with seeming unpredictably. Recent reviews of handedness in nonhuman primates by MacNeilage, Studdert-Kennedy, and Lindblom (1987, 1988) showed numerous systematic deviations from the symmetrical U hypothesis. And unlike the capricious disappearing act of the Chesire-Cat, asymmetrically lateralized reaching may, after all, be lawfully related to the demands placed upon the subject by the reaching. Briefly, McNeilage et al. (1987) suggested that visually guided reaching elicits left hand preferences in monkeys while increasing the amount of required manipulation increases right hand preference. At the very least, the review by MacNeilage et al. showed that the question of handedness in nonhuman primates is not, as previously assumed, a closed one. One basic question that should be answered involves the relationship between the prehensile requirements of a task and the type of manual laterality elicited.

Methodological Considerations

Humans are commonly classified as right or left handed, although the dichotomy has long been known to be an oversimplification since intersubject variability exists in the degree of right- or left-handedness across different manual tasks. In fact, human handedness may not even be unidimensional. Healey, Liederman, and Geschwind (1986) detected four factors across different manual preferences in humans. However, evidence on this question is not totally consistent (Porac, Coren, Steiger, & Duncan, 1980). These considerations notwithstanding, composite scores of human handedness are highly bimodal with most people displaying fairly consistent right or fairly consistent left hand preferences across multiple measures. Furthermore, human handedness is robust; it is reliable, stable over most of a person's life, and not highly dependent upon procedures used to measure it (Annett, 1985; Corballis, 1983). Human handedness can therefore be referred to as "strong handedness."

In contrast, manual preferences in nonhuman primates are far less robust than those in humans and can be described as expressions of "weak handedness". The "weak handedness" of nonhuman primates has implications for choice of testing procedures and statistical analysis of handedness data. The problem of detecting hand preferences in nonhumans is analogous to a signal detection problem (see Green & Swets, 1974) in which the handedness signal is weak and the noise sources that mask the handedness are strong. What are some of the noise sources that mask the appearance of weak handedness?

Hand preferences established by prior unequal reinforcement of responses by the two hands are one source of masking. If, for any reason, one hand is initially used more often than the other to reach for food, the greater frequency of reinforcement for that hand may occasion increasingly frequent use of that hand. If one hand is more dexterous or retrieves food more rapidly than the other hand, a similar effect may occur (see Warren, 1980). There is no way of totally eliminating this potential problem if more than one response is elicited from each subject. However, manual tasks on which subjects have extensive prior training should probably be avoided.

The location of the incentive or manipulated object relative to the animal's midline is more effective in determining hand choice in animals than in humans. Animals are less likely to reach across their bodies to retrieve an incentive with a preferred hand than are humans (Cronholm, Grodsky, & Behar, 1963; Harris & Carlson, 1988; Lehman, 1970). The response locus should be centered in the experimental apparatus, a consideration that is especially important if the animal's enclosure is small (Box, 1977). The problem of incentive location is potentially more serious in a free field situation in which food is thrown to the subjects. Since the trajectories of most thrown food items will deviate from the animals' midlines, a tendency for the animals to retrieve the items with the closest hand may obscure a less potent general hand preference (e.g. Itani, Tokuda, Furuya, Kano, & Shin, 1963) even though the procedure itself will cause no right or left bias over a sufficiently long series of trials. Fagot and Vauclair (1988) attempted to correct this problem in observations of reaching by gorillas by calculating handedness scores only from responses to the midline or across the body.

Behavioral stereotypies that primates often develop are also a possible masking source for hand preferences. The stereotypies are enhanced in small enclosures. Our observations of squirrel monkeys in small (31 x 40 x 36 cm) testing cages during memory testing showed that intertrial intervals were largely filled with stereotyped and highly idiosyncratic behaviors (Scott, 1990). Behaviors including repeated circling, somersaulting, rocking, or responding from a fixed location on one side of the manipulated object are common primate behaviors in small enclosures. If the stereotypy is invariant, it may strongly influence the hand used to manipulate an object or to retrieve food.

There is another systematic behavior that may possibly obscure manifestation of hand preferences; however, this source is more subtle and difficult to detect than the preceding sources. During visual discrimination and delayed response learning, animals display a variety of systematic error producing response sequences (Harlow, 1959; King & Michels, 1989; Levine, 1965; Lentz & King, 1981). One source of systematic errors is a position perserveration error (King & Michels, 1989). Position perseveration occurs when the animal makes an above chance frequency of responses to the position chosen on the previous trial irrespective of the outcome of that trial. Position perseveration responses are sequentially dependent and are different from position habits which are sequentially independent and occur when one position is preferred over the other independently of the response on the preceding trial.

An analogous sequential hand perseveration sequence may occur during long sequences of hand preference trials. It is expressed by runs of trials in which one hand is preferred, followed by runs in which the other hand is preferred. The stronger the tendency to shift hand preference the less detectable any overall preference for a particular hand would be. We will later describe a procedure we have used to separate sequential hand perseveration from handedness.

Statistical Considerations

The central problem in statistical analysis of handedness observations is that the distribution of hand preference scores across subjects may deviate markedly from normality. Therefore, nonparametric order statistics may often be the appropriate choice for an inferential test. Thus, if the null hypothesis is that median hand preferences do not differ across different tests, a Mann-Whitney or Kruskal-Wallis test could be used if independent groups of subjects were used and a Wilcoxon or Friedman test used if the same subjects were used in different conditions (Siegel, 1956).

The situation is more ambiguous, however, when a seemingly much simpler question is addressed: do handedness scores across a group of subjects differ significantly from a particular null value, usually equal use of each hand? The most common approach has been to classify subjects initially with respect to whether they meet some criterion for deviation from equal use of each hand. Usual criteria are attainment of some percentage use of one hand (e.g. 80%) or a statistically significant deviation from the 50% value (e.g. Brookshire & Warren, 1962; Lehman, 1970; Milner, 1969). A *chi-square* or binomial test using z scores can then be used to compare the number of subjects displaying strong right with the number displaying strong left preferences. Other related statistical strategies are also available. A *chi-square* test could be used to compare frequencies of three types of subjects: strong right handers, strong left handers, and those not showing a strong preference. Alternately, frequencies of right- and left-handed subjects could be individually compared with the frequency of subjects not showing a specified level of preference (see Cicchetti, 1987). The central question for each of these strategies is whether the null hypothesis being tested is the one intended by the experimenter. For example, comparing frequencies of significantly lateralized subjects with frequencies of nonlateralized subjects necessitates assuming that subjects not deviating by a specified amount from the null value contribute evidence against rejecting the null hypothesis. Thus, the procedure becomes a test of a null hypothesis about frequencies in the arbitrarily defined extremities of a population, not one about overall bias in that population. This approach seems to be based on an implicit tendency to transfer onto animals the typology of human hand preferences that is based on the feasibility of classifying most people as unambiguously right- or left-handed. Little evidence exists justifying a similar typology in nonhuman primates. Therefore, information about overall bias may be more important and appropriate than information about numbers of arbitrarily defined right- and left-handed subjects. Furthermore, when hypotheses about frequencies of extreme reactors are tested, the decision to reject the null hypothesis depends upon the arbitrarily determined stringency of the handedness criterion. As Calvin (1987) has noted, one should be highly skeptical when a decision favoring human uniqueness can always be made simply by demanding enough stringency in defining a characteristic.

An alternative strategy is simply to compare the number of subjects making a majority of right hand responses with the number making a majority of left hand responses, no matter how small the majority. A binomial test can then be used to test the significance of the difference. This is a perfectly legitimate procedure and is consistent with other typical uses of the binomial test. It should also be noted that a test of frequencies of highly lateralized subjects is not always simply a conservative version of a test for frequencies of subjects with simple majorities of left or right reaches. Depending upon the distribution of hand preferences, either test could be significant independently of the significance of the other. We believe that the best strategy is to present information on both types of tests.

Some Squirrel Monkey Natural History

Squirrel monkeys (*Saimiri sciureus*) are small, South American monkeys. They are highly saltatory in their arboreal habitat with over 40% of their locomotion accomplished by leaping, a higher percentage than that for other monkeys inhabiting similar habitats. Smaller but still substantial amounts of leaping occur during foraging and feeding (Fleagle, Mittermeir & Skopec, 1981).

Squirrel monkeys are omnivorous with most of their diet composed of fruit and insects. They are adept at capturing flying insects and seem to be attracted to them (Fleagle et al., 1981). Terborgh (1983) described squirrel monkeys as frequent, compulsive leaf foragers whose strategy for insect capture is searching instead of pursuing. Most insect prey are small and slow moving. Once a prey is located, usually on a leaf, it is quickly caught. Squirrel monkeys observed by Terborgh averaged 56.1 insect captures per hour during foraging.

Prehension During Fishing

Squirrel monkeys obviously do not catch fish in their natural arboreal habitat. However, we were intrigued by the possibility that a moving, brightly colored fish would elicit spontaneous capture responses similar to those elicited by flying or moving insects.

Our first study of prehension during fishing was a doctoral dissertation project by Virginia Landau (1987) that was designed to demonstrate the acquisition of fishing behavior in two groups of squirrel monkeys, each containing seven females and two males. The groups were housed in separate rooms measuring about 4.2 x 8.0 m. The rooms were equipped with a variety of racks, tree limbs, and ropes. Three weeks before the beginning of observations, the monkeys' regular diet was supplemented with ground sardines in order to accustom the monkeys to the taste of fish. Subjects in each room were then tested once a day for 14 consecutive days. At the start of each observation session, ten goldfish (*Carassius auratus*) were placed in a water filled ceramic bowl 40 cm in diameter. Behaviors of the monkeys directed toward the fish were then recorded on videotape for 20 minutes or until all ten fish had been caught. All attempts to capture fish by all monkeys were later scored from replays of the videotapes.

The second study was similarly conducted with the same subjects and in the same observation rooms. At the start of observations, however, the ten goldfish were placed in a square wading pool measuring 86 cm on each side and filled with water to a depth of about 11 cm. Testing was conducted daily for 22 days. Figure 4.1 shows a monkey about to make a left hand reach from the edge of the wading pool.

On the first test day in the first experiment, several monkeys were immediately attracted to the moving goldfish in the bowl. Frenzied attempts to catch the fish followed shortly. Two of the monkeys made only one attempt during the 14 observation days. The remaining 16 monkeys made 1,312 unimanual and bimanual attempts to capture fish. Figure 4.2A depicts proportions of left, right, and bimanual attempts to capture fish from bowls. About 60% of all attempts were left handed with right unimanual and bimanual attempts each comprising about 20% of the total. Frequencies of right and left attempts were compared with binomial tests ($p < .05$) for all 16 monkeys. Eleven monkeys made significantly more left than right responses and three made significantly more right than left responses. The two remaining monkeys made more left than right responses, but the differences were not significant. The total of 13 monkeys with predominant left hand responding significantly exceeded the three with significant right hand responding ($z = 2.25, p = .024$). The comparison involving monkeys with significant preferences (11 versus 3) was not significant.

In addition, frequencies of left and bimanual reaches were compared with binomial test for each monkey. Eleven monkeys made significantly more left

Figure 4.1. Monkey making a left hand reach from edge of wading pool.

than bimanual attempts; one made significantly more bimanual responses, and three made more left responses, the differences being nonsignificant. The number of monkeys (11) making significantly more left reaches significantly exceeded the number (1) making significantly more bimanual reaches (z = 2.60, p = .009).

Despite the monkeys' marked preference for using their left hands to capture moving fish, the success rate for the left hand was comparable to those for right-handed and bimanual attempts. A successful capture was defined as grasping a fish with one or both hands and removing it from the water even if the fish then slipped out of the monkey's hand and fell back into the water. Figure 4.2A shows that all three capture strategies had about a 30% success rate. Rates of successful capture were calculated separately for those monkeys displaying significant right and significant left hand reaching preferences. The mean success rate for the preferred hand was 34.8% and for the nonpreferred hand was 29.6%. This difference was not significant.

When the bowls were replaced by the wading pool in the second experiment, the monkeys maintained their left hand preference (See Figure 4.2B). Only 10 of 18 monkeys made more than 10 attempts, perhaps because

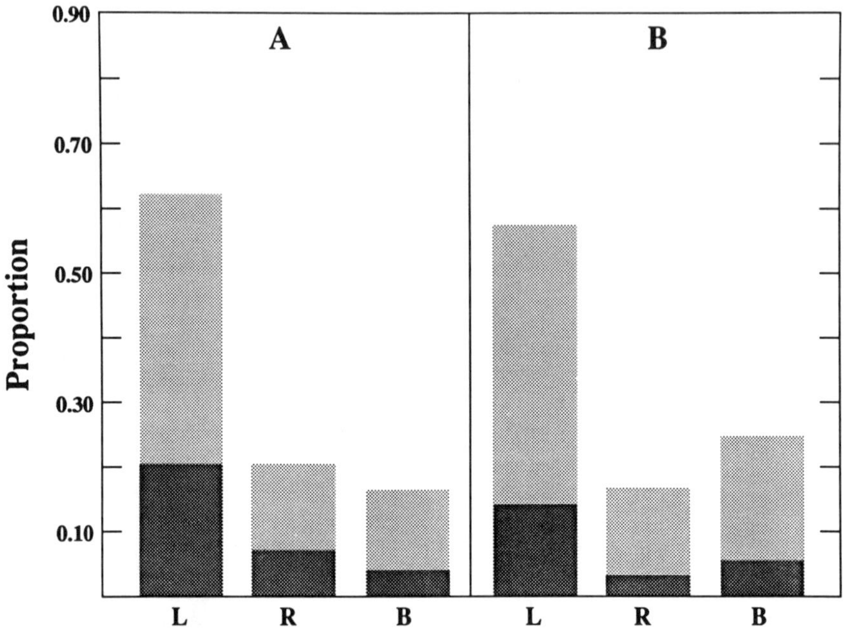

Figure 4.2. Proportions of fish capture attempts that were left (L) unimanual, right (R) unimanual, and bimanual (B). Attempts to catch fish from the bowls are shown in A and attempts from the wading pool in B. Dark shading shows proportions of attempts that were successful in removing a fish from the water.

of the increased difficulty of capturing fish in the wading pool. Number of left hand attempts significantly exceeded number of right hand attempts for eight monkeys. The other two monkeys displayed a nonsignificant left hand preference. The eight monkeys who displayed a significant left hand preference significantly exceeded the zero monkeys displaying a right hand preference ($z = 2.47, p = .013$). Comparison of left and bimanual responses yielded the same pattern of responses, with eight monkeys showing significantly more left than bimanual responses and the other two showing a nonsignificant difference in the same direction. As in the crock observations, the proportion of successful attempts was virtually identical for all three modes of attempted fish capture. As Figure 4.2 shows, however, proportion of successful captures was reduced to about 20%. The capture rate during use of the significantly preferred hand was 26% and for the other hand was 19.2%. As before, the difference was not significant. The correlation (Pearson's r) between proportions of left hand unimanual responses in the two sets of fishing observations was .70 ($p < .02$).

The top panel of Figure 4.3 shows the mean proportion of unimanual responses that were left handed for each subject in the fishing experiment.

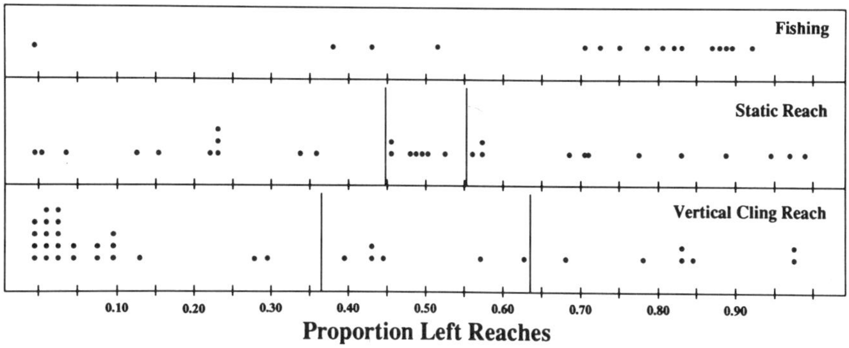

Figure 4.3. Proportion of left reaches by individual monkeys in the three experiments. Values for the fishing experiment are means from the ceramic bowl and wading pool experiments.

Values are means of combined data from both experiments. The left hand bias is clear and there is no evidence that the distribution is bimodal.

Static Reaching

Most handedness observations in laboratory settings have been conducted with simple reaching tasks in which the number of response loci was small and the subjects were confined to a small space (e.g. Lehman, 1970, 1978; Warren, 1977). As noted before, these confined environments may mask hand preferences that would otherwise be evidenced.

The static reach experiment was designed to examine handedness for simple food reaching in a relatively unrestricted environment. Subjects were 21 male and 9 female squirrel monkeys. The test enclosure was 112 cm long, 42 cm wide, and 51 cm high. Each of the two long walls were clear Plexiglas and contained a 3.2 cm slot running the entire 112 cm length of the wall. The slot in one wall was 10 cm from the floor thereby allowing the monkey to reach through from a quadrupedal or from a sitting posture. The slot on the opposite side was 41 cm from the floor, forcing the monkeys to assume a bipedal, upright posture in order to reach through. Heights of the slots on each wall were systematically varied during testing. An opaque wood barrier separated the tester from the monkey. The barrier had two rows of seven holes (2.5 cm diam.) positioned 13 cm apart. The two rows were at the levels of the high and the low slot. To present a trial, the tester placed a small piece of raisin or marshmallow in the end of a 2.5 cm diameter. PVC pipe. The pipe was then inserted through one of the holes and positioned so that the food was 5 cm from the slot. The monkey could then reach through the slot to grasp the food. The 30 monkeys were presented 42 trials a day, equally divided between high and low slot presentation, for eight days. The holes used for food presentation were systematically varied so that each hole was used equally often each day for each slot height.

In addition to the conventional data analysis in terms of proportions of left and right hand responses, another analysis was conducted in which responses were analyzed into three components; two components represented systematic responding and the third represented unsystematic responding. The technique is a simplified version of a procedure we have previously used to analyze delayed response in monkeys (King & Michels, 1989; Lentz & King, 1981; Scott, King, Michels, & Fobes, 1989). The analysis is calculated as if the subject is in some state (e.g. A) at the start of each trial and that state uniquely determines the response (e.g. response a) on that trial. It is assumed that if some alternative state is in effect the probability of response a, P(a), is 0.5. Thus, $P(a) = A + .5(1 - A)$ where A is the probability of state A and the summed probabilities of all states is 1. It then follows that $A = 2P(a) - 1$.

Simple hand preference is measured by state B and reflects an overall majority of responses by one hand. If the proportion of right hand responses, P(r), exceeds the proportion of left hand responses P(l), then $B = 2P(r) - 1$. If left hand responses predominate, $B = 2P(l) - 1$.

As noted before, another type of systematic perseveration may also occur in which the subject manifests an above chance use of the hand that was used on the preceding trial independently of whether that hand was the left or the right hand. The corresponding state is referred to as sequential hand perseveration, P, and the associated responses are perseverated right hand responses r_1 r_2 and perseverated left hand responses l_1 l_2 where subscripts 1 and 2 refer to successive pairs of trials. The observed proportion of perseverative responses P(p) is the unweighted mean of the conditional proportions of right and left perseverative sequences. Thus, $P(p) = [P(r_2/r_1) + P(l_2/l_1)] / 2$ and $P = 2P(p) - 1$.

Values for B and P are, of course, simply linear transformations of the original proportion of the most frequent hand response and the proportion of sequential hand perseveration responses and therefore add no new information to those original values. B and P do, however, estimate probabilities of response determinants corrected for chance. B and P will typically sum to less than one. The difference is assumed to measure effects of the monkeys' unsystematic selection of reaching hand. This residual unsystematic responding is designated as state R (random) and is defined as the proportion of responses remaining after those attributable to B and P have been subtracted from the total.

Partitioning of all responses into hand preference, sequential hand perseveration, and random components allows us to calculate a statistic K that measures hand preference independently of sequential hand perseveration effects. K is defined as B / (B + R). Since B + P + R = 1, K is a corrected value of B when sequential hand perseveration is zero. This use of K is roughly similar to use of bias free measures in signal detection theory (Green & Swets, 1974) to measure observer sensitivity independently of observer bias, although the underlying model is different in the two cases. Specifically, K is analogous to the nonparametric sensitivity measures A and A' in signal detection applications (Green, 1964; Grier, 1971). A and A' estimate proportions of correct responses in the absence of a simple response bias favoring one response (e.g. "yes") over the alternative response (e.g. "no"), whereas K estimates proportion of trials governed by hand preference in the absence of sequential hand perseveration.

The overall proportion of left hand responses was 0.49. Twelve monkeys made significantly more left than right responses, eleven made significantly more right than left responses, and seven showed no significant difference. Of the seven monkeys displaying no significant difference, five made more right and two made more left reaches. These data clearly show no evidence of a population level hand preference. The mean for males (0.46) was not significantly different from the mean for females (0.55). Since proportions of left hand reaches across monkeys had a reasonably normal distribution (See Figure 4.3), analysis of variance was used to test effects of posture (horizontal or vertical stance) and four-day trial blocks on proportion of left reaches, P (sequential hand perseveration), R (random), and K. K was signed positive when left reaches predominated and signed negative when right reaches predominated. Neither posture nor trial blocks had significant effects on any of the dependent measures. Mean values for P and R were 0.06 and 0.40 respectively. Both values were significantly greater than zero (t (29) = 3.50, p = .0015 for P; t (29) = 4.13, p = .0003 for R). Pearson's r between proportion of left hand responses on the first and second 4-day blocks was 0.78 ($p < .01$). The corresponding correlation between bipedal and quadrupedal left hand reaches was 0.93 ($p < .01$). These two high correlations indicate that hand preferences were reliable across trial blocks and postural conditions and that the almost equivalent right and left hand use across subjects was not simply a result of inconsistent preferences by individual monkeys.

The middle panel of Figure 4.3 depicts proportions of left reaches for each monkey. The distribution of scores is symmetrically centered around 0.50 and forms an approximately normal distribution. There is no evidence for the separate populations of strong right- and strong left-handed subjects.

Vertical Cling Reach

The static reach experiment revealed no evidence that a relatively unstable bipedal posture increased lateralized hand preferences even when the results were adjusted for perserverative hand choice sequences. However, we suspected that a more extreme increase in postural instability would elicit lateralized hand preferences. The vertical cling tests were conducted on 37 squirrel monkeys (24 males and 13 females) in their home cages. The cages measured 70 cm by 47 cm by 73 cm and were constructed from 2.54 cm sq. stainless steel mesh through which the monkeys could easily reach. Testing was conducted with a 2 cm diam wood dowel tapered at one end. A pin inserted into the tapered end held a small piece of marshmallow. The tester used the marshmallow to lure the monkey to either the side or the back of its cage at a position high enough to force it to cling with both hands and both feet. The tester then brought the dowel slowly towards the monkey at a point along the monkey's upper midline. To grasp the marshmallow, the monkey had to release its grasp of the mesh with one hand to reach toward the dowel while continuing to use the other hand to cling to the cage. Six trials were presented to each of the monkeys daily for ten days. Position of the food presentation was changed between the side and back of the cage after each trial so that the monkey had to descend from one surface and then ascend to the adjacent one after each response. These constant postural readjustments were imposed to discourage development of sequential perseverative reaching and other stereotyped reaching patterns. A monkey engaged in a clinging reach is shown in Figure 4.4.

The monkeys displayed surprisingly persistent right hand reaching preferences. Right hand responses accounted for 71.8% of all responses. Twenty-four monkeys made significantly more right than left responses ($p < .05$) and seven made significantly more left than right responses. Differences were not significant for the remaining six monkeys. Number of right hand reaches exceeded number of left hand reaches in four of the six. The total of 24 significantly right-handed monkeys significantly exceeded the total of 7 significantly left-handed monkeys ($z = 2.87, p = .004$) and the total of 28 monkeys with more right than left reaches significantly exceeded the total of 9 monkeys with the opposite preference ($z = 2.96, p = .003$). The difference between males and females was not significant. The distribution of hand preferences for individual monkeys is depicted at the bottom of Figure 4.3. As with the other two test conditions there is no indication of a bimodal distribution of scores.

Comparisons Among Reaching Conditions

There was a considerable overlap in monkeys serving in the three

Figure 4.4. Monkey reaching for incentive from vertical cling posture.

experiments just described. It was therefore possible to compare proportions of left hand responses in the same subjects across conditions. Fifteen monkeys, serving in the fishing and the vertical cling experiments, made significantly more left hand reaches in the former (69.7%) than in the latter (35.7%) condition ($p < .02$ for 2-tailed Wilcoxon test). Similarly, 29 monkeys made more left hand reaches in the static reach (47.5%) than in the vertical cling (28.6%) condition ($p < .01$). The difference in left hand reaches for the ten monkeys in both the fishing and static reach experiments was not significant.

Data from monkeys serving in multiple experiments were also used to calculate Spearman rank order correlations among the three conditions. The static and vertical cling scores were positively correlated ($r = .591$, $p < .001$). The correlations of reaches in these two conditions with reaches in the fishing condition were negative and nonsignificant ($r = -.084$ between fishing and vertical cling; $r = -.491$ between fishing and static reach).

Discussion

The observations described here support two important generalizations, one positive, one negative. First, the three experiments showed that hand preference in reaching can vary from significant right preference to no preference to significant left preference in the same group of squirrel monkeys. Manual reaching preferences in squirrel monkeys are not only "weak" but labile and highly modifiable by changes in the context of reaching. Clearly, the preponderant and "strong" right manual preference of humans is not paralleled in squirrel monkeys. Second, the data provide little support for the symmetrical U hypothesis. Only the data from the static reach experiment showed the approximately equal preferences for right and left hand reaching predicted by the hypothesis. The preference scores were, however, concentrated around 50% use of each hand with no evidence of a bimodal distribution.

Significant left preferences occurred during reaches for moving goldfish. Four characteristics of reaching for moving fish may have had significant roles in eliciting the monkeys' left hand preferences. First, capture attempts were quick, ballistic responses in which there was little modification of reaches once they were initiated. Second, tracking and grabbing the moving fish required a sustained, intense visual guidance. This was one of the main conditions hypothesized by MacNeilage et al. (1987) to facilitate left hand reaching. Third, tracking and capture of the fish necessitated constant movement and postural readjustment by the monkeys. In addition, particularly in the wading pool test, reaches were often made from a fairly unstable posture (see Figure 4.1). Sanford, Guin, and Ward (1984) observed that left hand reaching preferences in bushbabies occurred when the animals reached from an unstable vertical posture but not when they reached from a more stable quadrupedal posture. Similarly, significant left hand food reaching preferences in black-and-white ruffed lemurs occurred only when whole body postural adjustments preceded the reaching (Forsythe, Milliken, Stafford, & Ward, 1988).

Present data would therefore strongly suggest that substantial postural adjustment is a necessary if not sufficient condition for group reaching asymmetries. Fourth, the monkeys' fishing attempts were made in a context of intense social competition that caused them to become highly aroused. Limited evidence with human neonates suggests that right response biases are enhanced by increased alertness and irritability (Liederman & Kinsbourne, 1980; Michel & Goodwin, 1979; Reiser, Yonas, & Wikner, 1976). The possible relationship between manual laterality and arousal state clearly should be further investigated.

The similar rates of successful fish capture with preferred and nonpreferred hands was somewhat unexpected. However, the monkeys' almost equivalent ability to catch fish with either hand would not be unexpected if manual asymmetry arose from the problem of unifying the outputs of two separate motor systems. When this unification first develops, the relative skills of the two systems are not as important as the decision about which of the two equally capable systems is to be initiated first in a situation requiring quick responses (Kinsbourne, 1974). Manual preferences by squirrel monkeys catching live fish may reflect early development of motor organization where priority of movement is being established, but specialized skills for each hand have not yet developed.

The static reach experiment also required the monkeys to use visual guidance to reach for food items. The food was stationary and, unlike the fish, did not occasion any strong urgency for quick responding. The monkeys displayed no evidence of manual bias across the group even after the correction for sequential hand perseveration. Furthermore, the lack of bias was just as evident when the monkeys reached from a bipedal, upright posture as when they reached from a quadrupedal, horizontal posture. This result contrasts with the previously noted study by Sanford et al.(1984) who found a left hand preference in bushbabies reaching for stationary food from bipedal but not quadrupedal postures, a result they attributed to increased postural instability during vertical reaching. Our squirrel monkeys' lack of group hand preferences during the bipedal reaching condition may be attributable to their frequent use of the horizontal slot for manual support for one hand while the other hand reached for the food. Consequently, the bipedal posture was not especially unstable. The possibility that hand preferences are potentiated by postural instability remains highly tenable in light of other confirmatory evidence.

Some of the evidence that postural instability potentiates manual asymmetry came from the vertical cling reaching. The monkeys displayed pronounced right hand reaching preferences when clinging from vertical surfaces. The clinging posture was considerably less stable, and probably far less comfortable, than the bipedal posture in the static reach experiment. Therefore, the vertical cling condition probably elicited more hurried responses than did the static reach condition.

As MacNeilage et al.(1987, 1988) noted, right hand reaching preferences while the left hand is used for suspensory support may be related to lateral differences in bone and muscle weights in the upper limbs. Dhall and Singh (1977) found that muscle and bone weights on the upper limbs of 12 rhesus monkeys were generally significantly higher on the right side. Two notable exceptions were the pectoralis major, which was larger on the left side (although not quite statistically significant, $p < .1$) and the teres major (equal on both sides). A survey of upper limb sizes in 150 rhesus monkeys by Falk, Pyne, Helmkamp and DeRousseau (1988) also showed considerably larger sizes on the right side. However, a study of upper limb muscles of human fetuses showed heavier right side weights with the exception of the pectoralis major which was significantly heavier on the left side (Pande & Singh, 1971). In adult humans, however, the pectoralis major is heavier on the right side, probably as a result of more intensive use of the right arm (Chibber & Singh, 1972). The pectoralis major and teres major muscles are both important contributors to bodily suspension by the arms. Therefore, the large size of the left pectoralis major may be evidence of a primitive feeding pattern involving simultaneous suspension by the left arm and reaching by the right, exactly the pattern the monkeys preferred in the vertical cling reaching.

Finally, the correlation among hand preferences in the three reaching conditions suggests that two different manual reaching systems may have been evoked in these experiments. The static and clinging reaches were positively correlated; both conditions permitted reaching for a stable, stationary food source. In contrast, reaches in the fishing condition had negative and nonsignificant correlations with those in the other two conditions. These results suggest that visually guided reaching for stationary incentives evokes a motor system with right hand predominance, contingent on sufficient postural instability and urgency for rapid responding. In contrast, visually guided ballistic reaching for moving incentives may evoke a motor system with left hand predominance also contingent on sufficient postural instability and urgency for rapid responding. Confirmation of these two highly speculative hypotheses awaits further observations.

References

Annett, M. (1985). *Left, right, hand and brain: The right shift theory*. New York: Erlbaum.

Box, H.O. (1977). Observations on spontaneous hand use in the common marmoset (*Callithrix jacchus*). *Primates, 18*, 395-400.

Brookshire, K.H., & Warren, J.M. (1962). The generality and consistency of handedness in monkeys. *Animal Behaviour, 10*, 222-227.

Calvin, W.H. (1987). On evolutionary expectations of symmetry and toolmaking. *Behavioral and Brain Sciences, 10*, 267-268.

Chibber, S.R., & Singh, I. (1972). Asymmetry in muscle weight in human upper limbs. *Acta Anatomica, 81*, 462-465.

Cicchetti, D.V. (1987). On viewing the evidence for primate handedness: Some biostatistical considerations. *Behavioral and Brain Sciences, 10*, 368-369.

Cole, J. (1957). Laterality in the use of the hand, foot, and eye in monkeys. *Journal of Comparative and Physiological Psychology, 50*, 296-299.

Corballis, M.C. (1983). *Human laterality*. New York: Academic Press.

Cronholm, J.N., Grodsky, M., & Behar, I. (1963). Situational factors in the lateral preference of rhesus monkeys. *Journal of Genetic Psychology, 103*, 167-174.

Dhall, U., & Singh, I. (1977). Anatomical evidence of one-sided forelimb dominance in the rhesus monkey. *Anatomischer Anzeiger, 141*, 420-425.

Fagot, J., & Vauclair, J. (1988). Handedness and bimanual coordination in the lowland gorilla. *Brain, Behavior and Evolution, 32*, 89-95.

Falk, D., Pyne, L., Helmkamp, C., & DeRousseau, C.J. (1988). Directional asymmetry in forelimb of *Macaca mulatta*. *American Journal of Physical Anthropology, 77*, 1-6.

Fleagle, J.G., Mittermeir, R.A., & Skopec, A.L. (1981). Differential habitat use by *Cebus apella* and *Saimiri sciureus* in central Surinam. *Primates, 22*, 361-367.

Forsythe, C., Milliken, G.W., Stafford, D.K., & Ward, J.P. (1988). Posturally related variations in the hand preferences of the ruffed lemur (*Varecia variegata variegata*). *Journal of Comparative Psychology, 102*, 248-250.

Green, D.M. (1964). General prediction relating yes-no and forced choice results. *Journal of the Acoustical Society of America, 36*, 1042.

Green, J.B., & Swets, J.A. (1974). *Signal detection theory and psychophysics*. New York: Krieger.

Grier, J.B. (1971). Nonparametric indicies for sensitivity and bias: Computing formulas. *Psychological Bulletin, 75*, 424-429.

Hacaen, H., & Ajuriaguerra, J. de (1964). *Left handedness: Manual superiority and cerebral dominance*. New York: Grune and Stratton.

Harlow, H.F. (1959). Learning set and error factor theory. In S. Koch (Ed.), *Psychology: A study of a science, Vol. 2* (pp. 492-537). New York: McGraw Hill.

Harris, L.J., & Carlson, D.F. (1988). Hand preference for visually guided reaching in human infants and adults. *Behavioral and Brain Sciences, 11*, 738.

Healey, J.M., Liederman, J., & Geschwind, N. (1986). Handedness is not a unidimensional trait. *Cortex, 22*, 33-53.

Itani, J., Tokuda, K., Furuya, Y., Kano, K., & Shin Y. (1963). The social construction of natural troops of Japanese monkeys in Takasakiyama. *Primates, 4*, 1-42.

King, J.E., & Michels, R.R. (1989). Error analysis of delayed response in aged squirrel monkeys. *Animal Learning and Behavior, 17*, 157-162.

Kinsbourne, M. (1974). Lateral interactions in the brain. In M. Kinsbourne & W.L. Smith (Eds.), *Hemispheric disconnection and cerebral function* (pp.239-259). Springfield: Thomas.

Landau, V.I. (1987). *The adaptation of New World monkeys to new environmental situations: Food acquisition and food processing behaviors*. Unpublished doctoral dissertation, University of Arizona, Tucson.

Lehman, R.A.W. (1970). Hand preferences and cerebral dominance in 24 rhesus monkeys. *Journal of the Neurological Sciences, 10*, 185-192.

Lehman, R.A.W. (1978). The handedness of rhesus monkeys - I. Distribution. *Neuropsychologica, 16*, 32-42.

Lentz, J.L., & King, J.E. (1981). Sources of errors by capuchin monkeys on delayed response. *Animal Learning and Behavior, 9*, 183-188.

Levine, M. (1965). Hypothesis behavior. In A.M. Schrier, H.F. Harlow, and F. Stollnitz (Eds.), *Behavior of nonhuman primates, Vol. 1* (pp. 97-129). New York: Academic Press.

Liederman, J., & Kinsbourne, M. (1980). The mechanism of neonatal rightward turning bias: A sensory or motor asymmetry? *Infant Behavior and Development, 3*, 223-238.

MacNeilage, P.F. (1987). The evolution of hemispheric specialization for manual function and language. In S. Wise (Ed.), *Higher brain functions: Recent explorations of the brain's emergent properties* (pp. 285-308). New York: Wiley.

MacNeilage, P.F., Studdert-Kennedy, M.G., & Lindblom, B. (1987). Primate handedness reconsidered. *Behavioral and Brain Sciences, 10*, 247-303.

MacNeilage, P.F., Studdert-Kennedy, M.G., & Lindblom, B. (1988). Primate handedness: A foot in the door. *Behavioral and Brain Sciences, 11*, 748-758.

Michel, G., & Goodwin, R. (1979). Intrauterine birth position predicts newborn supine head position preferences. *Infant Behavior and Development, 2*, 29-38.

Milner, A.D. (1969). Distribution of hand preferences in monkeys. *Neuropsychologica, 7*, 375-377.

Napier, J.R., & Napier, P.H. (1967). *A handbook of living primates*. New York: Academic Press.

Pande, D.N., & Singh, I. (1971). One-sided dominance in the upper limbs of human fetuses as evidenced by asymmetry in muscle and bone weight. *Journal of Anatomy, 109*, 457-459.

Porac, C., Coren, S., Steiger, J.H., & Duncan, P. (1980). Human laterality: A multidimensional approach. *Canadian Journal of Psychology, 34*, 91-96.

Reiser, J., Yonas, A., & Wikner, K. (1976). Radial localization of odors by human newborns. *Child Development, 47*, 856-859.

Sanford, C., Guin, K., & Ward, J.P. (1984). Posture and laterality in the bushbaby (*Galago senegalensis*). *Brain, Behavior and Evolution, 25*, 217-224.

Scott, A.G. (1990). *Monkeys, movements, and memories: Behavioral sequences and short term memories in aged monkeys.* Unpublished doctoral dissertation, University of Arizona, Tucson.

Scott, A.G., King, J.E., Michels, R.R., & Fobes, J.L. (1989). Effects of $[D\text{-ala}]^2$ met-enkephalinamide, a met-enkephalin analog, on delayed response by squirrel monkeys. *Physiology and Behavior, 46*, 605-612.

Siegel, S. (1956). *Nonparametric Statistics for the Behavioral Sciences*. New York: McGraw-Hill.

Terborgh, J. (1983). *Five new world primates: A study in comparative ecology.* Princeton: Princeton University Press.

Warren, J.M. (1977). Handedness and cerebral dominance in monkeys. In S. Harnad, R.W. Doty, J. Jaynes, L. Goldstein & G. Krauthamer (Eds.), *Lateralization in the nervous system* (pp.151-172). New York: Academic Press.

Warren, J.M. (1980). Handedness and laterality in humans and other animals. *Physiological Psychology, 8*, 351-359.

5

Cerebral Asymmetry, Interhemispheric Interaction and Handedness: Second Thoughts About Comparative Laterality Research with Nonhuman Primates, About a Theory and Some Preliminary Results

Bruno Preilowski

The ultimate aim of comparative work on laterality is to contribute to an understanding of functional and structural lateral cerebral asymmetries in man and the principles underlying it. Off hand, nonhuman primates appear to be very suitable subjects for this type of research; our fellow primates are capable of learning and performing highly complex tasks, and some of the species which have been studied appear to use their brains and hands very much like humans. Nevertheless, I feel forced to conclude that in doing comparative research with nonhuman primates we have not yet made any significant contribution to the solution of the riddle of cerebral asymmetry.

Rather, it appears as if in using other animal models more substantial progress has been made. The work on laterality in birds is an outstanding example (see Rogers, 1986, for a recent review). Unfortunately, the insights gained into basic mechanisms of, for instance, asymmetric brain development, by way of these approaches are too far removed from the human situation, where some important questions urgently await answers. Take, for example, the relationship between handedness and functional as well as structural brain asymmetries. Without the use of invasive methods, like the Wada Test, we still cannot reliably indicate lateralization of speech functions in an individual patient who has to undergo brain surgery. We also do not know whether handedness is related to or influences the development of bi-manual coordination or cognitive functions in the growing child. So, especially — but not exclusively — with regard to questions concerning handedness and laterality, comparative research with nonhuman primates remains a necessary adjunct to work with other animal models.

It is the objective of this contribution to discuss possible reasons for the lack of progress in laterality research with nonhuman primates. A brief description of an attempt to learn about handedness and cerebral asymmetry by studying intermanual transfer of manual skills in rhesus monkeys and an outline of a theory will serve to illustrate some of the points I am trying to make.

Problems of Laterality Research with Non-human Primates

Laterality research with nonhuman primates is associated with two kinds of problems. Only one set of these difficulties is specific to comparative work with primates. The other kind appear to be more general problems of laterality research which are encountered in work with animals as well as human subjects. The existence of these general problems of laterality research is perhaps most clearly indicated by the fact that in the area of human experimental neuropsychology, despite enormous numbers of studies, many admittedly more imaginative than those conducted with non-human primates, progress also has been less than satisfactory.

Basically, these latter problems of laterality research appear to be related to the complexity of the questions that arise in its context. Thus, it has proven extremely difficult to formulate theories which allow the deduction of hypotheses specific enough to be tested and possibly discarded by decisive experiments. The problem of agreeing upon and adhering to precise definitions can also be largely attributed to the complexity of the phenomenon of laterality.

Comparative research with nonhuman primates has been plagued by two additional problems. One is the predominance of Scala Naturae thinking. The other problem, which is quite unique to primate research on laterality, is that of detrimental experimental compromises.

General Problems of Laterality Research

The Problem of Definition

It is not possible here to give a complete overview concerning the various definitions that have been proposed in the area of laterality research. I merely would like to describe how I shall be using certain terms and to detail the definitions of those concepts and their limitations, which play an important role in the following discussion of theoretical and methodological questions.

Laterality. I am using the term laterality as shorthand for Lateral Cerebral Asymmetry which describes a relatively permanent morphological or functional lateral asymmetry of the nervous system and behavioral asymmetries directly related to them.

Handedness. As may become more obvious after reading the next section in which the complexity of the various research questions is discussed, the problem of defining handedness is twofold. One part of the problem concerns finding a neutral description of the various behavioral expressions of manual asymmetries. The other part of the problem consists of defining the relationship of these handedness forms to laterality. This latter part is the difficult one. Here, it must be determined under what circumstances we accept inferences about laterality on the basis of behavioral measures. Such a definition may thus include additional, quite diverse arguments that incorporate logical, statistical, and neuroanatomical qualifications.

I shall first consider the easier part and distinguish two forms of manual asymmetries:

(1) *Qualitative handedness*: This kind of handedness can be defined as a difference between right and left hand in quality and speed of performance; it was previously called "performance handedness."

(2) *Quantitative handedness*: The other kind of handedness refers to a difference in the frequency of using right and left hand; it was called "preference handedness" before, since it was determined in situations in which the subject could choose to use one or the other hand.

In the course of our studies I have been impressed by the strong influence of habit formation in many testing situations so that the term "preference" — which would perhaps imply a conscious deliberate choice of hand — appeared to be inappropriate. Young, Corter, Segalowitz, and Trehub (1983) have proposed a similar distinction by differentiating between "handedness" (describing prevalence of hand usage), and "manual specialization" (indicating qualitative differences). As is the case with "qualitative handedness," Young et al. (1983) want to restrict "manual specialization" to a particular situation; nevertheless, I am uncomfortable with this term, probably because of its connotations of hemispheric specialization.

Although not totally satisfactory, the distinction of a quantitative and a qualitative form of handedness will serve well for the time being. It is important to note that the definition of these two kinds of handedness does not automatically imply a distinction between a superficial or acquired kind of handedness versus an innate or "real" handedness. Similarly, no inference is allowed from these preliminary definitions concerning laterality. As already indicated, any such inference would require a further refinement of the handedness definitions. Such a further differentiation presupposes answers to a whole complex of questions about laterality and handedness.

The Problem of the Complexity of Research Questions

The following four questions are probably the most important ones arising in the context of comparative research on laterality and handedness:

(1) Is handedness related to structural or permanent functional brain asymmetry(-ies) or to a temporary asymmetric brain state or brain states?

(2) Is handedness related to other forms of laterality (asymmetries in perception, information processing, communicative functions, etc.)? And if so, how is it related to these?

(3) Is handedness a product of the genes or of experience?

(4) With regard to the phylogenesis of handedness, has a continuous development with a common origin taken place, or independent species specific evolutions?

When I stated in the beginning that we haven't made much progress in the comparative study of laterality with nonhuman primates, it was because, among other reasons, we have no answers to any of these questions.

The complexity of this set of questions is obvious, as one must include the possibility that we are dealing with more than two definite alternative answers to any of these. Instead of being a matter of "either-or", it could well be in some cases "more-less" or "as well as".

To illustrate this point let us take a look at possible answers to the first question. Figure 5.1 summarizes some simple models of dual brain functions. However, it is entirely possible that laterality can be adequately described only by a model which combines the features of several of those simple ones listed in the table.

Further complications for a formulation of theories or models and subsequent clearcut research hypotheses arise from the fact that the questions outlined above are not independent of each other, in so far as possible answers to any of them may have important implications for research towards an attempt to answer the others. This in turn, however, does not mean that in the absence of satisfactory data concerning a particular question, results obtained with regard to any of the others could be taken as substitute answers or circumstantial evidence for arguments at a level other than that of the originally investigated question.

Constructions between different levels of investigation, for example, between the behavioral and the neuroanatomical level, are not uncommon in laterality research and quite problematic. Thus differences in reaction times between hands to lateralized stimuli are quite frequently interpreted to support a particular hypothesis of lateralized cognitive functioning. Often, this interpretation is heavily dependent upon the assumption of a certain neuroanatomical circuitry involved in cognitive-sensory-motor brain functions, which is itself quite hypothetical. Similarly, connections are frequently constructed between, for example, tentative answers to the first of the above formulated questions concerning a static or dynamic brain basis for handedness and suppositions about the third question concerning the nature-nurture issue.

The cumulative effects of these problems of laterality research are a lack of deductive approach, lack of adequate theories and precisely formulated hypotheses, of unambiguous definitions and decisive experiments, as well as the dangers of post-hoc additions of assumptions or changes in definitions.

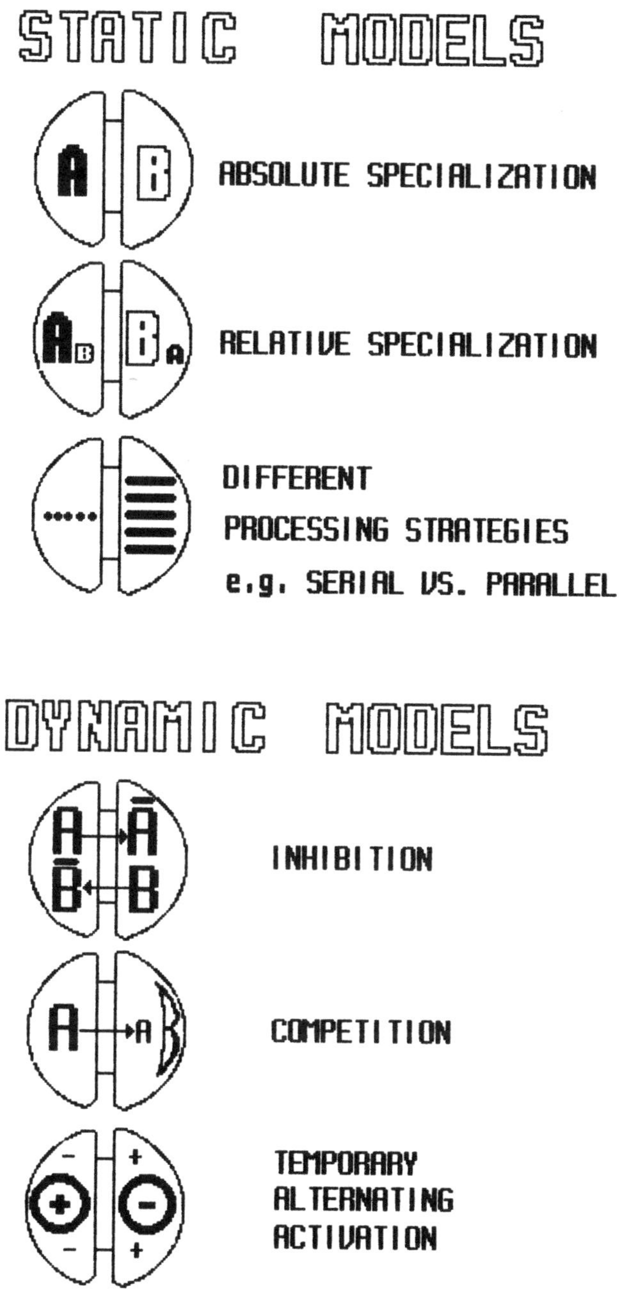

Figure 5.1. Schematic representation of the major dual brain function models of laterality. A and B represent functions like language, spatial perception, and so on.

These are most clearly demonstrated in human laterality research. Here some areas are characterized by huge accumulations of equivocal data which are of value only to questionable ex-post-facto models.

Problems of a Definition of Handedness as a Form of Laterality

Coming back now to our two forms of handedness, the quantitative and the qualitative one, what additional attributes are required to accept handedness as a form of laterality? How can one deal with the complexity of this question? The only sensible way, of course, is to use a theory to structure and simplify our task. Many such theories are possible. I shall discuss only the one we have followed. But before I do, I must complete the description of difficulties of comparative research with nonhuman primates. Whereas the former elaboration of general difficulties of laterality research may be seen as a pretext for introducing our own rather limited theoretical approach, this latter discussion of problems specific to research with nonhuman primates may be taken, if you like, as an attempt to excuse the rather modest rate of progress of our own experimental efforts.

Problems Particular to Comparative Research with Non-human Primates

The Scala Naturae Influence

Since various expressions of laterality have been documented and most extensively studied in man, it is legitimate to search for similar asymmetries in man's closest relatives. However, an exclusive orientation of comparative research that places man at the top of phylogenesis incorporating the final products of evolution and other lower primates carrying only precursors of human laterality is not acceptable. More than twenty years ago, Hodos and Campbell (1969) pointed out the detrimental effects of such Scala Naturae thinking on comparative research in general.

I have discussed elsewhere in more detail (Preilowski, 1978, 1983) how findings about human laterality and handedness in particular have been rather uncritically accepted as the standard with which methods and results of comparative work were compared. But it cannot be pointed out often enough that a Scala Naturae orientation prevents a truly comparative approach to laterality research in nonhuman primates. By excluding from consideration possible species-specific developments one foregoes the very advantages of comparative research. For instance, it precludes finding a more general biological mechanism possibly underlying laterality. It is in this sense that I have previously argued for a more biologically oriented evolutionary theory of laterality (Preilowski, 1972, 1978, 1983), not because I claim it to be the only correct one, but because it is the more open-minded one. Thus it could be profitable to look for dexterity in nonhuman primates before checking for right-handedness, or to test for cerebral asymmetry in the perception of facial grimaces as socially relevant signals rather than expecting an asymmetry in the perception of faces merely because faces are complex holistic stimuli.

Experimental Compromises

In a paper presented at a NATO Advanced Study Institute on Comparative Biology in Primates (Preilowski, 1972) on the question of whether there is laterality in nonhuman primates, I reviewed the literature and came to the conclusion that there is no convincing evidence for laterality, but neither is there sufficient evidence against it. To me this appeared to be due to a dearth of relevant behavioral, genetic, physiological, and anatomical studies. The existing behavioral studies up until that time had been restricted mainly to the investigation of handedness in monkeys.

Undaunted at that time by much experience of my own in working with nonhuman primates, I criticized these studies for using tasks which were not indicative of cerebral cortical functions. I also argued that the lateralization of such tasks would be irrelevant for the survival of the species, and therefore irrelevant to the study of laterality as a product of evolution. Furthermore, I criticized the fact that the monkeys studied were not selected for species, age, sex, and degree of experimental experience. I then proposed an evolutionary theory of laterality and boldly suggested a program of laboratory and field studies to remedy the situation.

I have stuck with the theory, and I think most of the critical remarks are still valid today as well, although at the time a certain degree of naivety probably helped to express them. It would certainly be more difficult now, as I find my own efforts subject to the same criticism and my original program of research changed beyond recognition, mostly because of external, nonscientific reasons. Exclusive dependence on short term grants often required contraction and selective thinning of designs. Lack of support for long-term animal care and a limited supply of monkeys frequently made it necessary to use whatever animals were available. Laterality and handedness not being "hot" topics until recently, behavioral work not considered quite scientific, and work with monkeys being rather expensive, sometimes meant agreeing to alternative work favored by funding agencies and thereby compromising the investigations on laterality research. I think it is not an unjustified generalization beyond my own situation to state that a lot of laterality research in nonhuman primates from other laboratories shows similar characteristics.

External or financial pressures were not always directly responsible for compromises in the experimental design. We sometimes ran out of support while following time consuming detours which had not been anticipated. At times we were forced to back up from a particular question and attack others first, as it turned out that certain assumptions, based on findings in the literature or on previous observations of our own, were untenable.

For example, while I was convinced that the investigation of lateralized motor functions indicative of lateralized cerebral control would need to involve individual fractionated finger movements, we were unable to train monkeys to perform such movements under controlled experimental conditions. Although we have found a report about an electrophysiological study in which monkeys were trained to press different buttons with different fingers (Schwartzman, 1978), I would conclude from our own substantial but unsuccessful efforts that it will take a separate research program to conclusively answer the question of whether or not nonhuman primates really are capable of controlled individual finger movements, such as lifting only one single finger by itself without moving the others of that hand.

Since we were unable to pursue such a program at the time, we had to rely on less clearly lateralized tasks, such as a force discrimination between two fingers of one hand (subsequently called Finger-Force-Discrimination or FFD-task). The currently used manipulanda are shown in Figure 5.2.

Early reports from animal studies and clinical studies indicated that such pincer grasp finger movements appear to be controlled only by the contralateral hemisphere (Brinkman & Kuypers, 1972). However, more recent electrophysiological data (M. C. Hepp-Reymond, personal communication, January 22, 1988) and observations following recovery from unilateral pyramidal damage (Kučera & Wiesendanger, 1985) as well as reports about ipsilateral control of finger movements in a split-brain patient (Trope, Fishman, Gur, Sussman, & Gur, 1987) suggest a reexamination of this issue. We tried to control for this as much as possible by including a control task in which whole arm movements constituted the response (the Arm-Force-Discrimination or AFD-task), and thus a greater degree of ipsilateral control was possible, in comparison with the more distal finger movements with presumably greater contralateral control.

Still other experimental compromises just cannot be avoided as they are the result of a dilemma for which no solution is yet available. This inescapable dilemma in handedness and laterality research lies in the fact that the more relevant we want a task to be in order to enable us to tax the capacity of each hemisphere and the more certain we want to be about exclusive unilateral cortical control, the more training is involved in the use of such tasks. However, this will then increase the risk of introducing a bias with regard to both qualitative and quantitative handedness measures.

At this point I would also have to talk about the enormous individual differences among monkeys as well as of changes over time in motivation and motivational effects on performance. These influences are complicated, and interactions between the consequences of performance changes of one hand on the other in an individual animal are difficult to assess.

I would also have to talk about the difficulties in comparing performance levels and capacities. This latter complex of problems can be exemplified by

Figure 5.2. Presently used manipulandum of the Finger-Force Discrimination Task. The animal has to reach into a slot of a loosely mounted cylinder and squeeze two metal tongues to produce a specified force.

the problem with which a coach is confronted when trying to nominate the best athlete for a meet: should he choose the one who has jumped a world record but only once or the one who has consistently demonstrated high performance although never quite achieving a top level result? Again, the operationalization of performance measures entails compromises. And, when working with highly developed animals like primates and complex tasks, the decisions required to reach a satisfactory compromise are much more debatable than when performance criteria are defined in simpler tasks for more uniformly behaving animals.

This is not the place for an extensive discussions of any of these or other related issues. But perhaps what has been discussed so far already suffices to make the point that comparative laterality research is very complex, time-consuming and expensive. It requires an all-out effort, or otherwise its results are likely to be disappointing.

About a Theory for the Study of Handedness and Laterality

The theoretical background of my work on handedness and laterality in nonhuman primates extends from a rather general theory of laterality to more restricted hypotheses about the role of the neo-commissures, specifically of the corpus callosum, in the development of asymmetrical functional states of the brain. The different parts of this framework are not new. If originality can be claimed at all, then only for the configuration as a whole.

If we consider laterality to be a general biological phenomenon, a kind of specialization as an evolutionary product, then this implies that it may assume a variety of adaptive forms in different species depending on particular, perhaps species-specific, selective pressures. Fairly convincing explanations for the increase in size and complexity of mammalian brains have been proposed in terms of the evolutionary advantages of a growing capacity to integrate more and more sensory systems and of an increased flexibility in the use of these systems. There is good evidence for the assumption that this development involved a move from the simple representation of the immediate environment in specific brain structures to a complex reconstruction of the world throughout the brain, a world which extends beyond the immediately perceived environment (i.e. beyond the present to include the past as well as expectations of the future). Similarly, it would entail a growing flexibility in reacting to environmental and internal stimuli.

Consideration of the metabolic cost of maintaining an organ like the brain makes it clear that its growth, in order to increase its power, must be limited. In addition, the problems of larger brains and heads at birth can only be partly circumvented by retarding prenatal development. Flexible usage of brain systems, of which an example of perhaps its highest form is "learning to learn", appears to be a reasonable solution which imparts obvious advantages with regard to biological fitness at a tolerable expense.

Lateral specialization by "de-duplication" has often been invoked as a logical step in this development to increase the efficient use of brain tissue. Many investigators, however, reserve this developmental step for man. For instance, Levy argues that de-duplication of functions and hemispheric specialization "arose in response to specifically human selective pressures, occurring in a socially organized species, the members of which were mutually interdependent and who each served to increase the fitness of the species as a whole by the contribution of his specific skills" (Levy, 1977, p. 271). But why should one decide a priori that laterality is restricted to man and identical with his individualized superior language and cognitive skills? Why assume the effectiveness of an evolutionary mechanism for man and deny its influence in other species which have had a separate, equally long or even longer, evolutionary history of their own? In the absence of any evidence to the contrary, I assume the working of a general biological — not specifically human — evolutionary mechanism of lateral specialization.

As the brain evolves from a representational to a reconstructional system and finally to a "de-duplication" of functions (i.e. functional differentiation of homotopical brain areas) there must be a transitional phase which I would like to characterize as a change from congruence to independence. In other words, systems in the right and left halves of the brain must first be able to function independently before they can assume different roles. If we look for a system which could be instrumental in bringing about this change from congruence to independence, the commissural system is an obvious candidate. However, neither our knowledge of its ontogenentic or phylogenetic development nor the existing neurophysiological, behavioral, or clinical data allow us as yet to say exactly what this development entails and how the commissures enable such development. The problem is that the neocortical commissures apparently do not have a unitary function and cannot be considered as an independent system of their own.

On the contrary, the neocortical commissures must be considered (1) part of an intricately and completely balanced organization including not only the neocortical hemispheres but subcortical structures as well and (2) consisting of various components, playing different and varying roles. The variability in the functioning of smaller or larger subsets of the commissures (a) extends over different time periods, for instance, as a function of maturation as well as experience (e.g. learning); (b) depends on the various dynamic processes involved in a particular task; for example, according to the level of learning which has been achieved or specific requirements of performance; and (c) depends on the changing contributions and states of various subsystems and structures of the brain.

We are just beginning to learn about the anatomical and electrophysiological complexity of the commissures and their potential for plastic changes. At the level of behavioral investigations it has long since become apparent that the commissures are not just direct connections for the simple exchange of information between the two halves of the brain. To give only a few examples, the use of negative and positive reinforcement has been shown to have different effects on interhemispheric transfer, and this in turn influences acquisition and retention in split-brain cats (Sechzer, 1970). Split-brain patients show more interlimb interaction than one would expect considering the extent of the disconnection of their hemispheres. For example, they show no obvious deficits in preoperatively overlearned bimanual sensory-motor skills, but they have even greater difficulties than normal in performing newly acquired interdependent as well as, as is especially important, independent arm, hand, and finger movements (Preilowski, 1975, 1987; Tuller & Kelso, 1989).

With regard to intermanual transfer of lateralized sensory-motor skills, which is of special significance to the discussion about handedness, the situation is equally complex: On one hand, there are reports about the possibility of transfer in split-brain subjects, and, on the other hand, about a lack of transfer with normal, intact interhemispheric commissures. While the positive transfer of learning, which has been observed in both human and animal split-brain subjects, could be due to a failure or incompleteness of lateralization of the skill during original learning or due to subcortical interhemispheric communication, the lack of transfer of learning we have seen in intact animals (Preilowski, 1977a, 1977b, 1979; Preilowski, Reger, & Engele, 1986) is not as easily explained.

My hypothesis is that the development of a delicate sensory-motor skill dependent on control of the distal extremities can remain lateralized to the opposite hemisphere despite intact commissures. As an additional hypothesis it might be proposed that, with extended training of only one hand, an active interhemispheric inhibition precludes the bilateral establishment of skill. Support for such a hypothesis comes from studies describing a lack of interocular transfer despite intact commissural connections in rats (Steele Russell & Ochs, 1963; Burešová, Bureš, & Nadel, 1972), rabbits (e.g. Van Hof, 1970), fish (Ingle, 1968), and pigeons (e.g. Palmers & Zeier, 1974). Suppression of interocular transfer appears to be dependent upon a certain degree of independence of the visual fields of both eyes. Here it may serve to prevent conflict in the event of different visual inputs to the two eyes. Similarly, an increased independence of the upper extremities may be correlated with an increased capacity for interhemispheric inhibitory actions with regard to the cerebral control mechanisms of both hands.

The idea of lateralized learning in subjects with intact neocortical commissures and of an active role of the corpus callosum in preventing interhemispheric transfer carries important implications for a possible theory of handedness and laterality, as well as for an understanding of cortical motor functions in general. The latter would be especially important for teaching bimanual skills as well as for the rehabilitation of patients with motor problems due to unilateral brain lesions.

A Summary of Our Results on Quantitative and Qualitative Handedness in Rhesus Monkeys and the Role of the Corpus Callosum

Together with a number of students, I have studied handedness and intermanual transfer in nonhuman primates for several years now. Almost all of the experimental work was done with rhesus monkeys. The aim was to learn about handedness by looking for quantitative and qualitative asymmetries in hand use, uninfluenced by cultural pressures and unconfounded by cognitive reflections about possible consequences of choosing a particular hand in a specific task. The latter can be expected to influence the choice of hand in man. And this, together with lifelong habit formation through the use of one hand in all sorts of situations, surely contributes to the extreme expressions of asymmetry in human hand preferences. For testing the monkeys several different tasks were used, ranging from simple food retrieval under various conditions to more demanding manipulations which either required more skill or larger efforts, and sometimes both, to obtain a reward.

Quantitative Handedness

With regard to quantitative aspects of handedness, we found that the asymmetry in hand usage increased with the amount of sensory-motor effort expended in performing the task. By sensory-motor effort I mean the amount of force that had to be produced as well as adjustments required in posture or stance. For example, in 14 animals the average degree of asymmetric hand usage increased from 62% when a piece of fruit had to be picked from a short tube to 66% when the animals had to reach into a long tube. The asymmetry increased further to 80% when the reward had to be plucked from a string dangling in front of the animal, and to 94% when the latter had to be accomplished with one eye peeping through a hole to control the act (equal use of right and left hand would be indicated by 50%, whereas exclusive use of one hand would result in a score of 100%). The highest consistency of hand usage was found in a task which required the animals to produce a specific force between the fingertips of one hand for a defined length of time. Here all animals used only one hand from the beginning and stuck with it for months and thousands of trials. Even when repeatedly failing with one hand we never observed the least attempt to use the other hand (Preilowski, 1977a).

Although the degree of asymmetry (i.e. the consistency with which only one hand was used) was influenced by task variables (complexity or required effort), no systematic effect was observed on the direction of the asymmetry (i.e. prevalence of right or left hand usage). It appeared as if a particular hand was used initially only by chance. Even in a situation in which an animal was successively tested with each hand separately and for a certain period performed consistently better with one hand, it was not possible to predict which hand it would use under conditions of free choice of hand.

Qualitative Handedness

Performance differences between hands were, of course, of greatest interest to us. It was for this reason that we had tried to find a task which would maximize the contribution of unilateral cortical processes and which would thus allow us to infer hemispheric from manual performance asymmetries. A task which had been successfully used in the study of pyramidal tract functions (Wiesendanger & Hepp-Reymond, 1970) appeared to fit our demands. This was the already mentioned FFD-task in which the animals had to produce a force between the fingertips and to maintain this force between specified lower and upper limits for a certain duration. To increase the difficulty of the task the required duration was increased up to 1.5 sec. of continuous pressure. Then the gap between the lower and upper limits of acceptable force was narrowed in successive steps. The titration procedure assured that performance at a certain level of difficulty had reached a specified criterion before the difficulty of the task was increased.

The experiments were run under two basically different conditions. In one series the animals were tested for up to an hour in a separate experimental cage within a sound attenuated room. Later, we incorporated our testing into the room in which the animals were housed, such that up to two animals could work whenever they felt like it, receiving their normal food as rewards from special feeding devices. In the earlier tests, series of trials of 10 sec. length were run, which allowed us to distinguish between trials without attempts as well as unsuccessfully and successfully completed trials. In the later experiments, when the animals had continuous access to the manipulanda from their home cage, instead of using a trial structure we recorded any attempt of the monkey at manipulation, thus giving us a measure of the working time they needed to reach a certain criterion of success.

When a trial was on, or, in the continuous access situation, when an animal touched the manipulanda, a tone (e.g. 350 Hz) was presented. This trial-on-tone was switched to a tone of a different frequency (e.g. 500 Hz) as long as the force produced by the animal was within the specified limits. This all-or-none feedback was chosen over continuous feedback in order to increase the probability that the lateralized tactile kinesthetic feedback would remain essential for monitoring the fine adjustment of force production between the fingers.

With the use of this task we observed consistent performance differences in several normal monkeys (Preilowski, 1979). Although the fact that all monkeys until that time had shown better performance with the right hand was found of greater interest (e.g. MacNeilage, Studdert-Kennedy, & Lindblom, 1987), I did not think then (and still don't now) that an asymmetry in direction could be meaningfully interpreted on the basis of only eight animals. The importance of the results in my view lay in the demonstration of performance differences between hands of normal animals. Such a difference would indicate, even if reliably present only in a single animal, the possibility of the development of asymmetric capacities of both hemispheres. The only problem was, how could we make sure that these performance differences were reliable and valid?

The Problem of Verifying Performance Differences Between Hands

It may seem an easy task to verify qualitative differences. However, I have to admit now, almost 10 years and several experiments later, that we still cannot be sure what these performance differences mean. The difficulties in evaluating the data from performance tests are manifold. One of the problems was already mentioned earlier, namely that of determining the criteria of better performance. Does a 25% success rate in producing a force within the range of 20 pond constitute a better performance than an 80% success rate within a 25 pond range? How should a measure of variability of performance enter into determining a performance score?

A further basic problem is that of statistical analysis and interpretation of dependent data, namely those from the two hands of the same animal. Somewhat related to this is the problem of the influence of motivation on the performance of right and left hand. As long as the average success rate of performance was relatively high, changes in motivation, which were indicated by changes in the number of attempts made by the animals per unit time, seemed to be influencing both hands in similar ways. As the average success rate dropped, we often saw that one hand became better and better just as the performance with the other hand was getting worse and worse.

Performance tests were run mainly under two conditions: Either both hands had to work at the same level of difficulty, which was the highest level reached by both hands, or the level of difficulty was individually adjusted for each hand depending on its running performance. In either case, consistent differences in performance between the hands appeared mainly at the higher levels of difficulty. This, of course, was what we had aimed at, because it was reasonable that in order to tax hemispheric capabilities one would have to get at the limits of both right and left hand performance. It was also to be expected that at these levels of difficulty the success rate would drop. As long as we tested animals only unimanually we were able to handle low success rates by titration of difficulty, (i.e. by lowering the level of difficulty after a series of failures and raising it again as performance improved). What we didn't realize was that such a procedure would have unpredictable effects on either hand when both hands were tested. In other words, we still cannot be sure whether or not the performance differences which we found were accentuated or even produced by motivational rather than skill factors.

Controlling Motivational Influences. Alternative methods of analysis, such as time series analysis or other cross-correlational analyses also produced no clear answers with regard to the validity of the observed performance differences or the motivation-performance interactions between hands and hemispheres (Preilowski et al., 1986). Therefore, one of my students undertook to test for qualitative differences in performance between hands under rigorous control of motivational influences (Engele, 1988). To consistently equate the number of successes with each hand, he allowed a switch of hands only after work with the previously used hand had been successful and rewarded. Whereas it was possible that in the long run giving more practice to the less perfect hand may eliminate possible existing subtle differences, such a conservative test for asymmetries in sensory-motor functions appeared to be necessary after previous tests had perhaps unduly favored the appearance of such asymmetries in performance.

The design of Engele's study was complex and I can report here only a small part of the results that were obtained. We look first at the data from nine normal intact animals which were trained with both hands and then tested for performance differences in two 5-day long sessions with an intervening pause of three to four weeks using the FDT-task. As a measure of quality of performance the quotient of "total time within the prescribed lower and upper limits" over "total working time" was calculated: In the first test three animals showed better performance with the left hand, four with the right hand, and two showed no significant differences. In the second test session the performance with the previously worse hand showed improvements in most animals while the better hand remained at the same level, so that only two animals now showed performance differences. This could not have come about because of a ceiling effect for the level of difficulty chosen for this test was below that reached during previous training.

The Effect of Callosotomy. It is also of interest to look at the performance of animals after their corpus callosum was sectioned. Three of the nine previously tested animals were given a complete callosotomy and re-tested, another three were tested after receiving a sham operation and once again after a callosotomy. Two additional animals that were naive to hand testing underwent callosotomy first and then were trained and tested. After the split-brain operation none of the animals showed a consistent asymmetry in performance. If a change in performance was observed at all after the operation, then that performance improved equally with both hands and the results became more similar for the two hands than they had been before the operation. Thus, there was no evidence for a qualitative handedness which would allow us to infer a unilateral specialization for sensory-motor skills in the rhesus monkey. We are presently analyzing the training data to check for asymmetries relative to those found in the earlier tests. However, even if such asymmetries would be found, they appear to be easily leveled by training.

Lack of Intermanual Transfer

Before I summarize and draw some conclusions with regard to monkey and human handedness, I would like to come back to a finding which has occupied our work for a considerable time. It concerns the fact that unimanually trained intact monkeys showed no intermanual transfer for the FFD-task. In a first experiment we had trained animals to the highest possible level of difficulty with their preferred hand. As I already mentioned, the animals, once they started to use one hand in this task, never switched although the manipulanda were within reach of both hands. When after extended training with one hand only the previously unused hand was given access to the manipulanda, the animals reached toward it subsequent to the tone signalling a trial, but appeared not to know what to do with their fingers once they touched the device. They had to be completely retrained with this hand. Although switching daily between the originally trained and the newly trained hand, the performance curves of both hands followed separate courses.

Figure 5.3 shows the data for an animal which, after the original learning (OL) and an unsuccessful transfer test, was tested each day for two blocks of trials with the previously trained hand (RL = relearning) and two trial blocks with the newly trained hand (TL = transfer learning). The trials needed to reach criterion on successively higher levels of difficulty as plotted in Figure 5.3 show TL on a course more similar to OL than to RL. There appear to be some savings for TL in comparison with OL which may be related to transfer of general experience with the testing situation.

When we observed this lack of transfer for the first time, we had used the original Wiesendanger and Hepp-Reymond manipulanda which required the animal to use a finger thumb opposition to reach the middle of a donut shaped metal device housing a tiny pressure transducer. This may have been too artificial or unnatural a task for the monkey to do; furthermore, the transducer was fastened to the end of a flexible goose neck to prevent the involvement of arm control in the production of a specific pressure. It may have been much more natural to pull the goose neck dangling in front of the cage, as most animals did at the beginning of training and transfer testing, than to apply a finger thumb pressure to the device attached at the end; thus the lack of transfer may have been the result of a stronger natural habit interfering with the previously learned and possibly transferred habit.

Several other possible explanations for the lack of transfer were

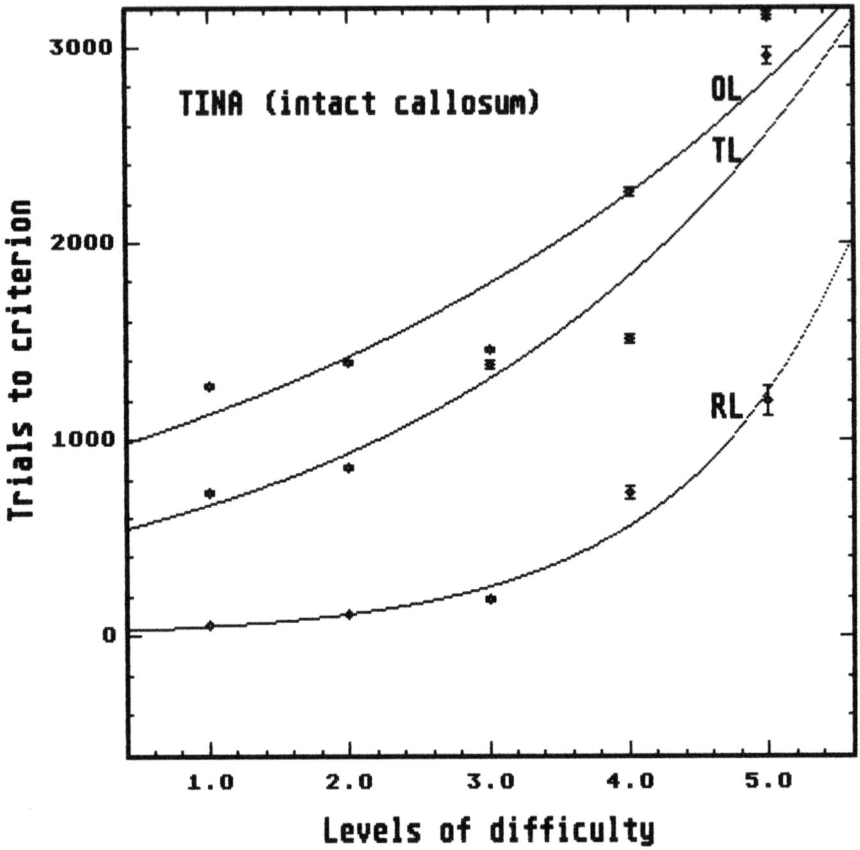

Figure 5.3. Exponential regression lines fitted to trials to criterion at successively higher levels of difficulty needed by an animal with intact commissures. OL = Original learning with the right hand; RL = Relearning with the originally trained right hand; TL = Transfer learning with the left hand.

considered, such as "situation specific learning", but the most intriguing possibility was that learning had remained lateralized within one hemisphere despite intact commissures. The results of several subsequent experiments with new manipulanda (Figure 5.2) and with different designs as well as control experiments which involved more proximal motor control (e.g. arm movements in the AFD-task) appear to support the latter explanation, although we have by no means eliminated all other ones. In favor of such an explanation of lateralized learning, for example, is the finding that animals trained with both hands on alternate days are slower in learning and their final performance remains far below that of animals trained with only one and the same hand. They do not seem to profit with one hand from the work done with the opposite hand. Animals which were tested with the arm-force-discriminationtest (AFD-task) by one of my students also showed no immediate but more extensive delayed transfer (Preilowski & Reger, 1987).

It appears then as if a number of variables, such as degree of anatomic lateralization with regard to motor and sensory control, duration of unimanual training, level of performance attained, the extent to which the task is similar to habits within the behavioral repertoire of the animals, as well as motivational factors determine the degree of intermanual and interhemispheric transfer. Although much remains to be done to elucidate the effects of these variables it can nevertheless be tentatively concluded that the development of a fine sensory-motor skill dependent on motor control of the distal extremities can remain lateralized to the opposite hemisphere despite intact commissures. As an additional hypothesis it might be proposed that with extended training of only one hand in such a task, an active interhemispheric inhibition might preclude the bilateral establishment of skill. I have already mentioned that such a possibility is not altogether out of bounds if the hands of primates are regarded as tools of the brain, the efficient use of which may require their independence as well as their coordination.

Does a Callosal Split Enhance Intermanual Transfer?

Supported by the already mentioned results of increased concordance of performance in the split-brain animals, we finally decided to test whether an animal which was trained with one hand after a callosotomy would perhaps show better transfer than the animals who had been trained and tested with intact callosum. As much as my theoretical deductions persuaded me to regard this as a reasonable thing to do, the long tradition of "disconnection-thinking" within split-brain research made such a hypothesis appear quite speculative. However, since we had one callosotomized animal that was still naive with regard to the finger discrimination task, we thought it was worth a try. Therefore, the monkey Nero was trained with one hand to the highest possible level of difficulty, which he reached within 12 days. During relearning with the same hand he again reached the same level within four days. When switching hands, however, Nero also showed no immediate transfer. He would not touch the manipulandum with the left hand and had to be tricked into reaching into the slot with the manipulanda. But then he worked at almost the same level as during the relearning with the right hand, reaching that same top level on day 5 of transfer learning.

Figure 5.4 shows his performance, which for TL looks much more like RL than OL, suggesting a substantial amount of delayed transfer. Lest I should be scolded that I am following one of the worst traditions of laterality research which reserves the most interesting data to anecdotes or single case

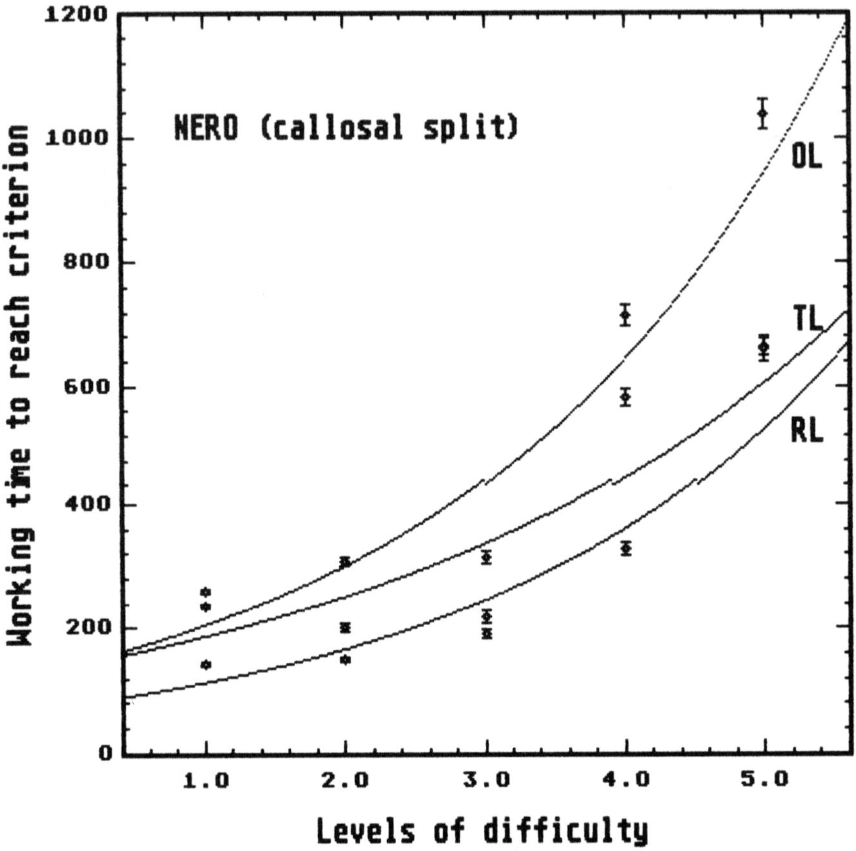

Figure 5.4. Exponential regression lines fitted to working time to reach criterion at successively higher levels of difficulty needed by an animal with split callosum. OL = Original learning with the right hand; RL = Relearning with the originally trained right hand; TL = Transfer learning with the left hand.

reports in conference volumes, never to be followed by completed research, I would like to emphasize the preliminary nature of these data.

It is clear that before we can make any definite statements a lot more research has to be done. I have already mentioned some of the variables to be investigated. In addition, we need much more basic research, such as research on functions of the commissures, especially the medial and frontal portions of the corpus callosum. We also need research about what the animals are actually doing and what they are learning as they are being trained and tested with the FDT- and ADT-tasks. As indicated above it will take an all-out effort on all levels of investigation if we want to make headway in understanding laterality. The research described here only points out some of the directions which have to be explored and the preliminary results are reported to indicate that it may be worth the effort.

Comparative Laterality Research: Human Handedness Reconsidered

Finally, I would like to add that while my own research with animals has not yet provided definite answers to the questions about handedness and laterality in monkeys, it has certainly raised some questions about handedness and laterality in humans which are worth reconsidering. As indicated by the investigations which included qualitative aspects of handedness (e.g. Annett, 1985), most of the handedness measures commonly applied in human research reflect the effects of learned expectancies of an asymmetry of performance rather than a performance asymmetry itself.

Figure 5.5 shows the hypothetical relationships between brain asymmetry, habit formation, and qualitative (performance handedness) as well as quantitative forms of handedness (preference handedness). The thin arrows indicate weak or only temporary active interactions, such as during early development of a child. The thick arrows stand for dominant persistent and causal relationships. Even very weak brain asymmetries could thus lead to manual preferences which would be strengthened by subsequent experience, as has often been suggested. What I would like to stress here is that habit formation may become more and more independent of the original brain asymmetry which started it off. The snow-ball effect is not only positively reinforced by experience and training but, with growing mental development, also by expectancies. These expectancies appear to be most influential in generalizing motor habits. This could explain why quantitative handedness and qualitative handedness show a fairly good correlation in humans, although only the qualitative handedness form may be directly related to laterality.

If I were forced to say what the difference between monkey and man with regard to handedness could be, I would propose that the main difference could be that they lack awareness of behavioral asymmetries and that they lack expectancies concerning the use of a particular hand. Therefore they are also unable to generalize their use of a one hand from one situation to the next. Thus, it is possible that their brains are lateralized with regard to motor functions much as ours are. However, resulting behavioral asymmetries would not be systematically reinforced nor generalized. In other words, monkeys could have handedness and laterality either if they could know or think that they were handed or if they could develop an equally

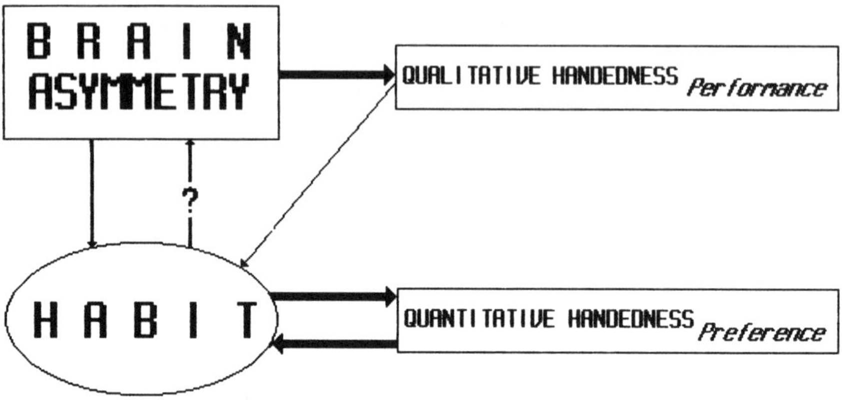

Figure 5.5. Schematic diagram indicating hypothetical relationships between brain asymmetries and qualitative as well as quantitative forms of handedness. Thick arrows indicate stable interactions, thin arrows only temporary or weak influences.

efficient mechanism providing for a generalization of hand use between different tasks.

Acknowledgements

The first drafts of this article were prepared while the author was on leave at the University of Auckland (New Zealand) and Monash University in Melbourne (Australia) thanks to the hospitality of Jim Wright, Mike Corballis, as well as John and Judy Bradshaw. The feedback received in the discussions with these colleagues and with Joël Fagot and Jacques Vauclair in Marseille is gratefully acknowledged. The research reported here has been supported by the Universities of Konstanz and Tübingen and the Deutsche Forschungsgemeinschaft.

References

Annett, M. (1985). *Left, right, hand and brain: The right shift theory*. London: Lawrence Erlbaum and Associates.

Brinkman, J., & Kuypers, H. G. J. M. (1972). Split-brain monkeys: cerebral control of ipsilateral and contralateral arm, hand and finger movements. *Science, 176*, 536-538.

Burešová, O., Bureš, J., & Nadel, L. (1972). The reversible split-brain and interhemispheric communication. In J. Cernacek & F. Podivinsky (Eds.), *Cerebral interhemispheric relations* (pp. 227-233). Bratislava: Publishing House of the Slovak Academy of Sciences.

Di Stefano, M., Morelli, M., Marzi, C. A., & Berlucchi, G. (1980). Hemispheric control of unilateral and bilateral movements of proximal and distal parts of the arm as inferred from simple reaction time to lateralized light stimuli in man. *Experimental Brain Research, 38*, 197-204.

Engele, H. C. (1988). *Split-brain und lateralität: Ein intermanueller Vergleich motorischer Leistungen bei intakten und callosotomierten Rhesusaffen (Macaca mulatta).* (Dissertation Universität Tübingen) Frankfurt/M.: Peter Lang.

Hodos, W., & Campbell, C. B. G. (1969). Scala naturae: Why there is no theory in comparative psychology. *Psychological Review, 76*, 337-350.

Ingle, D. J. (1968). Interocular integration of visual learning in the goldfish. *Brain, Behavior, and Evolution, 1*, 58-85.

Kučera, P., & Wiesendanger, M. (1985). Do ipsilateral corticospinal fibers participate in the functional recovery following unilateral pyramidal lesions in monkeys? *Brain Research, 348*, 297-303.

Levy, J. (1977). The mammalian brain and the adaptive advantage of cerebral asymmetry. *Annals of the New York Academy of Sciences, 299*, 264-272.

MacNeilage, P. F., Studdert-Kennedy, M. G., & Lindblom, B. (1987). Primate handedness reconsidered. *Behavioral and Brain Sciences, 10*, 247-303.

Palmers, C., & Zeier, H. (1974). Hemispheric dominance and transfer in the pigeon. *Brain Research, 76*, 537-541.

Preilowski, B. (1972, June). *Is there a specialization of the cerebral hemispheres in nonhuman primates ?* Paper presented at the NATO Advanced Study Institute on Comparative Biology of Primates, Torino (Italy).

Preilowski, B. (1975). Bilateral motor interaction: Perceptual-motor performance of partial and complete "split-brain" patients. In K. J. Zülch, O. Creutzfeldt, & G. C. Galbraith (Eds.), *Cerebral localization* (pp.115-132). Berlin: Springer-Verlag.

Preilowski, B. (1977a). *Projekt "Zerebrale Asymmetrie".Arbeitsbericht Nr. 2, DFG Pr 117/3.* Konstanz: Universität Konstanz.

Preilowski, B. (1977b, March). *Lack of intermanual transfer of a fine sensory-motor skill in normal rhesus monkeys.* European Brain and Behaviour Workshop on Callosal Functions, Rotterdam (Netherlands).

Preilowski, B. (1978). On the biological basis of human laterality: Problems with "grand schemes" and handedness. *Behavioral and Brain Sciences, 1*, 312-313.

Preilowski, B. (1979). Performance differences between hands and lack of transfer of finger posture and sensory-motor skill in intact rhesus monkeys: possible model for the origin of cerebral asymmetry. *Neuroscience Letters, 3*, S89.

Preilowski, B. (1983). Is there a cerebral hemispheric specialization in nonhuman primates?. In P. K. Seth (Ed.), *Perspectives in primate biology* (pp. 109-117). New Delhi: T & T Publishers.

Preilowski, B. (1987). The role of corollary discharges, the corpus callosum, and the supplementary cortices in bimanual coordination. *Behavioral and Brain Sciences, 10*, 322-323.

Preilowski, B., & Reger, M. (1987). Intermanueller Lerntransfer einer proximalen sensomotorischen Regelungsaufgabe bei Rhesusaffen (*Macaca mulatta*). In F. G. Barth & E.A. Seyfarth (Hg.), *Verhandlungen der Deutschen Zoologischen Gesellschaft, 80* (p.322). (Jahresversammlung 1987 in Ulm. S. 322). Stuttgart: Gustav Fischer Verlag.

Preilowski, B., Reger, M., & Engele, H. C. (1986). Handedness and cerebral asymmetry in nonhuman primates. In D. M. Taub & F. A. King (Eds.), *Current perspectives in primate biology* (pp. 270-282). New York: Van Nostrand Reinhold Company.

Rogers, L. J. (1986). Lateralization of learning in chicks. *Advances in the study of behavior, 16*, 147-189.

Schwartzman, R. J. (1978). A behavioral analysis of complete unilateral section of the pyramidal tract at the medullary level in *Macaca mulatta*. *Annals of Neurology, 4*, 234-244.

Sechzer, J. A. (1970). Prolonged learning and split-brain cats. *Science, 169*, 889-892.

Steele Russell, I., & Ochs, S. (1963). Localization of a memory trace in one cortical hemisphere and transfer to the other hemisphere. *Brain, 86*, 37-54.

Trope, I., Fishman, B., Gur, R.C., Sussman, N.M., & Gur, R.E. (1987). Contralateral and ipsilateral control of fingers following callosotomy. *Neuropsychologia, 25*, 287-291.

Tuller, B., & Kelso, J. A. S. (1989). Environmentally-specified patterns of movement coordination in normal and split-brain subjects. *Experimental Brain Research, 75*, 306.

Van Hof, M. W. (1970). Interocular transfer in the rabbit. *Experimental Neurology, 26*, 103-108.

Wiesendanger, M., & Hepp-Reymond, M. C. (1970). Cerebrale Organisation der Motorik und Fertigkeitsbewegungen. In K. Bättig (Hg.), *Psychologische Experimente* (pp. 110-116). Bern: Huber.

Young, G., Corter, C.M., Segalowitz, S.J., & Trehub, S. E. (1983). Manual specialization and the developing brain: An overview. In G. Young, S. J. Segalowitz, C. M. Corter, & S. E. Trehub (Eds.), *Manual specialization and the developing brain* (pp. 3-12). New York: Academic Press.

6
Manual Preference in Prosimians, Monkeys and Apes

Ralph A. W. Lehman

Bilaterally symmetrical organisms usually respond asymmetrically. Though their structural symmetry is imperfect, their behavioral asymmetry seems disproportionately great. Although most behavioral cues seem to have equal access to both sides of the brain, the two sides of the body do not compete with or duplicate each others' responses. This occurs despite the obvious mirror-image similarity of the two sides of the brain, each of which is concerned predominantly with the opposite side of the body. This review will focus upon studies of one example of this phenomenon, hand preference in nonhuman primates.

As early as 1871, Ogle (1871) discussed most of the issues. After a period of investigation at the London Zoological Gardens, Ogle became able to predict which hand each monkey would use when offered a morsel of food and he used this knowledge to entertain his friends. Aside from ascertaining the consistency of each individual's hand preference, he became convinced that monkeys, like humans, are generally right-handed. He also noted that hand usage was altered as a matter of convenience with a tendency to use the nearest hand when this was clearly advantageous. In relating his observations, Ogle commented upon the existence of asymmetrical hand usage, its predictability for individuals, alteration by circumstance, consistency between individuals, and its relationship to human handedness. He ascribed both monkey and human hand preference to "some common factor in their anatomical structure", but considered this to reside in bodily proportions rather than brain organization.

These same issues will be addressed in this review under four headings:

I. Asymmetry rather than duplication of hand usage: Why does not the other hand do the same thing at the same time?

II. Asymmetric hand usage by the individual: Does each individual display a consistent asymmetry?

III. Asymmetric hand usage within the population: Is there sufficient consistency between individuals to produce a population asymmetry?

IV. Laterality of hand usage and of other behaviors: Is asymmetry of other behaviors consistently similar or consistently opposite to that of hand preference?

Asymmetry Rather Than Duplication of Voluntary Hand Use

It is not surprising that most voluntary hand usage is asymmetrical. As often pointed out, when one hand plays a leading role, the other generally functions in a supportive or complementary fashion. This subordinate activity makes performance more effective. Thus both hands are often actively involved during behavior labeled as unilateral. One hand, the leading hand manipulates the object which the observer considers the goal of the activity being observed, while the other hand does not. When both hands are utilized sequentially or concurrently to gain access to and manipulate this object or portions of it, such activity is designated as bimanual and the definition of which hand has the leading role becomes more arbitrary (Beck & Barton, 1972; Vauclair & Fagot, 1987a; Warren 1977a).

It seems likely that asymmetry of hand usage is the expression of an asymmetry of cerebral activity. Evidence suggests that each choice of one or the other hand for the leading role is dependent upon the active contribution of the contralateral cerebral hemisphere. Anatomical and physiological studies indicate that motor control exercised by each cerebral hemisphere extends to both arms proximally but that distal fine movements are governed largely by the contralateral hemisphere (Bernhard & Bohm, 1954; Clough, Kernell, & Phillips, 1968; Lawrence & Kuypers, 1968; Liu & Chambers, 1964; Tanji, Okano, & Sato, 1988). There also appears to be little transfer of learned manual skill from one hand to the other, suggesting that the skills of each hand are controlled by different hemispheres (Preilowski, 1983; Preilowski, Reger, & Engele, 1986).

Particularly pertinent are studies of hand usage by animals with an intact but sensorially deprived hemisphere. If one cerebral hemisphere is "blinded" by sectioning its optic input pathway and also deprived of access to visual information from the other hemisphere by division of the interhemispheric commissures, the animal usually stops reaching for pieces of food with the hand controlled by the blind hemisphere unless it has no other choice (Bossom & Hamilton, 1963; Butler, 1968; Downer, 1959; Gazzaniga, 1964; Lehman, 1968; Sperry, 1961; Trevarthen, 1962, 1965a, b). Apparently choice of the hand selected to carry out voluntary manipulation is dependent upon the informed participation of the contralateral hemisphere. When this is not available, the other hemisphere displays no hesitancy in assuming the leading role.

Ordinarily, however, information is available to both hemispheres. What prevents them from competing in pursuit of the same goal? This question has two components: What determines which hemisphere takes the leading role, and how is it that the other hemisphere does not? Major candidates for a role in ensuring the coordination rather than independent activity of the brain halves in determining hand usage are the interhemispheric commissures including the corpus callosum. It might be expected that division of these commissures would lead to some concurrent duplication of effort by the two hands and even to a more equal frequency in their usage. Both of these predictions are fulfilled after section of the forebrain commissures, but only to a disappointingly meager degree (Brinkman & Kuypers, 1973; Kennard & Watts, 1934; Lehman, 1972; Trevarthan, 1965a, b, 1978). It may be concluded that these commissures serve as a minor pathway by which the hemisphere controlling the leading hand deters its mate from engaging the otherwise subordinate hand in independent actions. Apparently the commissures also help recruit the subordinate hand for supportive activity since such subordinate activity may be diminished after forebrain bisection (Trevarthen, 1978), and rarely will the two hands even compete with one another (Trevarthen, 1965b).

Other conclusions derive from a study of bimanual coordination after a unilateral ablation of the supplementary motor cortex. Following such a lesion, hand and finger posturing of the arm opposite the lesion tended to duplicate rather than complement the activity of the normal limb (Brinkman, 1984a). This deficit was abolished by callosal section, suggesting that the message to duplicate is transmitted transcallosally. Brinkman concluded that each supplementary motor area sends duplicate motor commands to its own and the opposite cerebral hemisphere and that normally the transcallosal messages are nullified by the supplementary motor cortex of the recipient hemisphere. The findings indicate the existence of a transcallosal mechanism by which the leading hemisphere could influence its mate to produce mirror-image responses. They are opposite to the findings noted earlier which provided evidence for a transcallosal mechanism which helps prevent duplication of activity.

The occurrence of opposing transcallosal messages may account in part for the relatively mild changes resulting from forebrain bisection. Nevertheless, the paucity of findings suggests that other mechanisms operating through neural pathways crossing the midline at levels below the interhemispheric commissures are probably those most important for the asymmetric use of the hands during voluntary activity (Lehman, 1972; Trevarthen, 1965a, 1978).

Supporting this conclusion are experiments which demonstrate that if separated by section of the optic chiasm and forebrain commissures, the two cerebral hemispheres can concurrently learn and perform different visual discriminations. When so engaged, each hemisphere responds to its task with the contralateral hand. Despite the concurrent presentation of both tasks, the two hands are used alternately rather than in unison (Gazzaniga & Young, 1967). This is another example in which despite section of the forebrain commissures the two hemispheres, each employing a different hand, do not act independently. Instead, each functions only in alternation as if they were allocated permission to do so by a single controlling mechanism.

Whereas normal adult monkeys rarely exhibit duplication of hand usage in carrying out a task, young individuals may do so with some frequency (Lehman, 1978b; Vauclair & Fagot, 1987a). This difference is presumably due to neural maturation, but the role of experience must also be considered.

Hand Preference of the Individual

Observation indicates that most individuals will use one hand in the leading role more frequently than the other. Whether the greater use of one hand is the result of a definite hand preference or not is a matter of definition. One of three criteria have been used:

1. The individual's use of the preferred hand during a series of trials upon a single task exceeds that expected by chance.
2. The same hand is preferred upon a second trial series on the same task in a greater proportion of animals than would have been expected by chance.
3. The hand preferred on different tasks is consistently the same or consistently opposite for a greater proportion of individuals than expected by chance.

To this a fourth criterion can be added - the maintenance of the original preference by a significant majority of individuals after the task is repeated under conditions disadvantageous to the previously preferred hand. This invariably involves testing on the same task before and after the handicap.

One other criterion should be mentioned — that the hand preferred for a task is the same for a greater number of individuals than expected by chance. This is discussed in Section III, which concerns the hand preference of the population.

Hand Preference for a Single Task

Definition

The hand used more often in the leading role during a series of trials on a particular task is termed the preferred hand. In order to determine whether the frequency of preferred hand use is indicative of hand preference, investigators have resorted to tests which permit comparison with chance performance.

Binomial or chi-square tests have been utilized (Bard, Hopkins, & Fort, 1990; Bresard & Bresson, 1983; Brookshire & Warren, 1962; Byrne & Byrne, 1991; Deuel & Dunlop, 1980; Forsythe, Milliken, Stafford, & Ward, 1988; Forsythe & Ward, 1988; Fragaszy & Mitchell, 1990; Kruper, Boyle, & Patton, 1966; Kubota, 1989, 1990; Larson, Dodson, & Ward, 1989; Milliken, Forsythe, & Ward, 1989; Milliken, Stafford, Dodson, Pinger, & Ward, 1991; Olson, Ellis, & Nadler, 1990; Rothe, 1973; Stafford, Milliken, & Ward, 1990; Tokuda, 1969; Ward, Milliken, Dodson, Stafford, & Wallace, 1990). Using these tests, individuals whose performance exceeds certain levels of improbability have been labeled right- or left-handed as the case may be. All others have been classified as ambidextrous (ambilateral). Such a statistical system of categorization is captive to several influences not directly related to an animal's preference. The number of reaches observed helps determine the percentage of hand usage classifiable as handed rather than ambidextrous. Consequently this system categorizes a number of animals as ambidextrous that would have proved to be classifiable as right- or left-handed had more observations been made. Furthermore, the level of improbability chosen to differentiate handedness from ambilaterality is arbitrary.

In a second method of defining hand preference, an arbitrary percentile has been used as the criterion level for dividing left- and right-handers from the ambidextrous group (Annett, 1972; Deuel & Dunlop, 1980; Deuel & Schaffer, 1987; Ettlinger, 1961; Ettlinger & Moffett, 1964; Finch, 1941; Fischer, Meunier, & White, 1982; Fragaszy, 1983; Garcha, Ettlinger, & Maccabe, 1982; Gautrin & Ettlinger, 1970; Horster & Ettlinger, 1985; Milner, 1969; Tokuda, 1969). Overall, these two methods should be considered conservative and probably underestimate the number of individuals who are classifiable as right- and left-handed.

An additional problem may arise if only a small number of trials are observed. There is evidence that in some circumstances hand preference is less pronounced during the early trials of a new task than it is later on in testing (Cole, 1957; Ettlinger, 1961; Lehman, 1978a, 1980b; Warren, 1958, 1977a). In such cases, if only a small number of trials on a new task are observed, performance will be more ambidextrous or less handed than if a larger series of trials is observed. Field studies of everyday behavior are probably not subject to such an effect.

Utilizing the binomial or chi-square method, a number of investigators have found many individual monkeys and some apes to display a statistically significant degree of hand preference (Bresard & Bresson, 1983, 1987; Brookshire & Warren, 1962; Byrne & Byrne, 1991; Fischer et al., 1982; Itani, Tokuda, Furuya, Kano, & Shin, 1963; Kawai, 1967; Kruper et al.,, 1966; Kubota, 1989, 1990; Marchant & Steklis, 1986; Olson et al., 1990; Rothe, 1973; Stafford et al., 1990; Steklis & Marchant, 1987; Tokuda 1969). Similar results have been reported for prosimians (Forsythe et al., 1988; Forsythe & Ward, 1988; Larson et al., 1989; Milliken et al., 1989; Milliken et al., 1991; Ward et al., 1990). Though often unremarked upon, the number of such strongly handed individuals always exceeds that expected by the statistical criterion ($p < .05$) chosen. Similar claims have been made or implied when a percentile criterion was employed (Deuel & Dunlop, 1980; Deuel & Schaffer, 1987; Horster & Ettlinger, 1985; Milner, 1969; Tokuda, 1969). In most instances, as many or more animals have been identified as ambidextrous as were classified as either right- or left-handed.

Alternatively, one can avoid the question of what degree of handedness is significant and content oneself with the consideration that handedness varies upon a continuum from 0 to 100% as a percentage right (or left) hand preference (Kounin, 1938). This avoids categorization of individuals as ambidextrous and classifies individuals as right- or left- handed according to whether or not they exceed the fiftieth percentile. When this is combined with the strength of handedness (percentage preferred handedness from 50 to 100%), each animal is fully characterized as having a specific laterality and degree of hand preference. Of course the first measure, percentage right hand preference, provides a measure of both. While the use of this measure to define laterality is rather liberal, no information regarding laterality is lost. This permits a more complete comparison of laterality during repetitions of the same task or between different ones.

Distribution

The hand preferences of monkeys and apes performing a single task are generally U-shaped or even J-shaped in distribution and not bell shaped as would be expected from chance alone (Brooker, Lehman, Heimbuch, & Kidd, 1981; Byrne & Byrne, 1991; Cole, 1957; Cronholm, Grodsky, & Behar, 1963; Finch, 1941; Horster & Ettlinger, 1985; Kounin, 1938; Lehman, 1978a, 1980b, c; Warren, 1953). Such U-shaped and J-shaped distributions contain more strongly handed individuals than expected by chance. Similar evidence has already been noted in the studies relying upon the chi-square classification system. This constitutes strong evidence that the hand preference of individual monkeys is not just the result of a random process. Nevertheless, for some tasks a relatively normal unimodal distribution has been evident (Brooker, Lehman, Heimbuch, & Kidd, 1981; Heestand, 1986).

Even though the population distribution of hand preferences for a single task is commonly U-shaped or even J-shaped, this does not imply that the distribution will retain this shape when each individual's preferences are averaged over several different tasks. The shape will only be maintained if there is a strong correlation between hand preferences on the different tasks. As discussed later, this correlation varies with the tasks considered. Consequently some data for hand preference distributions are poorly or not at all U-shaped (Vauclair & Fagot, 1987a; Warren, 1977a).

Hand Preference upon Task Repetition: the Role of Practice

Is the same hand preferred after an interval of time? Such a consistent bias would be another indication that hand preference is not a matter of chance. Because the distribution of hand preference indicates that more individuals are strongly handed than expected by chance, one may conclude that individual hand preference is biased and therefore likely to be maintained when a task is repeated after an interval of time. Indeed, studies of prosimians, monkeys, and apes have shown this to be the case (Beck & Barton, 1972; Bolser, Runfeldt, & Morris, 1988; Brookshire & Warren, 1962; Chivers, 1974; Fagot & Vauclair, 1988b; Finch, 1941; Forsythe & Ward, 1988; Kounin, 1938; Lehman, 1978a, 1980b; Ogle, 1871; Rothe, 1973; Sanford, Guin, & Ward, 1984; Stellar, 1960; Vauclair & Fagot, 1987b; Warren, 1958, 1977a; Warren, Abplanalp, & Warren, 1967; Warren & Nonneman, 1976). Sometimes preference for the same hand strengthens with task repetition (Lehman, 1978a, 1980b; Warren, 1958). The maintenance of hand preference upon task repetition confirms that individual hand preference for a task is not simply a matter of chance. In Warren's experience (1977a), consistency of hand preference upon task repetitions depends upon prolonged previous practice on that task. This has not been true in other studies (Lehman, 1978a, 1980b, c) in which ultimate preference has been evident upon the very first reach.

Training the non-preferred hand can alter preference (Deuel & Dunlop, 1980; Kempf, 1917; Lashley, 1917; McGonigle & Flook, 1978). Stamm, Gadotti, and Rosen (1975) demonstrated that such training produces electrophysiologic changes in the contralateral prefrontal cortex. But whether such changes are accompanied by a change of hand preference in favor of the trained hand was not determined.

Because practice on a task provides training, one might wonder whether the reinforcement provided by the food reward associated with tests of hand preferences produces those preferences. It could be argued that the greater number of reinforcements received by the preferred hand would bias future usage ever more strongly in the same direction. However, because the chance of receiving a reward for each instance of hand use is equal for each hand, this seems unlikely.

Nevertheless, the first trial of a new test has no distinct predecessor and the reward received for this initial effort might bias future usage in favor of the same hand. Thus, the first reach made by monkeys who had never received any other type of testing is predictive of the hand preferred over a series of 599 additional reaches (Lehman, 1978a, 1980b). In order to further study this possibility, a new group of animals were given the same test, but this time the hand chosen for the first reach was never rewarded until after the other hand had been chosen and rewarded for the first time. It was the first hand to be used, rather than the first to be rewarded, that proved to be predictive of hand preference over the ensuing trials (Lehman, 1980c). It was concluded that differences in the reward received are not important in determining the laterality of hand preference when tasks offer equal opportunity for reward for use of either hand.

Other factors must be important. These factors apparently antecede any experience with the task at hand since they are evident during the first reach. These data provide some refutation Warren's contention (1980) that for the most part hand preference on a task results from the progressive effects of repeatedly rewarding a slight initial asymmetry. It is more consistent with the suggestion that these asymmetries are established well before adulthood (Byrne & Byrne, 1991). These data (Lehman, 1978a, 1980b, c) also lead to the expectation that, irrespective of which hand is preferred for a given task, the same factors should predispose individuals to prefer the same hand for certain other tasks. Evidence for this is presented later.

The Effect of Brain Lesions

The tendency to prefer the same hand upon task repetition is evident even when the controlling contralateral cerebral hemisphere is placed at a disadvantage. This can be done by dividing its optic pathway. This hemisphere can then only receive visual information secondarily from its mate, which still enjoys the advantage of direct visual input through its optic pathway. Despite this, the hand preferred usually does not change (Downer, 1959; Lund, Downer, & Lumley, 1970; Trevarthan, 1965a) though it may do so transiently (Lehman, 1968) or perhaps longer (Ettlinger, 1961). Preference for the same hand continues even though it remains under the control of the now visually deprived hemisphere. This is proven by isolating this hemisphere from its major remaining source of visual information by sectioning the interhemispheric commissures. As would be expected if the deprived hemisphere had continued to control hand preference, this additional lesion usually causes hand preference to shift strongly to the other side (Bossom & Hamilton, 1963; Butler, 1968; Downer, 1959; Gazzaniga, 1964; Lehman, 1968; Sperry, 1961; Trevarthan, 1962, 1965a, b). Thus the hemisphere responsible for the preferred hand continues to exercise its control of hand preference if it can obtain relevant sensory information from its mate. In this sense it is the predominant hemisphere.

Furthermore, brain lesions within the hemisphere contralateral, but not that ipsilateral, to the preferred hand can produce a shift of hand preference toward the opposite side. This occurs after removal of the occipital cortex (Ettlinger, 1961; Ettlinger & Dawson, 1969; Franz, 1913; Jason, Cowey, & Weiskrantz, 1984; Lehman, 1968), and sometimes after lesions not affecting primary sensory or motor areas (Ettlinger & Dawson, 1969; Horster & Ettlinger, 1987). Similar changes have been noted by Moffett and Ettlinger (1970) after removal of posterior parietal cortex contralateral to the hand preferred for a tactile discrimination in four monkeys and ipsilaterally in three others. None of the latter changed hand preference postoperatively but at least two of the contralateral group did so. However, all the contralaterally operated and one ipsilaterally operated animals severely misreached with the hand opposite the lesion. Less pronounced changes have been observed after unilateral post-arcuate lesions. Slowness of movement with the hand opposite such lesions was accompanied by a diminished use of that hand (Rizzolatti, Matelli, & Pavesi, 1983). These results indicate that preference for one particular hand is dependent upon the intact function of a variety of regions in the contralateral cerebral hemisphere.

Lesions produce similar results when hand preference has been shaped by training. Contralateral cortical ablations in the periarcuate (Deuel & Dunlop, 1980) and the post- but not the pre-central (Semmes & Porter, 1972) cortex seem to reverse the hand preference favored by training.

Other Factors

Despite some evidence to the contrary (Byrne & Byrne, 1991; Forsythe & Ward, 1988; Warren et al., 1967), hand preference appears to be somewhat stronger in older animals than in younger ones (Brooker et al., 1981; Itani et al., 1963; Kruper et al., 1966; Kubota, 1989, 1990; Lehman, 1970, 1978a, 1980b; Preilowski, 1983; Stafford et al., 1990; Vauclair & Fagot, 1987a; Ward, 1991; Ward et al., 1990). Whether this is the result of practice and experience, maturation of bodily organs, genetics, or other factors is unknown. With the exception of a few studies showing stronger hand preference in females (Byrne & Byrne, 1991; Stafford et al., 1990; Ward, 1991), there does not appear to be any difference in the degree of hand preference manifested by the two sexes (Forsythe & Ward, 1988; Itani et al., 1963; Kubota, 1989, 1990; Lehman, 1978a, 1980b; Sanford et al., 1984; Vauclair & Fagot, 1987a; Ward et al., 1990).

Situational (Deuel & Dunlop, 1980; Deuel & Schaffer, 1987; Forsythe et al., 1988; Gautrin & Ettlinger, 1970; Kluver, 1933; Lashley, 1917; Milliken et al., 1989; Sanford et al., 1984; Ward, 1991; Warren, 1977a) and positional (Beck & Barton, 1972; Bresard & Bresson, 1983; Cronholm et al., 1963; Deuel & Dunlop, 1980; Finan, 1939; Fagot & Vauclair, 1988a; Grzimek, 1949; Lashley, 1917; Lehman 1978a, 1980b; Marchant & Steklis, 1986; Ogle, 1871; Schaller, 1963; Vauclair & Fagot, 1987a) factors affect the relative frequency with which each of the hands is chosen. The effect of position is readily ascertainable as an increase in preference for whichever hand has the closer or more convenient reach. In circumstances which do not automatically favor one hand in this fashion, animals most often position themselves so that their preferred hand has the shortest reach (Lashley, 1917; Lehman 1978a, 1980b). Both studies in which vertical and horizontal posture were compared revealed bush babies to be more strongly handed when reaching from bipedal rather than quadrupedal postures (Larson et al., 1989; Sanford et al., 1984). This may also be true for apes (Olson et al., 1990).

There is a distinct tendency for hand preference to be perseverative (Lehman, 1991). It is possible that such a tendency would magnify the use of the preferred hand and should be considered when quantifying degree of handedness.

Concern exists that hand preference may only be strongly expressed during "complex" manipulations. Complexity is rarely defined in a systematic fashion. Using their own criteria for complexity, a number of investigators have offered answers to this question. Negative results have been reported (Brookshire & Warren, 1962; Fagot & Vauclair, 1988b; Kluver, 1937; Marchant & Steklis, 1986; Olson et al., 1990; Rensch & Ducker, 1966-7; Rothe, 1973; Warren, 1958, 1977a) though some studies found handedness to be more pronounced during more demanding manipulations (Byrne & Byrne, 1991; Ettlinger, 1961; Kounin, 1938; Preilowski, 1983; Warren, 1958) including the components of bimanual tasks (Warren, 1977a).

Furthermore, relatively little has been done to determine whether the preferred hand is the more skillful one. Deuel and Schaffer (1987), Preilowski (1979, 1983, 1986) as well as Marchant and Steklis (1986) found no such relationship though Trevarthan (1978), Bolser et al. (1988), Hopkins et al. (1989), Fragaszy and Mitchell (1990) and, to some extent, Byrne and Byrne (1991) did. At times task difficulty is sufficiently great so that differences in the ability of the two hands will dictate preference (Bresard & Bresson, 1983). There was no evidence that this occurs frequently despite Warren's (1980) contention.

The converse possibility, that increased practice with the preferred hand accounts for its having the greater skill also deserves investigation. That more strongly handed animals perform better than less strongly handed ones when tested on other tasks seems unlikely (Fragaszy & Mitchell, 1990; Horster & Ettlinger, 1985; Sanford & Ward, 1986).

Hand Preference for Different Tasks

Is the hand preference for two different tasks related? Judgement of consistency depends upon finding within the sample population a sufficient proportion of individuals that are congruent (or incongruent) in hand use between tasks to permit the conclusion that hand usage is usually lateralized in the same (or opposite) direction. Some studies of this problem include bimanual tasks as defined earlier. Since the designation of one portion of such a task as defining the leading hand can be arbitrary, comparison of preference in such tasks with that in others has been excluded from most of the subsequent considerations.

Most reports of unimanual tasks fail to show a significant tendency for monkeys or apes to prefer the same hand on differing tasks (Bard et al.,

1990; Beck & Barton, 1972; Box, 1977; Cole, 1957; Deuel & Dunlop, 1980; Deuel & Schaffer, 1987; Ettlinger, 1961; Ettlinger & Dawson, 1969; Fagot & Vauclair, 1988a; Fragaszy & Mitchell, 1990; Haas, 1958; Hall & Mayer, 1966; Heestand, 1986; Hopkins, Washburn, & Rumbaugh, 1989; Kempf, 1917; Kluver, 1933; Kounin, 1938; Lashley, 1917; Lehman, 1980a; Marchant & Steklis, 1986; Moffett & Ettlinger, 1970; Olson et al., 1990; Rothe, 1973; Sanford et al., 1984; Stafford et al., 1990; Steklis & Marchant, 1987; Tokuda, 1969; Ward, 1991; Warren, 1977a). Nor do they reveal any tendency for individuals to consistently prefer opposite hands for different tasks. Nevertheless, when data is offered, it most frequently points toward a tendency for individuals to prefer the same hand for different tasks (Lehman, 1980a). In some studies this has been statistically significant (Brooker et al., 1981; Brookshire & Warren, 1962; Byrne & Byrne, 1991; Ettlinger, Blakemore & Milner, 1968; Hamilton & Vermeire, 1988; Larson et al., 1989; Masataka, 1989; Milliken et al., 1989; Rawlins, 1991; Warren, 1977a; Warren et al., 1967). Collectively these observations appear to justify the tentative conclusion that monkeys use the same hand for different tasks slightly more frequently than would be expected by chance.

While hand preferences are consistent across some tasks, they are not consistent for all tasks (Box, 1977; Byrne & Byrne, 1991; Deuel & Dunlop 1980; Lehman, 1989). Warren calculated the correlations between hand preferences on an extensive series of tests involving 14 monkeys. Hand preferences for these tasks and their components fell into 3 distinct groups, two of which correlated to a very high degree. Correlations across groups were much smaller. Warren (1977a) as well as Deuel and Dunlop (1980) concluded that hand preference is task-specific and thought that this reflected the effects of practice and experience rather than constitutional asymmetries.

While some studies indicate that the strength of originally preferred handedness increases with task repetition, usually this is not the case across tasks (Brookshire & Warren, 1962; Lehman, unpublished observations; Rothe, 1973) though there are documented exceptions (Ettlinger, 1961; Warren, 1977a).

It seems appropriate to discuss hand preference during multicomponent tasks at this point. In a sense, the multiple components of such tasks appear to offer opportunities for diversified hand usage similar to those present when differing tasks are performed. However, each trial of a multicomponent task contains both concurrent and sequential interdependent activities directed toward the same goal, whereas the trials of separate tasks are independent with respect to time and reward. The interdependence of the operations performed in a multicomponent task affords an advantage when different hands are used for different components. This would permit the two hands to be used in complementary fashion and might make performance quicker and easier. From studies of multiple component tasks (Beck & Barton, 1972; Brookshire & Warren, 1962; Cole & Glees, 1951; Ettlinger, 1961; Ettlinger & Dawson, 1969; Fagot & Vauclair, 1988a,b; Hamilton & Vermeire, 1988; Hauser et al., 1991; Marchant & Steklis, 1986; Rensch & Ducker, 1966-7; Warren & Nonneman, 1976) it is evident that primates often preferred to use different hands for the various manipulative (preparatory) and incentive (consummatory) components. As first suggested by Cole and Glees (1951) and confirmed by review (Lehman, 1987) of the reports cited above, when the same hand is preferred for all of the components of such a task, it is usually the one preferred for most other tasks as well.

Hand Preference of the Population

The hand preferences of humans are peculiar. Not only does each individual usually prefer to use the same hand for different tasks, but the hand which is preferred is usually the same for different individuals: The population is strongly skewed towards right-handedness. The hand preference of monkeys differs in both respects (Annett, 1972; Milner, 1969; Warren, 1980). The consistency with which monkeys prefer the same hand within and between tasks has been discussed; it is much less than that commonly attributed to man. Even more strikingly different is that prosimians, monkeys, and apes show little or no population bias toward right (or left) handedness. The latter has been a subject of much concern because the study of handedness in nonhuman primates has potential for shedding light upon the evolutionary precursors to the development of right-handedness in humans. Though Ogle (1871) accepted this premise, Steir (1911) did not. Steir found that a variety of primates preferred the right hand for a number of different tasks, but he considered this to be quite a different phenomenon from human hand preference.

Reports of unimanual hand usage in prosimians include a study of 25 bushbabies by Sanford et al. (1984). These were observed during two different tasks; one requiring a horizontal stance, the other demanding a vertical bipedal one. There were only seven left-handed individuals among the 12 classified as having a distinct hand preference when using a horizontal stance, whereas 14 of 19 were left-handed when responses were in a vertical position. The increased proportion of left-handers for the vertical bipedal task was offered as evidence that the bipedal position may favor a J-shaped population skew toward the left hand rather than the right. It should be noted that the frequency of left-handedness for the two stances is not significantly different (binomial test). Nor was it different in a separate study of 10 other bush babies (Larson et al., 1989) though this study also demonstrated a preponderance of 7 left-handed as compared to 3 right-handed individuals. Furthermore, the upright bipedal position was not associated with any decrease in tendency for right-handedness in the small-eared bushbaby (Milliken et al., 1991). Sanford and Ward (1986) reported all 8 of the bush babies they observed to be left-handed. Ward (1987) subsequently reported a left-handed trend in lemurs. Furthermore, Subramoniam (1957) observed eight lorises to be left-handed whereas Bishop (1963) reported a single loris to be ambidextrous. Among 33 black lemurs, 20 were considered left and 12 right-preferent (Forsythe & Ward, 1988). Even more strikingly, Masataka (1989) reported 20 of 22 members of five families of ring-tailed lemurs to be left-handed. In the largest study of prosimians, 194 lemurs of differing species (Ward et al., 1990), significantly more were considered left (46.9%) than right (33.5%) preferent. As the authors note, generalization of these findings might be limited by the relatively few matrilineages represented in many of the individuals studied. Nevertheless, taken together these reports raise the possibility of a population bias toward left-handedness among prosimians.

Studies of monkeys have sometimes suggested a preponderance of left-handedness (Beck & Barton, 1972; Ettlinger, 1961; Ettlinger & Moffett, 1964; Fagot & Vauclair, 1991; Hatta & Koike, 1991; Hauser et al., 1991; Itani et al., 1963; Kawai, 1967; King, Landau, & Scott, 1988; Kluver, 1933, 1937; Kruper et al., 1966; Kubota, 1989, 1990; Rothe, 1973; Stellar, 1960; Tokuda 1969), and occasionally a right-handed trend (Cole, 1957; Franz, 1913; Hall & Mayer, 1966; King & Landau, 1988; Kounin, 1938; Kuhl, 1988; Ogle, 1871; Post, Hausfater, & McCuskey, 1980; Rawlins, 1986; Ross, Wilczynski, & Albert, 1987; Vauclair & Fagot, 1987a, b; Yuanye, Yunfen, & Ziyun, 1986). However, a great number of studies find no trend in either direction for monkeys (Box, 1977; Brooker et al., 1981; Brookshire & Warren, 1962; Cole & Glees, 1951; Deuel & Dunlop, 1980; Deuel & Schaffer, 1987; Furuya, 1963; Gautrin & Ettlinger, 1970; Grzimek, 1949; Hamilton & Vermeire, 1988; Horster & Ettlinger, 1985; Kempf, 1917; Lashley, 1917; Lehman, 1978a, 1980b, 1980c; Ross et al., 1987; Warren, 1953). Interpretation of some of these studies is compromised by the classification of a very large proportion of the population as ambidextrous. The difficulties entailed by such a scheme have been described earlier. Of note is the unpublished study of Fagot and Vauclair (1991) demonstrating a statistically significant left hand bias in monkeys reaching for food while clinging vertically. However, King and Landau (1988) found a significant right hand preference in the 16 squirrel monkeys they observed under similar circumstances.

Reports of handedness in apes have been more limited in number and scope. Again, most of these include an ambidextrous category. As with monkeys, different tasks were studied by various observers. For chimpanzees, 44 individuals each had their data presented, and review suggests that 22 preferred the left hand, 20 the right hand and two were ambidextrous (Bolser et al., 1988; Bresard & Bresson, 1983; Chorazyna, 1976; Dimond &

Harries, 1984; Finch 1941). Steklis & Marchant (1987) note no tendency for right- or left-handedness among 26 chimpanzees, most of whom were considered to be ambidextrous. Similar findings were noted by Heestand (1986) for 20 other individuals, and also by Grzimek (1949). Bard et al. (1990) observed seven of ten infant chimpanzees to have a right hand bias for one of three measures of hand usage.

A tally of data for gorillas indicates that about 22 preferred the left hand, 19 the right hand and 11 were grouped as unclassifiable or ambidextrous (Annett & Annett, 1991; Dimond & Harries, 1984; Fischer et al., 1982; Haas, 1958; Lockard, 1984; Preilowski & Leder, 1985; Rensch & Ducker, 1966-7; Riess, Ross, Lyerly, & Birch, 1949; Yerkes, 1916). Additionally, Heestand (1986) noted 29 individuals and Schaller (1963) 72 others to be mostly ambidextrous for a variety of activities with no group tendency toward left or right except for a tendency to right handedness in chest beating. Olson et al. (1990) observed 12 gorillas for handedness on different tasks and found no clear group bias to the right or left, whereas Fagot & Vauclair (1988a) found a left hand bias on one of three tasks in seven of their group of eight. Most recently, Byrne and Byrne (1991) found no evidence of a population bias toward right or toward left handedness among 44 free-ranging mountain gorillas. Orang-utans have received less study. Data indicate eight left-handed, four right-handed and four ambidextrous individuals (Annett, 1985; Bresard & Bresson, 1983; Dimond & Harries, 1984; Finan, 1939; Rensch & Ducker, 1966-7). In addition, Heestand (1986) studied eight and Olson et al. (1990) studied 13 other individuals. Most tended toward ambidexterity and there was no trend to right or left. Similar findings were reported by Grzimek (1949). While all six gibbons studied by Olson et al. (1990) proved to be left-handed on one of the tasks studied and three of five siamangs observed by Chivers (1974) were left-handed, most of the 19 gibbons and siamangs studied by Stafford et al. (1990) were right preferent. There was no population bias among the 13 siamangs studied by Heestand (1986).

When comparing her observations of the various species of apes, Heestand noted no overall bias for any activity except locomotion, which was initiated preferentially with the right limb. An attempt to test for a population bias for handedness in the vertical (bipedal) position dictated one of the tasks given to the apes studied by Olson et al.(1990). On this task all six gibbons proved left-handed, 10 of 12 gorillas preferred the right, and the 12 orang-utans displayed no group bias.

It has been proposed that there is a left-hand specialization for reaching when clinging is required (MacNeilage, Studdert-Kennedy, & Lindblom, 1987, 1988, 1991) though the data of King and Landau (1988) contradict this. Taken together, the data related above from prosimians, monkeys, and apes barely hints at a bias toward left-handedness in the vertical (bipedal) position.

Arguments for the occurrence of population trends are based upon data that have usually been strikingly modest in the trends observed. Such arguments have generated post hoc qualifications of the conditions which determine whether the bias will be to the right or left. Overall, it seems that if any population asymmetry for right- or left-handedness exists among prosimians, monkeys, or apes, it is far less pronounced than in humans. This conclusion, in combination with the suggestion that there may be asymmetric brain structure and lateralization of communicative function in nonhuman primates, has led to the suggestion that cerebral dominance for speech and other cognitive functions may have had an earlier evolutionary origin than the population bias towards right-handedness characteristic of humans (Lehman, 1980b, Warren, 1977b).

Even if the population as a whole has no marked bias for the right or the left hand, there could be subpopulations with a strong lateral bias. The ability of environmental, experiential, and postural factors to bias hand usage toward one side has already been mentioned. However, it is difficult to imagine how they would be translated into a subpopulation or population bias. Whether there are population trends within rather than across different species must be regarded as a matter open for study.

Comparison of the sexes had usually revealed no significant differences until the more recent analyses by Ward of the studies performed by her group (Ward, 1991). These analyses suggest that some species of both the lemur and the bushbaby may show a tendency toward greater left-handedness in males than in females, with the right-handedness of the females becoming more prominent with age. However, this has not been true of all these studies (Forsythe & Ward, 1988; Larson et al., 1989) and with one exception (Stafford et al., 1990) apparently has not been noted in monkeys or apes.

The possibility of a hereditary or familial basis for a subpopulation bias has been investigated. While Brinkman (1984b, 1987) noted a significant tendency for offspring to have similar hand usage to their mothers, five other studies found no clear evidence to support this conclusion (Brooker et al., 1981; Byrne & Byrne, 1991; Forsythe & Ward, 1988; Kubota, 1989, 1990; Vauclair & Fagot, 1987a). This discrepancy may be an artifact of the criteria utilized to define hand preference. Whereas the studies with negative findings used the same concept of hand preference as in this review and determined hand preference from the relative frequency of right and left hand usage, Brinkman's abstract (1984b) indicates that measures of skill including speed, dexterity, and types of movement were used to define the preferred hand. Brinkman's findings could be interpreted as indicating that skill and idiosyncracies of motor coordination might be familial (genetic or learned) on the maternal side. This does not necessarily imply that hand preference (defined as the tendency to use one hand more frequently than the other) would have a similar familial tendency.

Finally, it should be mentioned that attempts have been made to correlate handedness with limb size (MacNeilage et al, 1988; Ogle, 1871; Steir, 1911).

Hand Preference in Relation to Other Lateralized Functions

Contrary to Ogle's surmise, the evidence reviewed indicates that hand preference in the monkey is predominantly an expression of contralateral brain activity. In this sense the laterality of hand preference is an expression of cerebral lateralization. The extent to which this laterality is consistently related to that of other functions is one measure of hemispheric specialization. Two approaches to this problem may be distinguished: within individuals and within populations. The first examines the tendency of individuals to display the same relationship (homolateral or contralateral) between the laterality of handedness and that of another function; the other is concerned with the possibility that the population's laterality (right or left) for handedness is skewed in the direction of the population's laterality for another aspect of the activity being performed. Investigations of other aspects of lateralized function in non-human primates which do not relate them to hand preference are not discussed here.

Hand Preference and Lateralization of Other Functions: Laterality at the Level of the Individual

Three approaches are possible and will be discussed in turn: The laterality of hand preference can be compared to that of other asymmetric behavior of the intact animal; it can be compared to the side of the brain exhibiting better performance after commissurotomy; and it can be compared to the side of the brain showing the greatest deficit after a unilateral lesion.

Only a few studies of this subject have concerned themselves with the relationship between the laterality of hand preference and that for other parts of the body. Except for Hall and Mayer (1966), none provide evidence for a relationship between the laterality or strength of hand preference and that of the sighting eye (Cole 1957; Hall & Mayer, 1966; Kounin 1938; Kruper et al., 1966). Cole (1957) found no relationship between handedness and footedness; while a total of 34 monkeys were studied for sighting preference, only six had their foot preference reported. An initial investigation of intact animals suggesting a relationship between the laterality of hand preference and that for lateral mirror-image shapes (Lehman, 1978c) has not been confirmed when the study was repeated (Lehman, unpublished research). The preferred direction of turning has been studied in the bush baby and found to bear no relationship to the side of the preferred hand (Larson et al., 1989).

Other studies have involved cognitive tasks. The format has been to present both cerebral hemispheres with a discrimination task and to note the animal's hand preference. Following this, the interhemispheric commissures are divided. This permits the animal to be tested one hemisphere at a time to determine whether or not the hemisphere with the better recollection of the discrimination is also the one which controlled the preferred hand. Underlying these studies is the assumption that the hemisphere more effectively learning the task is also the one more frequently controlling the choice of hand usage.

In its least rigorous form, such an experiment was carried out by Lehman and Spencer (1972). They trained optic chiasm sectioned monkeys on a visual discrimination task while both eyes were open and tested them for retention with one eye at a time. This test allows assessment of hemispheric recall under conditions in which both hemispheres communicated normally with one another. Since the optic chiasm was already sectioned, monocular testing permits the visual discrimination to be presented directly to the ipsilateral hemisphere, but only by way of the forebrain commissures to the contralateral one. As a result, it was thought that recall under monocular testing would more greatly reflect the recall of the ipsilateral hemisphere and that comparison of recall with the right eye to that with the left might indicate the hemisphere which had learned best. The negative results of this experiment are hard to interpret because this assumption may be unjustified (Hamilton, 1977a).

A more rigorous approach was pursued by Hamilton (1977a). He taught his chiasm sectioned monkeys a variety of visual discriminations before sectioning the interhemispheric commissures. Subsequently, each animal was tested one eye (hemisphere) at a time for recall. Hand preference was determined separately prior to both surgery and training. Monocular testing failed to reveal a correlation between the laterality of the preferred hand and that of the hemisphere with greater recall.

These results differ from earlier and more limited experiments by Trevarthen (1962, 1965a). He trained several monkeys after the optic chiasm and forebrain commissures had been sectioned. During binocular (and hence bihemispheric) training, one hand was preferred. Afterwards, testing was carried out with one eye (hemisphere) at a time. These monocular tests revealed that the hand preferred during binocular learning was contralateral to, and hence the one controlled by, the cerebral hemisphere which had learned best. Trevarthen suggested that better learning by one hemisphere might be related to an asymmetry of attention as reflected in the larger frequency of hand choices controlled by the superior hemisphere.

A more extensive series of similar experiments studied learning rather than recall (Hamilton, 1977b, 1983; Hamilton & Vermeire, 1982). Monkeys were tested for hand preference prior to division of their forebrain commissures and then trained monocularly afterwards. The rate of learning with the right eye (left hemisphere) was compared to that of the left eye (right hemisphere). For a variety of tasks there was judged to be no consistent correlation between the originally preferred hand and the learning of the contralateral hemisphere (Hamilton, 1977b, 1983).

However, the correlation between sequentially presented stimuli in a same-different go/no-go paradigm was somewhat stronger than the others. Encouraged by these results, more animals were added to the study so as to give a total of 12. The results, which embraced both the old and the new animals, showed a significant linear correlation between the initial hand preference and learning by the contralateral hemisphere (r=0.768) for the sequential but not (r=0.428) for the simultaneous task (Hamilton & Vermeire, 1982). However, the authors' suggestion that these two performances were different in their correlation with handedness is questionable since the two coefficients do not differ significantly from one another (t=0.97, df=10). It is noteworthy that the laterality index (-100 to +100 possible) of individuals for the discriminations tended to be unimodal and varied only between -19 to +55 on the sequential and -41 to +52 on the simultaneous tasks, whereas handedness was bimodal (U-shaped distribution) and ranged from -97 to +93, suggesting that laterality on the former tasks is significantly less marked than for handedness.

The ultimate issue is whether these experiments demonstrating a contralateral relationship between preoperatively preferred hand and the postoperatively more competent cerebral hemisphere will stand the test of time. While Trevarthen's (1962,1965a) and Hamilton and Vermeire's (1982) findings appear to support one another, Hamilton's other studies were negative. Some of those utilizing recall of go/no-go and other forms of visual discriminations failed to show any relationship in the laterality of postoperative hemispheric recall and preoperative hand preference (Hamilton, 1977a, b). It is noteworthy that Hamilton determined hand preference during tasks other than those upon which the animal was trained and tested for cognitive recall. This could account for his negative results; it is not hard to imagine that it is the hand preferred during discrimination learning rather than that preferred during some irrelevant task which provides the most valid clue to the hemisphere which learned most effectively. Although the data is limited, confirmation of this concept may be found in the reports of Trevarthen (1962, 1965a). However, he determined both hand preference and learning after section of the commissures so that the relevance of these observations to hand preference in the intact individual is uncertain. The ablation studies which are discussed next offer further support for the relationship between the preferred hand and learning.

Such an approach has been pursued by Ettlinger and his colleagues. After observing the hand preferred by each monkey during mastery of a discrimination task, lesions were made in an area of the brain thought to be critical for the performance of the task. In some animals these were produced on the side contralateral to the preferred hand, in others ipsilateral to it. One could then determine which of the brain lesions was more apt to interfere with performance.

Data of Moffett and Ettlinger's studies (1970) are pertinent. Unilateral lesions of the posterior parietal cortex were made contralateral or ipsilateral to the hand previously preferred for tactile discriminations. Postoperatively, the three ipsilateral animals showed little or no impairment in tactile performance whereas the four contralaterally lesioned ones had moderate deficits for both recall and new learning. Part of the deficit exhibited by the contralateral lesion group might be attributed to problems with sensory perception or observed changes in motor coordination of the originally preferred hand. Interestingly, the incoordination was noted to affect both hands in animals with contralateral lesions but only one or none in those with ipsilateral ones.

Garcha et al. (1982) gave their monkeys tactile discriminations before and after ablations of SII sensory cortex (often with additional bilateral removals of other brain regions). Removals of SII contralateral to the hand originally preferred during learning impaired performance with each hand whereas ipsilateral ablation had little or no effect upon either. It appears that the SII cortex contralateral to the hand preferred for the tactile task is critical for task performance irrespective of the hand employed. When this experiment was repeated with resection of the insula instead of SII, the outcome was similar but the decrement in performance also extended to "visual/auditory tasks" (Ettlinger, 1987). These extended observations have since been reported as a new series of animals undergoing contralateral and ipsilateral ablations of SII or the posterior insula (Horster & Ettlinger, 1987). Ettlinger concluded that specific portions of the "same neural system" within the cerebral hemisphere contralateral to the preferred hand may be specialized to different degrees.

Contrary results were obtained by Warren, Cornwell, and Warren (1969) and by Warren and Nonneman (1976) when prefrontal or prestriate cortex was removed unilaterally following a determination of hand preference on manual tasks. Postoperative testing included delayed response, position reversal, visual, and other discriminations. These tasks are known to require an intact prefrontal or prestriate cortex. Animals receiving contralateral lesions did no better or worse than their cohorts with lesions ipsilateral to the preoperatively preferred hand. It is possible that this failure to duplicate the Ettlinger group's demonstration of greater cognitive competence by the hemisphere opposite the preferred hand is due to differences in the lesion sites, cognitive tasks, or the determination of hand preference. It is noteworthy that Warren's group, like Hamilton's, determined hand preference preoperatively during a task unrelated to the cognitive task whereas Ettlinger's group not only measured hand preference preoperatively but did so upon the cognitive task in question. The study by Sutton, Trachy, and Lindeman (1981) is subject to similar criticism. A relationship between the preferred hand and loss of vocal agility after unilateral cingulate lesions was claimed, but only 4 animals were studied and one had a hand deformity.

Even if one accepts those split-brain and unilateral lesion studies which demonstrate superiority of cognitive function in the hemisphere opposite the preferred hand, an unresolved issue remains. The preferred hand was necessarily (Ettlinger, 1987; Garcha et al., 1982; Moffett & Ettlinger, 1970; Trevarthen, 1962, 1965a) or possibly (Hamilton & Vermeire, 1982) the one used most often while learning the task (or a similar one) utilized for comparing the cognitive competence of the two cerebral hemispheres. Consequently, it is possible that the demonstrated superiority of the hemisphere contralateral to the preferred hand might be the result of greater training of that hand-hemisphere combination rather than the expression of a preexisting asymmetry of cerebral function. Consistent with the former possibility, though far from conclusive in this regard, are the experiments of Ettlinger and Kalsbeck (1962). Monkeys forced to train preoperatively with one particular hand (left) on a tactile discrimination had worse postoperative deficits when their lesions were contralateral to the hand with the only preoperative practice. Electrophysiologic recording and stimulation studies also support the concept that it is the hand which has received training upon a task that determines the hemisphere in which greater learning resides (Stamm, Gadotti, & Rosen, 1975; Stamm, Rosen, & Gadotti, 1977).

Hand Preference and Task Requirement: Laterality at the Population Level

If the same hand is consistently preferred for a particular task, it is reasonable to assume that the controlling contralateral hemisphere is also the one more actively engaged in carrying out the other mental activities involved in performing that task. Should a significant majority of animals display a consistent preference for the same hand when performing that task, one might conclude that there is some degree of lateralization of cerebral function (hemispheric specialization, cerebral dominance) for that task in the population. This chain of reasoning underlies most of the experiments which have attempted to use studies of hand preference to demonstrate cerebral dominance at the population level in monkeys. Caution is necessary in interpreting such an experiment; because any task makes multiple demands, it may be difficult to decide which hemisphere should be most actively involved (Ettlinger, 1987).

Commonly superimposed upon this line of reasoning is an evolutionary retrospective. This argues that the relative dominance of the human right cerebral hemisphere for visuo-spatial or other functions and the dominance of the left for other functions such as language, tactile discrimination, and manipulation might be present to a lesser degree in nonhuman primates. These suppositions have led to the expectation that hemispheric specialization in monkeys should express itself as a tendency for more animals to prefer the right hand during tasks which place their demands upon left hemisphere functions and to prefer the left during those more demanding of right hemisphere activity. This rationale is contrary to the underlying assumption that hemispheric specialization in the monkey is a weak version of that in humans. Right hand preference is the rule in humans and hence is usually a left hemisphere function. Human hand preference on a task does not vary with the cognitive or perceptual demands of that task and there is no reason why it should be singled out to function differently in the monkey brain than in the human brain.

Nevertheless, MacNeilage et al., (1987) utilized such an evolutionary argument to rationalize their conclusions that most monkeys use the right hand for manipulation and the left for visually demanding tasks. The evidence provided was considered tenuous by most reviewers and their appended comments and the authors' rebuttal deserve careful review by the concerned reader. In particular, part of the data supporting this conclusion is based on the study of Beck and Barton (1972) which demonstrated a tendency to right-handedness for manipulation on bimanual tasks. However, only a limited number of animals were studied and some of the conclusions are based upon median rather than mean performance scores (Jerison, 1987). Alternatively, these results are explicable as another example of the tendency to use the less preferred hand for the manipulatory portion of multicomponent tasks while reserving the preferred hand for the consummatory portion (Lehman, 1987; but see Hamilton & Vermeire, 1988). Most of Beck and Barton's animals preferred the left hand on single component tasks and therefore would have been expected to use the right for the preliminary component of multiple component tasks. Furthermore, MacNeilage et al. (1987) disregard studies which fail to indicate right-handedness for manipulation in the monkey population on the grounds that these were not field observations and often included tasks of simple reaching. Such reasoning involves arbitrary criteria, and this latter argument certainly goes contrary to the wellspring of their argument which is based upon the specialization of the human brain. Additional critique is provided by Fagot and Vauclair (1991).

This evolutionary hypothesis has also been stated with respect to other types of tasks. Thus it has been presumed that tactile discriminations might favor right hand (left hemisphere) usage whereas visual ones would promote left-handedness (right hemisphere) (Ettlinger et al., 1968; MacNeilage et al., 1987). The evidence favoring right-handedness during tactile tasks (Beck & Barton, 1972; Ettlinger, 1961; Ettlinger & Moffett, 1964; Gautrin &

Ettlinger, 1970; Milner, 1969) and left-handedness for visual discriminations (Ettlinger, 1961; Ettlinger et al., 1968; Gautrin & Ettlinger, 1970; Lehman, 1987) or visually guided tasks (Deuel & Schaffer, 1987; Fagot & Vauclair, 1988a, b; Fragaszy & Mitchell, 1990) has usually been equivocal despite the compilation of some of the results into a more compelling form (see Table 2, MacNeilage et al., 1987; Fagot & Vauclair, 1991).

The perception that these results fail to provide convincing support for the initial hypothesis is reinforced by those studies in which the same animals were trained upon both visual and tactile discriminations. No consistent difference in hand preference was found for the groups of animals given these tasks (Ettlinger, 1961; Ettlinger et al., 1968; Ettlinger & Dawson, 1969) and sometimes the results were opposite to those expected (Gautrin & Ettlinger, 1970). Difficulties in interpreting some of these studies are compounded by the classification of a majority of animals as ambidextrous. An intriguing study (Kuhl, 1988) of 30 monkeys revealed all individuals to rely exclusively upon the right hand when performing an auditory discrimination task but to vary their hand choice during other activities.

Fagot and Vauclair (1991) have added another element to the debate over hemispheric specialization for handedness in non-human primates. From their observations (1988a, b) and the literature, they suggest that left and right hand preferences in non-human primates only manifest during tasks that are novel and complex; simple, highly practiced activities do not evidence a population bias for either hand. Many of the studies which Fagot and Vauclair do and do not cite appear to differ with respect to whether the right or the left hand was preferred even though comparable tasks were involved. Furthermore, at least in the study of Kuhl (1988), increased practice was associated with the manifestation of a decided preference for the right hand. Additional critique is offered by Annett and Annett (1991).

An alternative approach with a similar rationale has been to investigate whether the learning of left or right-handed animals might be different. In this case the hypothesis investigated was whether right-handed animals learned tactile discriminations better than left-handed ones. One might similarly attempt to discover whether left-handed animals learned visual discriminations better than right-handed ones. The tactile hypothesis was explored by Horster and Ettlinger (1985). Contrary to expectation, left-handed animals solved tactile problems significantly more quickly than right-handed ones and, even more surprisingly, ambidextrous ($< 90\%$ handed) individuals did so even faster. The results of six animals studied preoperatively by Ettlinger and Dawson (1969) do not give support to either the tactile or the visual hypothesis.

Finally, the studies of McGonigle and Flook (1978) deserve attention. Five squirrel monkeys were taught to associate one color as a cue that the right hand was required to obtain reward and that a second color required a left hand response. Learning occurred equally quickly for both hands (colors). Taken as a whole, these inquiries do not seem to provide any convincing evidence for a relationship between right or left hand usage and rapidity of learning.

Controversy continues as to whether non-human primates have a population bias toward right- or toward left-handedness for certain tasks (MacNeilage et al., 1987, 1988; Warren, 1980; Fagot & Vauclair, 1991). The beguiling aspect of this debate arises from those observations which demonstrate a species preference for one particular hand during a specific activity; the bewildering aspect of this work lies in the frequent failure to replicate the same result in the same or another species with the same or a related task. Attempts to reconcile these discrepancies have produced a plethora of hypotheses. Studies with operational definitions of task context, components, and requirements, as well as their integration into task performance articulated a priori in terms of hypothesized right and left hemisphere functions are prerequisite to a satisfactory conceptual synthesis in this and related areas.

Conclusions: Manual Preference in Prosimians, Monkeys and Apes

1. Voluntary hand usage is usually asymmetric, with one hand playing a leading role and the other a subordinate one. This distinction is less clear for multicomponent tasks.

    a. The ability of either hand to play a leading role in voluntary activity is dependent upon the informed activity of the contralateral cerebral hemisphere. It may be disrupted if this hemisphere is deprived of relevant sensory information or injured.

    b. Subordinate and coordinate activity of the opposite hand and the cerebral hemisphere contralateral to it is dictated in small part by messages from the leading hemisphere which pass through the corpus callosum. However, some messages passing through the callosum also promote a duplication of effort by the two hands. This duplication is normally suppressed by the supplementary motor cortex of the subordinate hemisphere.

    c. Major pathways integrating the activities of the two hands/cerebral hemispheres probably originate and certainly cross the midline at subhemispheric levels.

    d. Duplication of hand usage is unusual but occurs more frequently in young individuals.

2. Hand preference appears to be a reproducible phenomenon at the level of the individual (within-tasks/ within-individuals and across-some-tasks/ within-individuals) but weakly, if at all, at the population level (within-or across-tasks/ across-individuals).

a. Hand preference is not due to chance alone: The population distribution of preference from 0 to 100% right hand preference is usually U-shaped or even J-shaped rather than bell-shaped for any one task; preference for the preferred hand tends to remain the same or strengthen upon repetitions of the same task; there is often a significant correlation between the hand preferences exhibited by individuals for different tasks.

b. The task-specific nature of hand preference, its variation with situational factors, and occasional strengthening with practice suggest that hand preference is of multifactorial origin. It probably has no genetic or familial basis and is evident prior to training on a specific task.

c. When a task has several components involving the object which contains the goal of voluntary activity, there is a tendency to switch hands between components. The hand preferred for the consummatory component probably tends to be that preferred during other tasks. When one hand is preferred throughout, this is usually the same as that preferred for single component tasks.

d. The possible correlation between the degree of preference for one hand and the skill or strength of that hand relative to the non-preferred hand has received too little attention for even tentative conclusions. Evidence that hand preference is more marked on tasks demanding more complex manipulation is mixed.

e. Hand preference can be altered in favor of the non-preferred hand by ablations of the contralateral cerebral hemisphere or lesions which deprive it of relevant sensory information. Preference for one hand appears to depend upon the active, informed participation of the cerebral hemisphere contralateral to the preferred hand and will be maintained even if this information is only available from the opposite hemisphere.

3. There is probably little if any similarity between the cerebral organization responsible for hand preference in monkeys and the hemispheric specialization of the human population. Data from studies of apes is limited but seems similar to that of monkeys.

a. Evidence from prosimians rather tentatively suggests a weak population trend toward left-handedness. However, among the evolutionarily more advanced monkeys and apes there is no or extremely little population bias toward one hand. Overall, studies of monkeys indicate no bias, or at most a minimal one to the left.

b. The data available offers little suggestion of any correlation between the preferred hand and the preferred foot or sighting eye.

c. Evidence favoring the use of the right hand for one type of task (e.g., tactile or manipulative) or the left for others (e.g., visual) seems inconsistent and limited. Evidence regarding auditory tasks is limited.

d. Evidence favoring better learning by right-handed individuals for one type of task or by left-handed individuals on others also seems unconvincing.

e. The most provocative results suggest that the hemisphere opposite the hand preferred for a cognitive or perceptual task is also the one which learns to perform that task best. It is not clear that this hemispheric superiority for task performance is the expression of a preexisting superiority. It may very well be the product of an attentional or other benefit accruing to one hemisphere in association with the greater use of the contralateral hand during training.

f. The evolution of cerebral hemispheric specialization for handedness in the population may not have preceded that for speech and cognition.

References

Annett, M. (1972). The distribution of manual asymmetry. *British Journal of Psychology, 63*, 343-358.

Annett, M. (1985). *Left, right, hand and brain: The right shift theory*. London: Lawrence Erlbaum.

Annett, M., & Annett, J. (1991). Handedness for eating in gorillas. *Cortex, 27*, 269-275.

Bard, K. A., Hopkins, W. D., & Fort, C. L. (1990). Lateral bias in infant chimpanzees (*Pan troglodytes*). *Journal of Comparative Psychology, 4*, 309-321.

Beck, C. H. M., & Barton, R. L. (1972). Deviation and laterality of hand preference in monkeys. *Cortex, 8*, 339-363.

Bernhard, C. G., & Bohm, E. (1954). Cortical representation and functional significance of the corticomotoneural system. *Archives of Neurology and Psychiatry, 72*, 473-502.

Bishop, A. (1963). Use of the hand in lower primates. In J. Buettner-Janusch (Ed.), *Evolutionary and genetic biology of primates*. Vol. 2. (pp.133-225). New York: Academic Press.

Bolser, L.A., Runfeldt, S., & Morris, R.A. (1988). Handedness in language-trained chimpanzees (*Pan troglodytes*) in daily activities and assessment tasks. *Journal of Clinical and Experimental Neuropsychology, 10*, 40-41.

Bossom, J., & Hamilton, C. R. (1963). Interocular transfer of prism-altered coordinations in split-brain monkeys. *Journal of Comparative and Physiological Psychology, 56*, 769-774.

Box, H. O. (1977). Observations on spontaneous hand use in the common marmoset (*Callithrix jacchus*). *Primates, 18*, 395-400.

Bresard, B., & Bresson, F. (1983). Handedness in *Pongo pygmaeus* and *Pan troglodytes*. *Journal of Human Evolution, 12*, 659-666.

Bresard, B., & Bresson, F. (1987). Reaching or manipulation: left or right. *Behavioral and Brain Sciences, 10*, 265-266.

Brinkman, C. (1984a). Supplementary motor area of the monkey's cerebral cortex: Short- and long-term deficits after unilateral ablation and the effects of subsequent callosal section. *Journal of Neuroscience, 4*, 918-929.

Brinkman, C. (1984b). Determinants of hand preference in *Macaca fascicularis*. *International Journal of Primatology, 5*, 325.

Brinkman, C. (1987). Personal communication to P. F. MacNeilage, M. G. Studdert-Kennedy & B. Lindblom. Primate handedness reconsidered. *Behavioral and Brain Sciences, 10*, 247-303.

Brinkman, J., & Kuypers, H. J. M. (1973). Cerebral control of contralateral and ipsilateral arm, hand and finger movements in the split-brain rhesus monkey. *Brain, 96*, 653-674.

Brooker, R. J., Lehman, R. A. W., Heimbuch, R. C., & Kidd, K. K. (1981). Hand usage in a colony of bonnett monkeys. *Behavioral Genetics, 11*, 49-56.

Brookshire, K. H., & Warren, J. M. (1962). The generality and consistency of handedness in monkeys. *Animal Behavior, 10*, 222-227.

Butler, C. R. (1968). A memory-record for visual discrimination in habits produced in both cerebral hemispheres of monkey when only one hemisphere has received direct visual information. *Brain, 10*, 152-167.

Byrne, R. W., & Byrne, J. M. (1991). Hand preferences in the skilled gathering tasks of mountain gorillas (*Gorilla g. berengei*). *Cortex, 27*, 521-546.

Chivers, D. J. (1974). *The Siamang in Malaya*. Basel: S. Karger.

Chorazyna, H. (1976). Shifts in laterality in a baby chimpanzee. *Neuropsychologia, 14*, 381-384.

Clough, J. F .M., Kernell, D., & Phillips, C. G. (1968). The distribution of monosynaptic excitation from the pyramidal tract and from primary spindle afferent motoneurones of the baboon's hand and forearm. *Journal of Physiology (London), 198*, 145-166.

Cole, J. (1957). Laterality in the use of the hand, foot, and eye in monkeys. *Journal of Comparative and Physiological Psychology, 50*, 296-299.

Cole, J., & Glees, P. (1951). Handedness in monkeys. *Experientia, 7*, 224-225.

Cronholm, J. N., Grodsky, M., & Behar, I. (1963). Situational factors in the lateral preference of rhesus monkeys. *Journal of Genetic Psychology, 103*, 167-174.

Deuel, R. K., & Dunlop, N. L. (1980). Hand preferences in the rhesus monkey. *Archives of Neurology, 37*, 217-221.

Deuel, R. K., & Schaffer, S. P. (1987). Patterns of hand preference in monkeys. *Behavioral and Brain Sciences, 10*, 270-271.

Dimond, S., & Harries, R. (1984). Face touching in monkeys, apes and man: Evolutionary origins and cerebral asymmetry. *Neuropsychologia, 22*, 227-233.

Downer, J. L. (1959). Changes in visually guided behavior following midsagittal division of optic chiasm and corpus callosum in monkeys (*Macaca mulatta*). *Brain, 82*, 251-259.

Ettlinger, G. (1961). Lateral preferences in monkeys. *Behaviour, 17*, 275-287.

Ettlinger, G. (1987). Primate handedness: How nice if it were so. *Behavioral and Brain Sciences, 10*, 271-273.

Ettlinger, G., Blakemore, C. B., & Milner, A.D. (1968) Opposite hand preferences in two sense-modalities. *Nature, 218*, 1276.

Ettlinger, G., & Dawson, R. F. (1969). Hand preferences in the monkey: Effect of unilateral cortical removals. *Neuropsychologia, 7*, 161-166.

Ettlinger, G., & Kalsbeck, J. E. (1962). Changes in tactile discrimination and in visual reaching after successive and simultaneous bilateral posterior parietal ablations in the monkey. *Journal of Neurology, Neurosurgery, and Psychiatry, 25*, 256-268.

Ettlinger, G., & Moffett, A. (1964). Lateral preferences in the monkey. *Nature, 204*, 606.

Fagot, J., & Vauclair, J. (1988a). Handedness and bimanual coordination in the lowland gorilla. *Brain, Behavior and Evolution, 32*, 89-95.

Fagot, J., & Vauclair, J. (1988b). Handedness and manual specialization in the baboon. *Neuropsychologia, 26*, 795-804.

Fagot, J., & Vauclair, J. (1991). Manual laterality in nonhuman primates: A distinction between handedness and manual specialization. *Psychological Bulletin, 109*, 76-89.

Finan, J. L. (1939). A note on prehension in a captive orang-utang. *Journal of Genetic Psychology, 55*, 435-437.

Finch, G. (1941). Chimpanzee handedness. *Science, 94*, 117-118.

Fischer, R. B., Meunier, G. F., & White, P. J. (1982). Evidence of laterality in the lowland gorilla. *Perceptual and Motor Skills, 54*, 1093-1094.

Forsythe, C., Milliken, G. W., Stafford, D. K., & Ward, J. P. (1988). Posturally related variations in hand preferences of the ruffed lemur (*Varecia variegata*). *Journal of Comparative Psychology, 102*, 248-250.

Forsythe, C., & Ward, J. P. (1988). Black lemur (*Lemur macaco*) hand preference in food reaching. *Primates, 29*, 369-374.

Fragaszy, D. M. (1983). Preliminary quantitative studies of prehension in squirrel monkeys (*Saimiri sciureus*). *Brain, Behavior and Evolution, 23*, 81-92.

Fragaszy, D. M., & Mitchell, S. A. (1990). Hand preference and performance on unimanual and bimanual tasks in capuchin monkeys. *Journal of Comparative Psychology, 104*, 275-282.

Franz, S. I. (1913). Observations on the preferential use of the right and left hands by monkeys. *Animal Behavior, 3*, 140-144.

Furuya, Y. (1963). quoted in Itani, J., Tokuda, K., Furuya, Y., Kano, K. & Shin, Y. The social construction of natural troops of Japanese monkeys in Takasakiyama. *Primates, 4*, 1-42.

Garcha, H.S., Ettlinger, G., & Maccabe, J. J. (1982). Unilateral removal of the second somatosensory projection cortex in the monkey: Evidence for cerebral dominance. *Brain, 105*, 787-810.

Gautrin, D., & Ettlinger, G. (1970). Lateral preferences in the monkey. *Cortex, 6*, 287-292.

Gazzaniga, M. S. (1964). Cerebral mechanisms involved in ipsilateral eye-hand use in split-brain monkeys. *Experimental Neurology, 10*, 148-155.

Gazzaniga, M. S., & Young, E. D. (1967). Effects of commissurotomy on the process of increasing visual information. *Experimental Brain Research, 3*, 368-371.

Grzimek, B. (1949). Rechts und linkshandigkeit bei pferden, papageien und affen. *Zeitschrift fur Tierpsychologie, 6*, 406-432.

Haas, G. (1958). Handigskeitbeobachtungen bei gorillas. *Saugetierkundiche mitteilungen, 6*, 59-62.

Hall, K. R. L., & Mayer, B. (1966). Hand preferences and dexterities of captive patas monkeys. *Folia Primatologica, 4*, 169-185.

Hamilton, C. R. (1977a). Investigations of perceptual and mnemonic lateralization in monkeys. In S.Harnad, R. W. Doty, L. Goldstein, J. Jaynes & G. Krauthamer (Eds), *Lateralization in the nervous system* (pp. 45-62). New York: Academic Press.

Hamilton, C. R. (1977b). An assessment of hemispheric specialization in monkeys. *Annals of the New York Academy of Sciences, 299*, 222-232.

Hamilton, C. R. (1983). Lateralization for orientation in split-brain monkeys. *Behavioral Brain Research, 10*, 399-403.

Hamilton, C. R., & Vermeire, B. A. (1982). Hemispheric differences in split-brain monkeys learning sequential comparisons. *Neuropsychologica, 20*, 691-698.

Hamilton, C. R., & Vermeire, B. A. (1988). Cognition, not handedness, is lateralized in monkeys. *Behavioral and Brain Sciences, 11*, 723-725.

Hatta, T., & Koike, M. (1991). Left-hand preference in frightened mother monkeys in taking up their babies. *Neuropsychologia, 29*, 207-209.

Hauser, M., Perry, S., Manson, J.H., Ball, H., Williams, M., Pearson, E., & Berard, J. (1991). It's all in the hands of the beholder: New data on free-ranging rhesus monkeys. *Brain and Behavioral Sciences, 14*, 342-344.

Heestand, J. E. (1986). *Behavioral lateralization in four species of apes*. Unpublished doctoral dissertation, University of Washington, Seattle.

Hopkins, W. D., Washburn, D. A., & Rumbaugh, D. M. (1989). Note on hand use in the manipulation of joysticks by rhesus monkeys (*Macaca mulatta*) and chimpanzees (*Pan troglodytes*). *Journal of Comparative Psychology, 103*, 91-94.

Horster, W., & Ettlinger, G. (1985). An association between hand performance and tactile discrimination performance in the rhesus monkey. *Neuropsychologia, 23*, 411-413.

Horster, W., & Ettlinger, G. (1987). Unilateral removal of the posterior insula or of area SII: Inconsistent effects on tactile, visual and auditory performance in the monkey. *Behavioral Brain Research, 26*, 1-17.

Itani, J., Tokuda, K., Furuya, Y., Kano, K., & Shin, Y. (1963). The social construction of natural troops of Japanese monkeys in Takasakiyama. *Primates, 4*, 1-42.

Jason, G. W., Cowey, A., & Weiskrantz, L. (1984). Hemispheric asymmetry for a visuo-spatial task in monkeys. *Neuropsychologia, 22*, 777-784.

Jerison, H. J. (1987). Which hand lost its cunning? *Behavioral and Brain Sciences, 10*, 278-279.

Kawai, M. (1962). quoted by Itani, J., Tokuda, K., Furuya, Y., Kano, K., & Shin, Y. (1963). The social construction of natural troops of Japanese monkeys in Takasakiyama. *Primates, 4*, 1-42.

Kawai, M. (1967). Catching behavior observed in the Koshima troop - a case of newly acquired behavior. *Primates, 8*, 181-186.

Kempf, H. D. (1917). The preferential use of the hands in monkeys with modification by training and retention of the new habit. *Psychological Bulletin, 14*, 297-301.

Kennard, M. A., & Watts, J. W. (1934). The effect of section of the corpus callosum on the motor performance of monkeys. *Journal of Nervous Mental Disorders, 79*, 159-169.

King, J. E., & Landau, V. I. (1988). Quoted by MacNeilage, P. F., Studdert-Kennedy, M. G. & Lindblom, B. Primate handedness: A foot in the door. *Behavioral and Brain Sciences, 11*, 738.

King, J. E., Landau, V. I., & Scott, A. (1988). Quoted in Fagot, J. & Vauclair, J. (1991). Manual laterality in nonhuman primates: A distinction between handedness and manual specialization. *Psychological Bulletin, 109*, 76-89.

Kluver, H. (1933). *Behavior mechanisms in monkeys*. Chicago: University of Chicago Press.

Kluver, H. (1937). Re-examination of implement-using behavior in a Cebus monkey after an interval of three years. *Acta Psychologica, 2*, 347-397.

Kounin, J. S. (1938). Laterality in monkeys. *Journal of Genetic Psychology, 52*, 375-393.

Kruper, D. C., Boyle, B. E., & Patton, R. A. (1966). Eye and hand preference in rhesus monkeys. *Psychonomic Science, 5*, 277-278.

Kubota, K. (1989). Hand use preference of Japanese macaque monkeys raised in an open compound. *Neuroscience Research*, Suppl. *9*, S29.

Kubota, K. (1990). Preferred hand use in the Japanese macaque troop, Arashiyama--R, during visually guided reaching for food pellets. *Primates, 31*, 393-406.

Kuhl, P. (1988). On handedness in primates and human infants. *Behavioral and Brain Sciences, 11*, 727-729.

Larson, C. F., Dodson, D. L., & Ward, J. P. (1989). Hand preferences and whole-body turning biases of lesser bushbabies (*Galago senegalensis*). *Brain, Behavior and Evolution, 33*, 261-267.

Lashley, K. S. (1917). Modifiability of the preferential use of the hands in the rhesus monkey. *Journal of Animal Behavior, 7*, 178-186.

Lawrence, D. G., & Kuypers, H. G. J. M. (1968). The functional organization of the motor system in the monkey. I. The effects of bilateral pyramidal lesions. *Brain, 91*, 1-14.

Lehman, R. A. W. (1968). Motor coordination and hand preference after lesions of the visual pathway and corpus callosum. *Brain, 91*, 525-538.

Lehman, R .A .W. (1970). Hand preference and cerebral dominance in 24 rhesus monkeys. *Journal of Neurological Science, 10*, 185-192.

Lehman, R. A. W. (1972). Constancy of hand preference following forebrain bisection in monkeys. *Neurology, 22*, 763-768.

Lehman, R. A. W. (1978a). The handedness of rhesus monkeys I. Distribution. *Neuropsychologia, 16*, 33-42.

Lehman, R. A. W. (1978b). The handedness of rhesus monkeys. II. Concurrent reaching. *Cortex, 14*, 190-196.

Lehman, R. A. W. (1978c). Lateralized preference and right-left discrimination in the monkey. *Cortex, 14*, 530-539.

Lehman, R. A. W. (1980a). The handedness of rhesus monkeys. III. Consistency within and across activities. *Cortex, 16*, 197-204.

Lehman, R. A. W. (1980b). Distribution and changes in strength of hand preferences of cynomolgus monkeys. *Brain, Behavior, and Evolution, 17*, 209-217.

Lehman, R. A. W. (1980c). Persistence of primate hand preference despite initial training to the contrary. *Behavioral Brain Research, 1*, 547-551.

Lehman, R. A. W. (1987). On the other hand.... *Behavioral and Brain Sciences, 10*, 280-281.

Lehman, R. A. W. (1989). Hand preferences of rhesus monkeys on differing tasks. *Neuropsychologia, 27*, 1193-1196.

Lehman, R. A. W. (1991). Hand choice by monkeys is governed not only by preference but also by perserveration. In A. Ehara et al. (Eds.), *Primatology today* (pp. 265-268). Amsterdam: Elsevier.

Lehman, R. A. W., & Spencer, D. D. (1972). Hand preference and hemispheric learning in the monkey. *Experimental Neurology, 36*, 88-100.

Liu, C. N., & Chambers, W. W. (1964). An experimental study of the corticospinal system in the monkey (*Macaca mulatta*). The spinal pathways and preterminal distribution of degenerating fibers following discrete lesions of the pre- and postcentral gyri and bulbar pyramid. *Journal of Comparative Neurology, 123*, 257-284.

Lockard, J. S. (1984). Handedness in a captive group of lowland gorillas. *International Journal of Primatology, 5*, 356.

Lund, J. S., Downer, J. L., & Lumley, J. S. P. (1970). Visual control of limb movement following section of optic chiasm and corpus callosum in the monkey. *Cortex, 6*, 323-346.

MacNeilage, P. F., Studdert-Kennedy, M. G., & Lindblom, B. (1987). Primate handedness reconsidered. *Behavioral and Brain Sciences, 10*, 247-303.

MacNeilage, P. F., Studdert-Kennedy, M.G., & Lindblom, B. (1988). Primate handedness: A foot in the door. *Behavioral and Brain Sciences, 11*, 737-746.

MacNeilage, P. F., Studdert-Kennedy, M.G., & Lindblom, B. (1991). Primate handedness: The other theory, the other hand and the other attitude. *Behavioral and Brain Science, 14*, 344-349.

Marchant, L. F., & Steklis, H. D. (1986). Hand preference in a captive island group of chimpanzees (*Pan troglodytes*). *American Journal of Primatology, 10*, 301-313.

Masataka, N. (1989). Population-level asymmetry of hand preference in lemurs. *Behavior, 110*, 244-247.

McGonigle, B. O., & Flook, J. (1978). The learning of hand preferences by squirrel monkey. *Psychological Research, 40*, 93-98.

Milliken, G. W., Forsythe, C., & Ward, J. P. (1989). Multiple measures of hand-use lateralization in the ring-tailed lemur (*Lemur catta*). *Journal of Comparative Psychology, 103*, 262-268.

Milliken, G. W., Stafford, D. K., Dotson, D. L., Pinger, C. D., & Ward, J. P. (1991). Analysis of feeding lateralization in the small-eared bushbaby (*Otolemur garnettii*): A comparison with the ring-tailed lemur (*Lemur catta*). *Journal of Comparative Psychology, 105*, 274-285.

Milner, A. D. (1969). Distribution of hand preferences in monkeys. *Neuropsychologia, 7*, 375-377.

Moffett, A. M., & Ettlinger, G. (1970). Tactile discrimination performance in the monkey: The effect of unilateral posterior parietal ablations. *Cortex, 6*, 47-67.

Ogle, W. (1871). On dextral pre-eminence. *Transactions of the Royal Medical and Chirurgical Society of London, 54*, 279-301.

Olson, D. A., Ellis, J. E., & Nadler, R. D. (1990). Hand preferences in captive gorillas, orang-utans and gibbons. *American Journal of Primatology, 20*, 83-94.

Post, D. G., Hausfater, G., & McCuskey, S. A. (1980). Feeding behavior of yellow baboons (*Papio cynocephalus*): Relationship to age, gender and dominance rank. *Folia primatologica, 34*, 170-195.

Preilowski, B. (1979). Performance differences between hands and lack of transfer of finger posture and sensory motor skill in intact rhesus monkeys: Possible model for the origin of cerebral asymmetry. *Neuroscience Letters, 3*, S89.

Preilowski, B. (1983). Is there a cerebral specialization in nonhuman primates? In P. K. Seth (Ed.), *Perspectives in primate biology* (pp.109-117). New Delhi: Today and Tomorrow Printers and Publishers.

Preilowski, B., & Leder, F. (1985). Comparative studies of laterality: Hand use of a captive group of lowland gorillas. In A. Albert (Ed), *Bericht uber den Kongres des Deutschen Gessellschaft fur Psychologie* (pp. 550-552). Gottingen: Hogoefe.

Preilowski, B., Reger, M., & Engele, H.C. (1986). Handedness and cerebral asymmetry in non-human primates. In D. T. Taub & F. A. King (Eds.), *Current Perspectives in Primate Biology* (pp.270-282). New York: Van Nostrand Reinhold.

Rawlins, R. G. (1986). Distribution of manual asymmetry among Cayo Santiago macaques (*M. mulatta*). *Primate Report, 14*, 173-174.

Rawlins, R. G., quoted by Hauser, et al., (1991). *Behavioral and Brain Sciences, 14*, 343.

Rensch, B., & Ducker, G. (1966-7). Manipulierfahigkeit eines jungen Orang-utans und eines jungen Gorillas. *Zeitschrift fur Tierpsychologie, 23*, 874-892.

Riess, B. F., Ross, S., Lyerly, S. B., & Birch, H. G. (1949). The behavior of two captive specimens of the lowland gorilla, *Gorilla gorilla gorilla* (Savage and Wyman). *Zoologica, 34*, 111-118.

Rizzolatti, G., Matelli, M., & Pavesi, G. (1983). Deficits in attention and movement following the removal of postarcuate (area 6) and prearcuate (area 8) cortex in macaque monkeys. *Brain, 106*, 655-673.

Ross, K. C., Wilczynski, W., & Albert, J. W. (1987). Hand preference and cerebral asymmetry in Savannah baboons. *American Zoologist, 27*, 16A.

Rothe, H. (1973). Handedness in the common marmoset (*Callithrix jacchus*). *American Journal of Physical Anthropology, 38*, 561-566.

Sanford, C., Guin, K., & Ward, J. P. (1984). Posture and laterality in the bushbaby (*Galago senegalensis*). *Brain, Behavior and Evolution, 25*, 217-224.

Sanford, C., & Ward, J. P. (1986). Mirror image discrimination and hand preference in the bushbaby (*Galago senegalensis*). *Psychological Record, 36*, 439-449.

Schaller, G. B. (1963). *The mountain gorilla: Ecology and behavior.* Chicago: University of Chicago Press.

Semmes, J., & Porter, L. (1972). A comparison of precentral and postcentral cortical lesions on somatosensory discrimination in the monkey. *Cortex, 8*, 249-264.

Settlage, P. H. (1939). The effect of occipital lesions on visually-guided behavior in the monkey. *Journal of Comparative Psychology, 27*, 93-131.

Sperry, R. W. (1961). Cerebral organization and behavior. *Science, 133*, 1749-1757.

Stafford, D. K., Milliken, G. W., & Ward, J. P. (1990). Lateral bias in feeding and brachiation in *Hylobates*. *Primates, 31*, 407-414.

Stamm, J., Gadotti, A., & Rosen, S. (1975). Interhemispheric functional differences in prefrontal cortex of monkey. *Journal of Neurobiology, 6*, 31-39.

Stamm, J. S., Rosen, S. C., & Gadotti, A. (1977). Lateralization of functions in the monkey's frontal cortex. In S. Harnad, R. W. Doty, L. Goldstein, J. Jaynes & G. Krauthamer (Eds), *Lateralization in the nervous system* (pp. 385-402). New York: Academic Press.

Steir, E. (1911). *Untersuchungen uber Linkshandigkeit.* (pp.86-89). Jena: Gustav Fischer.

Steklis, H. D., & Marchant, L. F. (1987). Primate handedness: Reaching and grasping for straws. *Behavioral and Brain Sciences, 10*, 285-286.

Stellar, E. (1960). The marmoset as a laboratory animal. *Journal of Comparative and Physiological Psychology, 53*, 1-10.

Subramoniam, S. (1957). Some observations on the habits of the slender loris, *Loris tardigradus L. Journal of the Bombay Natural History Society, 54*, 386-398.

Sutton, D., Trachy, R. E., & Lindeman, R. C. (1981). Primate phonation: Unilateral and bilateral cingulate lesion effects. *Behavioural Brain Research, 3*, 99-114.

Tanji, J., Okano, K., & Sato, K. C. (1988). Neuronal activity in cortical motor areas related to ipsilateral, contralateral and bilateral digit movements of the monkey. *Journal of Neurophysiology, 60*, 325-343.

Tokuda, K. (1969). On the handedness of Japanese monkeys. *Primates, 10*, 41-46.

Trevarthen, C. B. (1962). Double visual learning in split-brain monkeys. *Science, 136*, 258-259.

Trevarthen, C. B. (1965a). Functional interactions between the cerebral hemispheres of the split-brain monkey. *Ciba Foundation Study Group, 20*, 24-41.

Trevarthen, C. B. (1965b). Motor responses in split-brain animals. *Ciba Foundation Study Group, 20*, 103-106.

Trevarthen, C. B. (1978). Manipulative strategies of baboons and origins of cerebral asymmetry. In M. Kinsbourne (Ed), *Asymmetrical function of the brain* (pp. 329-404). Cambridge, MA: Cambridge University Press.

Vauclair, J., & Fagot, J. (1987a). Spontaneous hand usage and handedness in a troop of baboons. *Cortex, 23*, 265-274.

Vauclair, J., & Fagot, J. (1987b). Visually guided reaching in adult baboons. *Behavioral and Brain Sciences, 10*, 287.

Ward, J. P. (1987). quoted by MacNeilage, P. F., Studdert-Kennedy, M. G., & Lindblom, B. Primate handedness reconsidered. *Behavioral and Brain Sciences, 10*, 293.

Ward, J. P. (1991). Patterns of food reaching and handling in prosimian species: Lateral hand preference as a function of age, sex, species, posture, and food handling specializations. In A. Ehara et al. (Eds.) *Primatology today* (pp. 261-264). Elsevier Science Publishers, B.P.

Ward, J. P., Milliken, G. W., Dodson, D. L., Stafford, D. K., & Wallace, M. (1990). Handedness as a function of sex and age in a large population of *Lemur. Journal of Comparative Psychology, 106*, 167-173.

Warren, J. M. (1953). Handedness in the rhesus monkey. *Science, 118*, 622-623.

Warren, J. M. (1958). The development of paw preferences in cats and monkeys. *Journal of Genetic Psychology, 93*, 229-236.

Warren, J. M. (1977a). Investigations of perceptual and mnemonic lateralization in monkeys. In S. Harnad, R. W. Doty, L. Goldstein, J. Jaynes, & G. Krauthamer (Eds), *Lateralization in the nervous system* (pp. 151-172). New York: Academic Press.

Warren, J. M. (1977b). Functional lateralization of the brain. *Annals of the New York Academy of Sciences, 299*, 273-280.

Warren, J. M. (1980). Handedness and laterality in humans and other animals. *Physiological Psychology, 8*, 351-359.

Warren, J. M., Abplanalp, J. M., & Warren, H. B. (1967). The development of handedness in cats and rhesus monkeys. In H. W. Stephenson, E. H. Hess & H. L. Rheingold (Eds), *Early behavior* (pp. 73-101). New York: John Wiley.

Warren J. M., Cornwell, P.R., & Warren, H. B. (1969). Unilateral frontal lesions and learning by rhesus monkeys. *Journal of Comparative and Physiological Psychology, 69*, 498-505.

Warren, J. M., & Nonneman, P. (1976). The search for cerebral dominance in monkeys. *Annals of the New York Academy of Sciences, 280*, 732-744.

Yerkes, R. M. (1916). The mental life of monkeys and apes: A study of ideational behavior. *Behavioral Monographs, 3*, 1-145.

Yuanye, M., Yunfen, T., & Ziyun, D. (1986). Handedness of golden monkeys and leaf monkeys in picking up food. *Primate Report, 14*, 248.

7 Lateralized Hand Use in the Precultural Behavior of the Koshima Monkeys (*Macaca fuscata*)

Kunio Watanabe and Masao Kawai

Earlier studies on Japanese monkeys indicated that when monkeys picked up various foods (Itani, 1957; Itani, Tokuda, Furuya, Kano & Shin, 1963; Tokuda, 1969) or caught food which was thrown to them (Kawai, 1967) some hand preferences were evident. It appeared that left-handed monkeys outnumbered right-handed ones in most cases. However, there were a number of monkeys which appeared to be ambidextrous in these groups. Recently, we obtained clear evidence of lateralized hand use by Koshima monkeys during the course of studies on precultural behavior,(i.e., wheat and sweet potato washing; Kawai, 1965). The wheat washing behavior in particular appeared to be established, more or less, by fixed behavioral patterns in the monkeys, and hand preferences were clearly apparent when they were engaged in this behavior. This chapter describes the behavioral situations in which hand preferences were observed. For comparative purposes, other data are reported on hand preference when picking up wheat grain dispersed on the beach. The relation between hand preferences for these two behaviors are also discussed.

Materials and Methods

The Koshima monkeys, which live on a small island off the east coast of Kyushu, Japan, have been provisioned since 1952, and observations on them have continued for the past 37 years. During this period, the monkeys have developed or acquired a number of precultural behaviors such as sweet potato and wheat washing, swimming in the sea, and "give-me-some" postures (see Kawai, 1965 for description). The behavior of washing food items was developed primarily for the purpose of cleaning soiled sweet potatoes or removing sand from handfuls of wheat grains.

Washing behaviors have undergone changes over the years as the environmental situation and management policies of the monkeys have changed. Variations in the amount of food supplied to the monkeys have, in particular, affected them in many respects. Provisions for the monkeys were restricted severely after 1972 because of a change in management policy toward feeding the monkeys. Only 15 monkeys frequently performed the process of washing wheat grains in 1983 and 1984 (Kawai, Watanabe, & Mori, in press). The number of monkeys who washed sweet potatoes did not actually decrease.

The wheat and sweet potato washing behaviors were surveyed intensively by the authors in the summers of 1983 and 1984. In washing wheat grain, the monkeys rake up grains which are scattered on the beach using either hand (occasionally each hand alternately). The monkeys then carry the handfuls of wheat grain for several meters to wash them in the water. In carrying the sweet potatoes, they may simply grasp one or more of the sweet potatoes and carry them to the water. On such occasions, all monkeys were assessed for the hand used to carry the wheat or sweet potatoes. In addition, the methods used for washing the food items and the approximate distances they were carried were concurrently recorded. Both behaviors were recorded as many times as they occurred during observation periods lasting 27 days for sweet potato washing and 15 days for wheat washing behavior.

When the Koshima monkeys were engaged in picking up wheat grains which had been scattered on the sandy beach, their hand preferences appeared to be clearly pronounced. Data on this behavior were recorded for 12 days in May of 1988. The data were analyzed in terms of the degree of lateralization, the situations or opportunities in which hand preferences could be recorded, and the stability of hand preference. Previous reports give more details on the Koshima monkeys themselves (Mori, 1979; Watanabe, Mori, & Kawai, in press) and their washing behaviors (Kawai, 1965; Kawai et al., in press).

Results

Carrying Wheat to the Sea

In 1983 and 1984, the wheat washing behavior was employed frequently by only 15 of the approximately 70 monkeys who belonged to the main troop on Koshima (these 15 monkeys accounted for 99.3% of all observed cases; N = 2,237). Most of these monkeys revealed a very clear preference trend in the hand with which they carried the wheat grains mixed with sand. These data can be found in Table 7.1. Eight monkeys preferred to use their left hand to carry the wheat grains while another six preferred to use their right hand. It proved difficult to distinguish the hand preference in one female (Cha) who exhibited the behavior infrequently. Concerning the 14 who exhibited hand preference, the data were significantly different from chance

Table 7.1. Hand Preferences in Koshima Monkeys for Each Behavior

Subject	Sex	Birth	Wheat Washing Hand R	L	B	Potato Washing Hand R	L	B
Geshi	M	1966	75	5	1 R	11	6	0 r
Geba	M	1969	27	113	29 L	9	12	7
Noso	M	1967	20	148	3 L	8	13	36 l
Bohze	M	1966	2	55	10 L	7	2	6 r
Kani	M	1970	1	1	0	8	5	2
Mebaru	M	1979	1	1	0	11	14	20
Gazami	M	1967	0	1	0	2	11	7 l
Yotaka	M	1980	0	0	0	4	1	0 r
Ume	F	1971	26	114	0 L	1	17	20 L
Uri	F	1978	11	112	0 L	0	6	13 l
Utsugi	F	1977	19	193	58 L	5	9	29 l
Kuri	F	1967	111	19	0 R	4	4	30
Kurumi	F	1970	12	195	1 L	4	3	26
Isu	F	1965	161	10	0 R	11	5	32 r
Ichou	F	1968	71	6	7 R	7	14	25 l
Sanoho	F	1970	10	105	1 L	6	16	24 l
Zakuro	F	1967	220	8	7 R	6	12	31 l
Sakaki	F	1961	20	1	0 R	1	4	8 l
Cha	F	1976	9	7	4	5	5	13
Sasage	F	1969	1	0	0	10	9	22
Gaku	F	1969	1	0	0	6	14	13
Sarubia	F	1971	-	-	-	6	17	9 l
Zassou	F	1969	0	1	0	1	7	12 l
Zarame	F	1970	-	-	-	6	1	6 r
Udo	F	1980	-	-	-	7	5	1
Hie	F	1977	-	-	-	3	9	4 l
Take	F	1977	-	-	-	3	8	2 l
Soba	F	1976	-	-	-	9	4	4 r
Nemu	F	1969	-	-	-	2	2	6
Totals				Right = 6 Left = 8		Right = 0(+6) Left = 1(+14)		

Note. R - indicates significant right hand preference $p<.01$; L - indicates significant left hand preference $p<.01$; l & r - indicate trends for either left or right handedness.

($p<.01$, chi-square test). For five of these monkeys (monkeys with significant hand preferences) data for their hand use had been recorded in 1974 and 1975 by Mori (personal communication). Although his data were fragmental, the direction of hand preference in these monkeys was similar to our observations and therefore appeared to have been maintained over a period of ten years.

Carrying Sweet Potatoes

We next examined the hand use for carrying sweet potatoes. For this behavior, only one monkey (Ume) revealed a significant hand preference (to the left) at a probability level of $p<.01$. No other monkeys showed preferences in their hand use at significant levels, although many of them did display, to a greater or lesser extent, some tendency to use either their left or right hand (6 monkeys to the right; 13 monkeys to the left). These data are also depicted in Table 7.1. Furthermore, many monkeys were rather fond of carrying potatoes with both hands (the cases included all examples in which the monkeys carried one or more potatoes in each hand). It appears that hand preferences in the monkeys were less pronounced for carrying sweet potatoes than for wheat-washing.

Reaching to Pick Up Wheat Grains

Wheat grains were usually presented to the monkeys scattered around on the sandy beach. Thus, the monkeys needed to sort the wheat grains out from the mixed sand. The most elaborate method was the wheat washing behavior described earlier. However, some monkeys did not carry the grains to the water to wash them, but ate the wheat grains directly after picking them up from the sand. Hand preferences were also noted in this condition and are depicted in Table 7.2(a & b).

As can be seen, some monkeys exclusively used the left hand to pick up the grains (e.g., Kera and Bara), while some others used the right (e.g., Yotaka and Hotaru). In the early period, just after the wheat grains had been scattered, most of the monkeys ate them directly with the mouth on the sandy ground without using their hand and/or used both hands alternately to pick up the grains. Therefore, we could not identify any lateralized hand use in most monkeys in this situation. After several minutes the grains became intermingled with the sand trodden under the feet of the monkeys. At this point the monkeys began to relax about taking the grains as they appeased their hunger. Their preferred hand then appeared gradually. Also, when the beach was wet the grains tended to mingle more with the sand, and the hand preference for picking up the wheat grains could be observed more clearly. Such a process was clearly apparent in the data for monkeys such as Geshi, Ume, Utsugi, Ichou, and Udo (see Table 7.2). Although they used both hands with similar frequencies in many cases, the frequencies became more asymmetrical as the number of responses increased. Continuation of the observations revealed that most monkeys displayed some trend in their hand use toward either side, although there were still many monkeys in which no preference was found. Thus, the difference between individual hand preferences were graded from strongly left-handed to strongly right-handed.

Table 7.2a. Hand Preferences in Male Koshima Monkeys for Picking Up Wheat

		Observation Period						
Subject	Birth	(1) R/L	(2) R/L	(3) R/L	(4) R/L	(5) R/L	(6) R/L	Hand
Geshi	1966	45/48	*51/12*	*52/15*	-/-	-/-	-/-	R
Geba*	1969	0/2	*1/55*	*0/46*	-/-	-/-	-/-	L
Noso	1967	65/61	33/38	61/50	-/-	-/-	-/-	
Bohze	1966	83/88	-/-	-/-	-/-	-/-	-/-	
Nosuri	1965	*66/0*	*40/0*	39/40	-/-	-/-	-/-	R
Kani	1970	48/56	14/15	-/-	-/-	-/-	-/-	
Gazami	1967	*1/87*	*17/88*	60/85	26/28	-/-	-/-	L
Yotaka	1980	*71/0*	*33/0*	*18/0*	-/-	-/-	-/-	R
Kuma	1974	50/29	63/57	16/8	*72/25*	-/-	-/-	R
Mimizu	1971	6/16	*7/37*	14/23	-/-	-/-	-/-	L
Kera	1972	*0/72*	*0/71*	*0/40*	-/-	-/-	-/-	L
Kogera	1974	*2/30*	*2/18*	*0/13*	-/-	-/-	-/-	L
Karasu	1977	*39/12*	*26/3*	37/21	75/52	-/-	-/-	R
Okoze	1981	*0/23*	*0/35*	-/-	-/-	-/-	-/-	L
Mutsu	1981	*0/90*	*0/45*	-/-	-/-	-/-	-/-	L
Biki	1982	44/43	*6/28*	-/-	-/-	-/-	-/-	L
Kobe	1983	56/47	*70/39*	14/14	3/0	*12/0*	-/-	R
Mozu	1983	49/61	38/57	21/25	-/-	-/-	-/-	l
Donko	1984	24/32	*39/73*	0/2	*31/72*	-/-	-/-	L
Hotaru	1984	*107/0*	*20/0*	-/-	-/-	-/-	-/-	R
Medaka	1984	14/10	66/45	*69/39*	-/-	-/-	-/-	R
Rakko	1984	5/7	6/5	56/52	-/-	-/-	-/-	
Yuda	1984	49/51	*19/65*	24/53	-/-	-/-	-/-	L
Ponih	1984	8/16	55/54	*37/69*	-/-	-/-	-/-	L
Zuaka	1985	20/30	*24/54*	-/-	-/-	-/-	-/-	L
Tai	1985	2/10	*6/21*	*10/43*	-/-	-/-	-/-	L
Zebura	1986	68/51	9/15	15/20	-/-	-/-	-/-	
Hokke	1986	1/11	*0/36*	-/-	-/-	-/-	-/-	L
Hera	1986	*25/79*	*0/15*	*3/23*	-/-	-/-	-/-	L
Beko	1986	*3/17*	*39/80*	14/15	-/-	-/-	-/-	L
Datsu	1986	54/59	-/-	-/-	-/-	-/-	-/-	
Chidai	1987	20/88	-/-	-/-	-/-	-/-	-/-	L
Tonbo	1987	18/15	51/59	-/-	-/-	-/-	-/-	
Keri	1987	1/27	12/46	-/-	-/-	-/-	-/-	L
Rama	1987	37/86	-/-	-/-	-/-	-/-	-/-	L
Nuh	1987	7/93	8/21	2/6	7/15	-/-	-/-	L
Buyo	1987	77/31	70/41	-/-	-/-	-/-	-/-	R
Poto	1987	30/72	6/9	-/-	-/-	-/-	-/-	L

Male Totals: Right = 9 (+0) Left = 21 (+1)

Note. L = left-handed classification; R = right-handed classification. Symbols l and r indicate trends for either left- or right-handedness. Absence of established hand preference or trend shown by blanks. Values in italics indicate observation periods wherein a significant lateral bias was observed ($p<.05$). *Data taken in the other small group.

Table 7.2b. Hand Preferences in Female Koshima Monkeys for Picking Up Wheat

		Observation Period						
Subject	Birth	(1) R/L	(2) R/L	(3) R/L	(4) R/L	(5) R/L	(6) R/L	Hand
Ume	1971	51/59	36/63	*64/100*	-/-	-/-	-/-	L
Uri	1968	*12/92*	23/40	22/86	-/-	-/-	-/-	L
Utsugi	1977	54/54	60/58	61/93	27/30	45/32	71/53	
Kuri	1967	51/53	*44/3*	-/-	-/-	-/-	-/-	R
Kurumi	1970	*1/84*	*0/44*	30/33	20/29	-/-	-/-	L
Ichou	1968	64/39	38/46	47/57	13/28	*9/83*	25\31	L
Nira	1974	0/3	*0/117*	*0/24*	-/-	-/-	-/-	L
Gaku	1969	52/56	53/54	8/11	12/6	-/-	-/-	
Fuki	1974	*42/101*	16/32	-/-	-/-	-/-	-/-	L
Tara	1979	9/23	27/28	*42/75*	45/59	*3/16*	*4/22*	L
Zarame	1970	47/49	55/58	27/28	-/-	-/-	-/-	
Udo	1980	33/32	30/24	24/23	56/37	*74/33*	*32/9*	R
Hie	1977	60/57	25/33	-/-	-/-	-/-	-/-	
Take	1977	25/28	21/20	30/28	25/25	18/27	*82/26*	R
Soba	1976	60/65	*41/77*	-/-	-/-	-/-	-/-	L
Nemu	1969	36/41	36/35	-/-	-/-	-/-	-/-	
Shion	1969	4/7	51/44	64/44	-/-	-/-	-/-	r
Shiran	1970	3/15	54/58	16/18	49/62	32/24	-/-	
Nigana	1978	9/10	52/57	52/52	26/25	-/-	-/-	
Boke	1970	63/55	48/52	66/40	8/6	-/-	-/-	r
Kaede	1973	*18/108*	52/56	*3/101*	53/54	*0/108*	-/-	L
Ai	1974	51/67	3/9	-/-	-/-	-/-	-/-	
Bara	1977	*0/44*	*0/74*	-/-	-/-	-/-	-/-	L
Momo	1978	*63/10*	*26/1*	18/7	-/-	-/-	-/-	R
Soyogo	1978	28/28	51/55	56/77	24/23	-/-	-/-	
Keshi	1981	21/21	32/42	*7/41*	*0/11*	-/-	-/-	L
Yuri	1981	40/40	67/52	*44/1*	-/-	-/-	-/-	R
Megi	1982	52/64	61/64	24/30	1/4	*5/32*	*1/41*	L
Chisha	1982	56/62	44/45	34/38	5/16	51/55	41/43	l
Tade	1983	46/84	41/48	42/60	*11/67*	-/-	-/-	L
Chigaya	1984	*6/25*	*0/18*	*2/48*	*5/44*	-/-	-/-	L
Paseri	1984	53/53	19/21	-/-	-/-	-/-	-/-	
Buna	1984	66/57	0/3	4/9	10/1	-/-	-/-	r
Momi	1985	44/65	9/8	23/36	17/13	-/-	-/-	
Nurude	1985	*0/57*	*0/26*	-/-	-/-	-/-	-/-	L
Meron	1986	12/2	15/5	3/4	53/51	-/-	-/-	r
Rohzu	1986	11/14	59/50	*55/16*	17/10	*42/9*	-/-	R
Gumi	1986	*2/40*	*0/22*	-/-	-/-	-/-	-/-	L
Pihman	1986	7/5	8/12	18/28	-/-	-/-	-/-	l
Moku	1987	*5/32*	0/9	*2/13*	*2/22*	*0/25*	*27/68*	L
Tuzu	1987	15/10	*15/0*	*34/2*	*42/0*	-/-	-/-	R
Zumi	1987	3/10	9/10	55/49	4/6	-/-	-/-	

Female Totals Right = 7(+4) Left = 17(+2)

Note. See Table 7.2a for codes.

Another small group of monkeys was formed by fission in 1977 and lived separately from the main troop on the island. We usually distributed wheat grains at particular spots on rocks for the monkeys of the small troop. In such cases, they took many grains at once by grasping them in one hand. The monkeys tended to continue using the same hand, but then shifted to the other hand, which continued to be used for a fairly long time. Furthermore, when we gave them wheat grains scattered on the rocks, most monkeys ate the grains directly with the mouth rather than using their hands. The hand preferences of the monkeys belonging to this small troop were thus often difficult to assign.

Among the total 80 monkeys (38 males and 42 females) whose handedness was assessed, 16 were classified as right-handed and another 38 as left-handed. In another 24 monkeys it proved difficult to judge their hand preferences consistently, despite the fact that sufficient data for them were collected. However, some of these monkeys did show a tendency toward either the right or left hand, although the differences did not reach a statistically significant level (four monkeys to the right and three monkeys to the left; see Table 7.2a & b).

The results obtained for each of the monkeys were arranged according to their kinship relations. The females appeared to produce both right- and left-handed offspring irrespective of their own handedness. Nine left-handed mothers had six left-handed and four right-handed offspring and three right-handed mothers had three left-handed and one right-handed offspring. Furthermore, both left- and right-handed monkeys occurred randomly for each sibling. Observations on the trends of hand preference between mothers and offspring failed to reveal any significant heritability effects. Knowledge of the fathers contribution to hand preference remains to be examined.

Discussion

When the present results are compared, it is clear that a relation generally exists in preferred hand use for different tasks carried out by particular monkeys. That is, seven monkeys showed the same hand preferences in two or more of the tasks, while only one showed the reverse (Ichou; see Tables 7.1 & 7.2b). Five other monkeys also appeared to use the same hand preferentially in these tasks while four showed a reversed preference. Zakuro and Bohze, who exhibited some reversal in hand preference, demonstrated a common feature in their method of wheat-washing. These two monkeys usually grasped wheat grains which were mixed with sand in one hand and did not release them in the water. Rather, they brushed down a small part of the grains and picked them up with the other hand. In contrast, the other monkeys usually just released the wheat grains in the water and then picked them up with both hands. Thus, while Zakuro and Bohze used their preferred hand to pick up grains (the other hand being occupied for grasping grains), the other monkeys could both carry and take grains in their preferred hand. Viewed in this light, there is no contradiction in the hand preferences of these two monkeys. We cannot explain the reason why Ichou employed different hand preferences for different tasks. However, this could imply that the task of carrying food items is different from that of reaching for foods and/or that hand use is determined independently for each task.

Why the hand preferences were less pronounced when the monkeys carried sweet potatoes than when washing wheat remains unclear. Lateralized hand use appeared more clearly in each of the monkeys when they were carrying wheat grains or picking them up on the sandy beach. This difference may reflect the difficulty in grasping and/or carrying these foods. Monkeys can carry a large number of potatoes at one time for their meals. Moreover, they have been observed to carry them long distances to the sea or away from other troop members to avoid plundering. On the other hand, the amount of wheat grains which can be carried by the monkeys is very small when compared to that of sweet potatoes. Furthermore, after the monkeys place the grains in the water, there is a possibility that dominant monkeys might start to plunder the grains. In fact, some monkeys did show specialized trends in their behavior toward plundering wheat grains which other monkeys had carried and placed in the water. The wastefulness of the task of wheat washing behavior could be one important reason for this behavior decreasing, while simplicity and effectiveness of operation may have reinforced the sweet-potato washing behavior (Kawai et al., in press). Thus, hand preferences appear more clearly in situations where the monkeys are required to perform more difficult and sophisticated tasks but is obscured when they are engaged in easier tasks. This conclusion is supported by the results on picking up wheat grains in the sand. Lateralized hand use was obscured when the monkeys took wheat grains which were not mixed with sand but became more pronounced after the grains had become mixed with the sand.

The motivation of the monkeys for food may represent another factor by which their hand preferences may be influenced. When monkeys are very hungry, most of them usually employ both hands to take foods, and washing behavior was observed infrequently. We believe that most monkeys have some general tendency to use either the left or right hand which may become more apparent according to the situation and/or just by chance in continued examinations. However, these preferences are reflected as a matter of degree rather than exclusive right- or left-handedness.

Despite the fact that a number of subjects show no significant hand preference, the present study confirmed earlier results showing that left-handed monkeys outnumbered right-handed (Itani, 1957; Itani et al., 1963;

Table 7.3. Hand Preferences in Picking Up Wheat Grains According to Age and Sex of Monkeys

	Left-Handed		Ambidextrous		Right-Handed		
Age(years)	Male	Female	Male	Female	Male	Female	Total
1 - 5	13(+1)	5(+1)	4	3	4	2(+2)	35
6 - 10	3	3(+1)	0	2	1	3	13
11 - 15	1	5	0	3	2	1	12
=>16	4	4	3	4	2	1(+2)	20
Totals	21(+1)	17(+1)	7	12	9	7(+4)	80

Kawai, 1967; Tokuda, 1969). Although the differences are apparently ones of gradient, there were roughly twice as many left-handed as right-handed monkeys.

Furthermore, it is worth noting that hand preferences were seen in monkeys from a variety of different age classes, from very young juveniles of 1 to 2 years of age to monkeys more than 20 years of age. These data are presented in Table 7.3. Concerning the hand used to pick up wheat grains, there was no difference according to the age class of the monkeys, although some older monkeys did exhibit a somewhat more ambidextrous trend. It also appears that the preferred hand of each monkey does not change with age.

Although detailed quantitative data are not available, the hand preferences of some monkeys had been ascertained as early as 1971 by Mr. N. Yamaguchi through similar observations. Among four monkeys (Noso, Gasami, Kuri, and Uri) whose handedness was estimated both by Mr. Yamaguchi (personal communication) and by us, no individual monkey changed their hand preference between these years. The exception was Noso, who had been classified as a left-handed monkey by Mr. Yamaguchi, whereas our data suggested that Noso picked up wheat grains with both hands equivalently. However, the middle finger of his left hand now cannot be bent, possibly because of the aftereffects of an injury, and this may account for the ambidextrous trend in his behavior. Further, Mori's (1979) data on the hand preferences of the monkeys in wheat washing in 1973 and 1974 also confirmed that the preferred hand of each monkey has remained unchanged over the years of study.

In conclusion, our data suggest that feral monkeys display a general left hand preference in picking up wheat grains, washing wheat grains, and washing sweet potatoes. For a limited set of subjects, there appears to be some consistency in hand preference across these different tasks. Age does not appear to affect the distribution of left- or right-handed subjects and there was no evidence of heritability of hand preference in this population of subjects.

Acknowledgements

We are very grateful to Dr. A. Mori and Mr. N. Yamaguchi who kindly allowed us to use their first-hand data in our analysis. Our thanks are also due to Drs. Y. Higuchi and U. Mori and Mr. F. Kanchi for their assistance in collecting data in 1983 and 1984.

References

Itani, J., 1957. Personality of Japanese monkeys. *Iden, 11,* 29-33. (in Japanese)

Itani, J., Tokuda, K., Furuya, Y., Kano, K., & Shin, Y. (1963). The social construction of natural troops of Japanese monkeys in Takasakiyama. *Primates, 4,* 1-42.

Kawai, M. (1965). Newly-acquired pre-cultural behavior of the natural troop of Japanese monkeys on Koshima Islet. *Primates, 6,* 1-30.

Kawai, M. (1967). Catching behavior observed in the Koshima troop, a case of newly acquired behavior. *Primates, 8,* 181-186.

Kawai, M., Watanabe, K., & Mori, A. (In press). Pre-cultural behaviors observed in free-ranging Japanese monkeys on Koshima Islet over the past 25 years. *Primate Report.*

Mori, A. (1979). Analysis of population changes by measurement of body weight in the Koshima troop of Japanese monkeys. *Primates, 20,* 371-397.

Tokuda, K. (1969). On the handedness of Japanese monkeys. *Primates, 10,* 41-46.

Watanabe, K., Mori, A., & Kawai, M. (In press). Characteristic features of the reproduction in Koshima monkeys, *Macaca fuscata fuscata*: A thirty-four year summary. *Primates.*

8
Manual Specialization in Gorillas and Baboons

Jacques Vauclair and Joël Fagot

At the start of our studies on hand use in nonhuman primates we took for granted, as did the vast majority of other investigators, the theory that nonhuman primates did not show any bias in manual laterality at the population level, even though some individuals would demonstrate a clear preference for one hand (see Warren, 1980 for a relatively recent example of this position). One ape species, the gorilla (*Gorilla gorilla*), has sometimes been singled out within this commonly held view (e.g., Fischer, Meunier, & White 1982) because of its supposed right preference, which could be similar to the well established right bias of humans. The case of the gorilla was intriguing and challenging enough for us to undertake a study in which we wanted, among other things, to evaluate their manual preferences. Our study on gorillas followed the observation of the spontaneous hand usage of a group of eighteen Guinea baboons (*Papio papio*) living in semi-natural conditions (Vauclair & Fagot, 1987a, 1987b). With respect to the distribution of hand preferences, the results obtained in these works were in accordance with the view of a lack of population-level asymmetry in nonhuman primates.

Our work on gorillas used two kinds of approaches. The first relied on the measurement of hand preferences in reaching for food. The second examined manual preferences and hand collaboration in tasks involving precise visuo-spatial adjustments. For the sake of the present discussion, and in anticipation of the results which will be presented in the next section of this chapter, we were surprised to find, instead of the expected right hand advantage reported in previous studies, some apparently contradictory results. On the one hand, simple reaching tasks showed a symmetrical distribution of hand preferences for the group of 10 gorillas, but, on the other hand, a left hand preference appeared at the group level for solving a task that was complex both in terms of cognitive processing and of the hand movements involved. As is also illustrated below, similar results were later obtained with a group of six baboons.

While we were trying to understand and interpret our data, it became clear to us that the available frame of reference (i.e., the lack of population-level asymmetry in nonhuman primates) was theoretically insufficient. Though this theory could account for the distribution of hand preferences in simple reaching activities, it was unable to explain the asymmetry we found for the more refined spatial tasks. It appears that the literature on nonhuman primate laterality is unfortunately characterized by reliance on anectodal reports and on studies of manual activities mostly concerned with reaching for food. With few exceptions (e.g., Beck & Barton, 1972; MacNeilage, Studdert-Kennedy, & Lindblom, 1987), reports usually have not attempted to categorize manual activities as a function of their complexity.

It occurred to us that a simple and rather straightforward distinction had, however, been proposed by Young, Segalowitz, Corter, and Trehub (1983) in the field of human laterality. This approach distinguished "handedness", defined by a consistent lateralized usage in familiar and highly practiced tasks, from "manual specialization," which concerns lateralized hand usage on novel and relatively complex tasks. We applied this distinction to the tasks we had studied and tried to specify their complexity with reference to their spatial and temporal constraints. It became evident to us that simple reachings would fall into the category of "handedness", whereas the visuo-spatial tasks employed in our studies could easily be categorized as tasks of "manual specialization". Thus, this distinction was sufficient to explain that two types of preference could coexist within the same individual, and that these preferences did not contradict each other.

In order to evaluate the generality of the above distinction, we have reviewed a large body of data regarding manual lateralization among nonhuman primates (Fagot & Vauclair, 1991). This review has brought about several important conclusions that can be summarized as follows. "Handedness" expressed in highly practiced, routine activities usually leads to symmetrical hand biases. This type of laterality is not necessarily correlated with hemispheric specialization. In contrast, behaviors expressed in novel and highly controlled tasks suggestive of "manual specialization" showed asymmetrical distributions of hand biases. Moreover, these behaviors were indicative of a specialization of the contralateral hemisphere.

Such a distinction emphasizes the *tasks demands* and the variables of *novelty and practice* in the analysis of lateral biases. We thus proposed to examine these factors in the sample of gorillas and baboons we have investigated. In addition, our analysis will also include recent data gathered on experimentally naive baboons.

Study with Gorillas

We studied 10 gorillas (*Gorilla gorilla*) of the Barcelona Zoo (6 females and 4 males, mean age 13.7 years, s.d. = 8.3). Each ape was observed in a simple reaching task and in a panel displacement task. On average, we observed 120 quadrupedal reachings per subject. Only those acts which reflect a real choice between hands have been considered for analysis, namely reachings performed when the object was located in the midsagittal area of the animal. For the panel displacement task, the apparatus consisted of a transparent horizontal plexiglas sliding panel (H.S.P.) with two symmetrical windows which could slide toward the right or left in front of a vertical metal board with a central aperture. The subject had to operate a metal handle attached to the panel in order to adjust one of the two windows in front of the central aperture where a hazelnut was located. Thirty trials per subject were run in this situation. The hand (left or right) used to displace the panel was recorded.

All individual biases reported in this chapter were calculated with a two-tailed *Chi-square* test ($p<.05$) of the number of left versus right responses. Table 8.1 presents the results for simple reaching and the horizontal panel displacement tasks. As can be seen, the distribution of hand use between these two tasks are clearly different. Simple reaching revealed a symmetrical distribution of hand preferences (three left, three right and four subjects without statistically significant preference), whereas seven out of eight gorillas tested on H.S.P. were left-handed. The remaining subject preferentially used its right hand. The latter results correspond to a left hand preference for the group (Wilcoxon two-tailed test, p =.02). Interestingly, the asymmetry that appeared for H.S.P. was due to a shift of preference from the right to the left side for one gorilla and to the fact that three other subjects without preference in simple reaching became left-handed for sliding the panel. This study failed to confirm the claim frequently made of a right hand bias for the gorilla, at least concerning the simple reaching tested here and the horizontal sliding of the panel. The detailed report of these results can be found in Fagot and Vauclair (1988a).

Studies with Baboons

We will report here data gathered on six Guinea baboons, *Papio papio* (MAC, GRE, BIA, POM, YEL, RUM). The results concerning these first six baboons have been published elsewhere (Fagot & Vauclair, 1988b). Additionally, we report data on four previously unstudied baboons (REX, JIM, ZEU, EGA). The baboons ranged in age from 3 years to 8 years (mean 5.7; s.d = 2.4). Subjects (9 males and 1 female) were selected on the basis of their readiness to enter the test cage. The monkeys have been tested

Table 8.1. Hand Bias and Corresponding Percentage Hand Use for Each Gorilla in the Simple Reaching and Horizontal Sliding Panel Tasks

Subject	Reach	H.S.P.
YUM	Right 78	Left 100
NDE	Left 70	Left 100
BIM	No Preference	[-]
MAK	Left 78	Left 97
NTA	Left 75	Left 100
VIR	Right 67	[-]
KEN	Right 64	Right 90
COP	No Preference	Left 97
URK	No Preference	Left 100
BIN	No Preference	Left 97
Total right	[3]	[1]
Total left	3	7
Binomial	n.s.	[p =.07]
Wilcoxon	untestable	[T=1, p=.02]

Note. A binomial two-tailed test (p<.05) was used to compare the number of left vs. right handers for the group. The Wilcoxon two-tailed test (p<.05) was employed to compare individual numbers of left vs. right hand uses for the group.

in a cage specially adapted for that species (Trevarthen, 1978). This cage permits standardization of the posture of the subject. Two large symmetrical openings and a face mask with two eye-holes permitted free manipulation of the different apparati with one or both hands (cf. Figure 8.1).

Lateral preferences were measured through the hand used when reaching for a hazelnut located on a horizontal board. Two hundred trials were run in this task for six baboons and 100 trials for the remaining four baboons. Three versions of the sliding panel were also used. The first one was the horizontal sliding panel identical to the one used with gorillas. The second version consisted of a vertical sliding panel (V.S.P.) with only one window. The transparent panel could move vertically up and down in front of a central aperture where a hazelnut was located. The subject had to lift a metal handle attached on the plexiglas in order to align the window in front of the aperture. This apparatus was designed to force bimanual solutions by virtue of the fact that it was impossible to adjust the panel and simultaneously take the hazelnut with the same hand.

The third version (delayed vertical sliding panel: D.V.S.P.) was similar to the V.S.P. except that a second plexiglas prevented access to the hazelnut. In this case, the incentive fell into a pipe and was delivered on the floor of

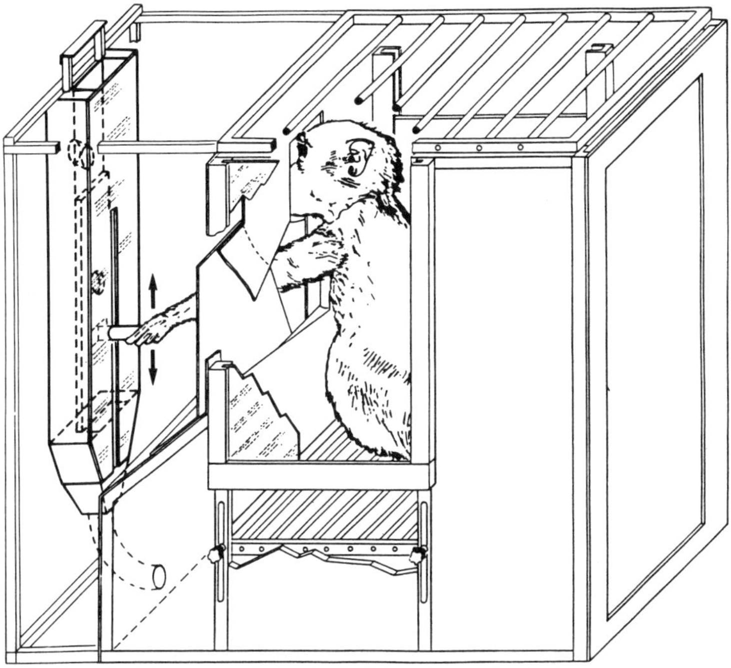

Figure 8.1. Illustration of the experimental setup used to test baboons with the Delayed Vertical Sliding Panel task.

the cage. As the food reward was delayed and could not be directly taken, the panel adjustment action can be considered as relatively independent of the food taking action. Unlike the V.S.P. task, this task could thus be solved with either one or two hands (see Figure 8.1).

Six baboons performed all three panel tasks. The other four were only tested with D.V.S.P. A similar number of trials were run in all situations: 4 sessions of 25 trials over 4 consecutive days. The analysis distinguished the hand used to initiate the displacement of the panel ("move") from the hand used to adjust it ("adjust") in the vicinity of the aperture. This distinction was applied for both vertical panel test conditions (V.S.P. and D.V.S.P.) but not for H.S.P., because in this horizontal panel condition subjects systematically used the same hand. Further information concerning the procedure are provided in Fagot and Vauclair (1988b).

The analysis will focus on two main aspects of manual preferences with respect to the distinction between "handedness" and "manual specialization", namely complexity and novelty. Complexity was investigated in terms of inter-task consistency of hand biases for only the first 25 trials of each sliding panel task in order to neutralize the factor of novelty. In contrast, novelty was examined through a comparison between the initial biases (trials 1 through 25) and the final biases (trials 75 through 100). For reasons of homogeneity, for the six baboons (BIA, GRE, MAC, POM, RUM, and YEL) only the first 100 trials were used in the analysis of simple reaching.

Inter-task Complexity

Table 8.2 summarizes the results for each subject on the first 25 trials of each task. Population-level hand lateralization was not found in the simple reaching task. In contrast, all three of the panel tasks elicited a left hand preference. This left hand preference approached significance ($p<.06$) for H.S.P. and was statistically significant for both the adjustment of V.S.P. and for the moving and adjusting actions on D.V.S.P. For the two tasks which were common to gorillas and baboons (reaching and H.S.P.), hand preference distributions were nearly identical. A consideration of hand

Table 8.2. Initial Bias (Trials 1 to 25) and Corresponding Percentage Hand Use for Each Baboon in the Simple Reaching and Sliding Panel Tasks

Subject	Reach	H.S.P.	V.S.P.		D.V.S.P.	
			Move	Adjust	Move	Adjust
BIA	L 100	L 100	L 92	L 100	L 100	L 100
GRE	R 100	L 88	L 79	L 100	NP	NP
RUM	L 100	L 100	L 100	L 100	L 100	L 100
MAC	R 100	L 84	NP	L 88	NP	L 70
POM	R 100	L 92	L 91	L 95	NP	L 86
YEL	L 100	NP	NP	L 92	L 100	L 100
ZEU	L 92	-	-	-	L 79	NP
REX	L 92	-	-	-	L 100	L 100
JIM	L 88	-	-	-	L 84	L 84
EGA	R 96	-	-	-	R 80	NP
Totals	6L/4R	5L/0R	4L/0R	5L/0R	6L/1R	7L/0R
Binomial	n.s.	p=.06	*	p=.06	n.s.	p=.02
Wilcoxon	T=24 n.s.	T=1 n.s.	T=2 n.s.	T=0 p=.05	T=5.5 p=.05	T=5 p=.02

Note. "Move" corresponds to the initiation of the displacement of the panel. "Adjust" refers to the final positioning of the panel. H.S.P.= horizontal sliding panel task; V.S.P.= vertical sliding panel task; D.V.S.P.= delayed vertical sliding panel task. L = "left hand" R = "right hand". * untestable. See legend of Table 1 for statistics.

preference between the reaching and the panel tasks indicates that inversions occurred. These inversions mostly appeared either from a right to a left hand preference or from a right to an ambipreference. Interestingly, we have never observed an inversion from a left to right hand preference.

Novelty Effects

The analysis of the effect of novelty is based on the comparison between individual biases expressed in each condition for the first 25 trials and last 25 trials. Data and statistical analyses are reported in Table 8.3. In contrast to the findings depicted in Table 8.2, the intra-task comparisons did not show any inversion of hand preference (either from the left to the right side or vice versa). However, several cases of changes can be seen in Table 8.3. Those changes are of three kinds and can be exemplified by the results of three subjects: GRE, on H.S.P. "move", changed from an initial left bias to an absence of preference (NP); MAC, in V.S.P. "move", showed an initial lack of preference and a left hand preference in the last trials; EGA, in D.V.S.P. "adjust", expressed no initial preference and a later right bias.

Intra-task comparisons showed an effect of practice on the distribution of hand biases for the group, in that an initial left preference had vanished in the last trials. Only two comparisons (Reaching and "Adjust" in V.S.P.) did not reveal any significant change over time for the group. For V.S.P. "Adjust", the strong left bias for the group was resistant to change despite the practice the subjects had with the apparatus. It was not suprising to find that preferences expressed in reaching were not influenced by the repetitions of the test. This general stability was observed for all baboons except one, which went from an initial left preference to an absence of preference. Reaching represents a highly familiar activity. Since it belongs to the minimal repertoire of hand use, it is hard to conceive the first 25 reachings as purely novel activities. In summary, the data and analyses offered in this section stress the necessity of integrating novelty factors in the assessment of lateral preferences.

Discussion

Methodological Remarks

Our gorilla sample was well-balanced with respect to the reaching preferences (3 right-handers and 3 left-handers). A left preference to displace the horizontal panel was observed. For baboons, the lateralized subjects in the reaching were slightly biased in favor of the left side (6 left- and 4 right-handers). If this group's composition can be the source of a possible artifact toward finding left preferences for tasks other than reaching, it can in no way explain the fact that three right-handers out of four inverted their preferences for the sliding panel task (cf. Table 8.2). Moreover, none among the left-handers in reaching became right-handers in the panel tasks. To us, the left preference we observed for baboons in panel tasks is thus directly related to the task employed and its requirements and not to a sample bias.

Under the hypothesis that nonhuman primates' laterality is a phenomenon as complex and multiple as it appears to be for humans (Healey, Liederman, & Geschwind, 1986), we have avoided a priori data summations over tasks. We have thus been attentive to analyze separately each task and its corresponding individual bias in terms of percentage of hand use. The analysis of our data has shown that different individual and group patterns appeared for different tasks and for the same task at different levels of practice. Thus, calculating average preferences on different tasks without taking into consideration the nature and features of the task can lead to wrong conclusions. That is the reason why this analytical procedure should be used to report and discuss manual preferences.

Inter-task Differences

In the simple reaching task, no asymmetrical population distribution of hand preference was noticed for either of the two species groups under study. This finding is in agreement with other reports in the literature. In a survey of 41 samples of nonhuman primates that were tested for food reaching, we found that only 3 studies described an asymmetry in the left-right distribution, whereas the vast majority of studies have found a symmetrical distribution of hand biases in the populations studied (Fagot & Vauclair, 1991). Nine out of ten baboons expressed a strong consistency between their first and last reaching trial preferences. In fact, our data confirm previous work on food reaching (e.g. Lehman, 1980), where preferences in reaching tasks did not change following extensive training. We have already noted that reaching is not in itself a novel activity. The stability for this kind of behavior can thus be easily understood.

With baboons and gorillas, distributions of reaching preferences can be clearly differentiated from other tasks' preferences. Sanford, Guin, and Ward (1984) and Ward, Milliken, and Stafford (Ch. 2 this volume) have observed a change of laterality in galagos tested from a quadrupedal stance and an erect posture. King and Landau (Ch. 4 this volume) have found a left preference for a group of 18 squirrel monkeys in live fish catching and no bias for simple reachings. These examples are convergent with our own findings, since they indicate that the distribution of hand biases is dependent upon the nature of the tasks and postural constraints.

Table 8.3. Hand Bias and Corresponding Percentage for Each Baboon in Initial Trials (1 to 25) and Last Trials (75 to 100) in Reaching and Sliding Panel Tasks (See legend of Table 8.1 for statistics)

	REACHING		H.S.P.		V.S.P.				D.V.S.P			
					MOVE		ADJUST		MOVE		ADJUST	
SUBJECT	INIT.	LAST	INIT.	LAST	INIT.	LAST	INIT.	LAST	INIT.	LAST	INIT.	LAST
BIA	L100	L100	L100	L100	L 92	L 95	L100	L100	L100	L100	L100	L100
GRE	R100	R100	L 88	L100	L 79	NP	L100	L 87	NP	R100	NP	R100
RUM	L100	L100	L100	L100	L100	L100	L100	L100	L100	L100	L100	L100
MAC	R100	R100	L 84	NP	NP	L 96	L 88	L 92	NP	NP	L 70	L 80
POM	R100	R100	L 92	L 96	L 91	NP	L 95	L 75	NP	NP	L 86	L 77
YEL	L100	L100	NP	R100	NP	L100	L 92	L100	L100	L100	L100	L100
ZEU	L 92	L 96	-	-	-	-	-	-	L 79	NP	NP	R100
REX	L 92	NP	-	-	-	-	-	-	L100	L100	L100	L100
JIM	L 88	L 88	-	-	-	-	-	-	L 84	L 88	L 84	L 88
EGA	R 96	R100	-	-	-	-	-	-	R 80	R100	NP	R100
TOTALS	6L/4R	5L/4R	5L	4L/1R	4L	4L	6L	6L	6L/1R	5L/2R	7L	7L/3R
BINOMIAL	n.s.	n.s.	.06	n.s.	untestable		.03	.03	n.s.	n.s.	.02	n.s.
WILCOXON	T=24 n.s.	T=28 n.s.	T=1 n.s.	T=4.5 n.s.	T=0 .05	T=1 n.s.	T=0 .05	T=0 .05	T=5.5 .05	T=16.5 n.s.	T=5 .02	T=16.5 n.s.

Manual Specialization and Novelty

The intra-task analyses (Table 8.3) have provided identification of individual shifts through practice for the three panel tasks. These shifts were twofold: First, left-handers lost their initial preference to become ambipreferent; secondly, ambipreferent subjects became either right- or left-handers. In haptic discrimination tasks and in the dark, Ettlinger (1961) and Milner (1969) found in macaques a greater amount of left hand usage compared to right hand usage to reach training criterion. The difference between the amount of left versus right hand use vanished when the training criterion was attained. In addition, left-handers needed fewer trials than right-handers to attain criterion (Hoerster & Ettlinger, 1985). An important question is how such changes occur through practice. In order to explain these changes, it is useful to review here the principles we have expounded elsewhere (Fagot & Vauclair, 1991). Under the assumption that in "manual specialization" only one hemisphere is competent to solve the task, it follows that the active hand is necessarily contralateral to the competent hemisphere, at least for movements of the distal portions of the upper limbs (Brinkman & Kuypers, 1972). For "handedness", we assume that the two hemispheres do not differ drastically with respect to their abilities in the control of manual activities. Our results suggest that a transient practice with the initially non-preferred hand could be sufficient to attenuate interhemispheric differences. We could then reasonably expect that an increased number of trials would lead to true inversions of laterality, both at the behavioral and neuro-functional level. Except for the work of Ettlinger and his colleagues (op.cit.), to our knowledge little attention has been paid to the distribution of preferences as a function of initial and final trials on nonreaching tasks.

In Ettlinger's studies and in our own, practice has been shown to affect the distribution of manual preferences as characterized by individual shifts of hand preference. It remains possible, however, that some highly specialized tasks might be more resistant to inter-manual transfer, as it is exemplified in Preilowski, Reger, and Engele's (1986) data obtained with an apparatus requiring an isometric finger pressure.

In brief, all data cited above and our own data emphasize the importance of the demands of the task and, especially, the analysis of the aspects of novelty and familiarity for an appropriate understanding of manual preferences and their proper relationships to the cerebral hemispheres. It seems to us that the distinction between the two types of laterality we have advanced has been helpful in interpreting the available data and could offer an adequate framework for the design of new experiments.

Direction of Hand Preferences

When tasks have been identified that consistently evoke reliable lateral biases, the bias must be characterized with respect to the nature of the task concerned. Although the frame of reference proposed by MacNeilage and his colleagues (MacNeilage et al., 1987) was unknown to us at the onset of our investigations with gorillas and baboons, we feel compelled to discuss our data and interpretations within the perspectives advocated by these authors. Their idea of a left hand preference for certain visually-guided activities could explain the left preferences in our panel tasks, as well as preferences obtained in the case of extreme prehension (Forsythe, Milliken, Stafford, & Ward, 1988) and in the hand fishing of squirrel monkeys (King and Landau, Ch. 4 this volume). But it is important to underline that a visually-guided task (Hopkins, Washburn, & Rumbaugh, 1989; Morris, Hopkins, Bolser-Gilmore, & Washburn, Ch. 9 this volume) revealed a right hand preference in three chimpanzees and two rhesus monkeys. Moreover, it appears that haptic discrimination tasks have revealed a left hand advantage (Hoerster & Ettlinger, 1985), although the task was performed in the dark.

Several simple conceptual frameworks could be borrowed from the human literature to explain these biases. One could propose, for example, a right hemisphere advantage in spatial functions. This relationship could explain the left preferences obtained in Ettlinger's studies, in King and Landau's study, and in our own investigations. Further, the concordance between the preferences found in haptic tasks with those found in visually guided tasks suggests that the right hemisphere specialization for spatial processing could be modality free. However, such an interpretation cannot be applied to Hopkins et al.'s (1989) finding of a right hand advantage in a joystick manipulation nor to Preilowski's finding of a right hand preference in the finger pressure task (Ch. 5 this volume). Moreover, lines differing in orientation are better discriminated by the left hemisphere of split-brain macaques. However, this neuro-functional asymmetry was not related to hand asymmetry in a set of simple manipulations (Hamilton, 1990). In fact, it appears that no classification can currently account for the disparate evidence gathered so far on manual asymmetries in nonhuman primates.

With respect to the direction of preferences, it is probably too early to draw a clear-cut pattern of relations between hand asymmetry and hemispheric specialization. But it is our firm belief that such attempts should soon be successful if researchers concentrate their efforts on the study of cognitively and motorically complex manual specialization tasks.

Acknowledgement

We thank V. Ginouvier for the drawing of Figure 8.1.

References

Beck, C. H. M., & Barton, R. L. (1972). Deviation and laterality of hand preference in monkeys. *Cortex, 8*, 339-363.

Brinkman, C., & Kuypers, H. H. J. M. (1972). Split brain monkeys: Cerebral control of ipsilateral and contralateral arm, hand and finger movement. *Science, 176*, 536-539.

Ettlinger, G. (1961). Lateral preferences in monkeys. *Behaviour, 17*, 275-287.

Fagot, J., & Vauclair, J. (1988a). Handedness and bimanual coordination in the lowland gorilla. *Brain, Behavior and Evolution, 32*, 89-95.

Fagot, J., & Vauclair, J. (1988b). Handedness and manual specialization in the baboon. *Neuropsychologia, 26*, 795-804.

Fagot, J., & Vauclair, J. (1991). Manual laterality in nonhuman primates: A distinction between handedness and manual specialization. *Psychological Bulletin, 109*, 76-89.

Fischer, R. B., Meunier, G. F., & White, P. J. (1982). Evidence of laterality in the lowland gorilla. *Perceptual and Motor Skills, 54*, 1093-1094.

Forsythe, C., Milliken, G. W., Stafford, D. K., & Ward, J. P. (1988). Posturally related variations in the hand preferences of the ruffed lemur (*Varecia variegata variegata*). *Journal of Comparative Psychology, 102*, 248-250.

Hamilton, C. (1990). Hemispheric specialization in monkeys. In C. B. Trevarthen (Ed.), *Brain circuits and functions of the mind: Festschrift for R.W. Sperry* (pp.181-195). Cambridge, M.A.: Cambridge University Press.

Healey, J. M., Liederman, J., & Geschwind, N. (1986). Handedness is not a unidimensional trait. *Cortex, 22*, 33-53.

Hoerster, W., & Ettlinger, G. (1985). An association between hand preference and tactile discrimination performance in the rhesus monkey. *Neuropsychologia, 23*, 411-413.

Hopkins, W. D., Washburn, D. A., & Rumbaugh, D. A. (1989). Note on hand use in the manipulation of joysticks by chimpanzees (*Pan troglodytes*) and monkeys (*Macaca mulatta*). *Journal of Comparative Psychology, 104*, 91-94

Lehman, R. A. W. (1980). Distribution and changes in strength of hand preference of cynomolgus monkeys. *Brain, Behavior and Evolution, 17*, 209-217.

MacNeilage, P. F., Studdert-Kennedy, M. G., & Lindblom, B. (1987). Primate handedness reconsidered. *Behavioral and Brain Sciences, 10*, 247-303.

Milner, A. D. (1969). Distribution of hand preferences in monkeys. *Neuropsychologia, 7*, 375-377.

Preilowski, B., Reger, M., & Engele, H. C. (1986). Handedness and cerebral asymmetry in nonhuman primates. In D. M. Taub & F. A. King (Eds.), *Current perspectives in primate biology* (pp. 270- 282). New York: Van Nostrand Reinhold.

Sanford, C., Guin, C. G., & Ward, J. P. (1984). Posture and laterality in the bushbaby (*Galago senegalensis*). *Brain, Behavior and Evolution, 25*, 217-224.

Trevarthen, C. B. (1978). Manipulative strategies of baboons and the origins of cerebral asymmetry. In M. Kinsbourne (Ed.) *Asymmetrical functions of the brain* (pp.329-389). Cambridge, M.A.: Cambridge University Press.

Vauclair, J., & Fagot, J. (1987a). Spontaneous hand usage and handedness in a troop of baboons. *Cortex, 23*, 265-274.

Vauclair, J., & Fagot, J. (1987b). Visually guided reaching in baboons. *Behavioral and Brain Sciences, 10*, 287.

Warren, J. M. (1980). Handedness and laterality in humans and other animals. *Physiological Psychology, 8*, 351-359.

Young, G., Segalowitz, S.J., Corter, C. M., & Trehub, S.E., (1983). Manual specialization and the developing brain: an overview. In G. Young, S. J. Segalowitz, C. M. Corter & S.E. Trehub (Eds.) *Manual specialization and the developing brain* (pp. 3-12). New York: Academic Press.

9
Behavioral Lateralization in Language-Trained Chimpanzees

Robin D. Morris, William D. Hopkins, Linda Bolser Gilmore, and David A. Washburn

From an evolutionary perspective, some have argued that the conspicuous lack of cerebral asymmetries in nonhuman primates is functionally related to the lack of "language" observed in these species (Annett, 1985; Beaton, 1986; Warren, 1977; Warren, 1980). Recently, with the emergence of "language-using" apes (Gardner & Gardner, 1968; Laidler, 1978; Patterson, 1978; Premack, 1970; Rumbaugh, 1977; Savage-Rumbaugh, 1986; Savage-Rumbaugh, McDonald, Sevcik, Hopkins & Rubert, 1986), our developing views on the evolution of language and related cognitive abilities have begun to challenge man's alleged uniqueness from other apes (Ettlinger, 1984; Menzel, 1974; Rumbaugh, Savage-Rumbaugh & Scanlon, 1982; Savage-Rumbaugh, 1986). From a biobehavioral or evolutionary perspective, the relationship between these cognitive processes and their supporting neurological systems is not well understood. The fact that many human studies of laterality employ complex linguistic or spatial stimuli, or utilize linguistic instructions to communicate the task requirements has limited our views of how laterality has evolved, what such linguistic instructions represent neurocognitively, or how laterality can be studied in nonhuman primates. We suggest that important questions regarding the evolution of laterality can be addressed by using nonhuman animals, particularly primates. Without theoretically and methodologically sophisticated comparable studies of laterality in nonhuman primates, particularly great apes, no theory on the basis or evolution of these processes in humans is complete. This chapter briefly reviews the current literature on anatomical asymmetries and functional lateralization in nonhuman primates in order to place our recent laterality studies with language-trained chimpanzees into perspective. After reviewing our studies, theoretical and methodological considerations relevant to the study of primate lateralization are discussed.

Neuroanatomical Foundations

There is a growing body of literature which suggests that adult humans, neonates (Geschwind & Levitsky, 1968; Kimura, 1979; Wada, Clarke, & Hamm, 1975; Witelson, 1977, 1987; Witelson & Pallie, 1973), and, to a certain extent, nonhuman primates possess neuroanatomical asymmetries which may be related to lateralized cognitive functioning (Falk, 1978, Falk, Cheverud, Vannier, & Conroy, 1986; Heilbroner & Holloway, 1988; Holloway & de la Coste-lareymondie, 1982; LeMay, 1985; LeMay & Geschwind, 1975; Yeni-Komshian & Benson, 1976). Neuroanatomical asymmetries have been found in apes and in lower primates, though typically the degree of asymmetry in apes is greater than in monkeys but less than that found in humans. In general, the Sylvian fissure has been found to be longer in the left hemisphere than the right hemisphere in humans and chimpanzees (*Pan troglodytes*) (Yeni-Komshian & Benson, 1976). These anatomical asymmetries were not found in rhesus monkeys (*Macaca mulatta*). However, another study, employing different procedures and fewer subjects, found the Sylvian fissure to be longer in the left hemisphere than in the right hemisphere in rhesus monkeys (*Macaca mulatta*) (Falk et al., 1986). Additionally, the height of the Sylvian fissure was located posteriorly higher in the right hemisphere than in the left hemisphere in the orangutan (*Pongo pygmeaus*) and chimpanzee (*Pan troglodytes*) (LeMay & Geschwind, 1975). Such anatomical asymmetries were not found in a variety of monkey and lesser ape species. In regards to petalia patterns, *Pongoids* show notable variability and the number of subjects showing no asymmetry is higher than seen in human data (Holloway & de la Coste-lareymondie, 1982). However, where asymmetries do exist they tend to follow a human pattern with greater extension of the left occipital lobe and right frontal lobe. The extension of the right frontal lobe, at least tentatively, seems greater in apes than in monkeys (LeMay, Bellig, & Geschwind, 1982). When considering both frontal and occipital petalia together, 66% of the apes evidenced right frontal extension while 58% showed a simultaneous left occipital extension.

Overall, these patterns of asymmetry are similar to those found in humans (i.e. the same hemispheric asymmetries are found in each species). In addition, the asymmetries are located in the superior temporal gyri and sulci, neocortical areas considered primary for auditory perception and language comprehension in humans. Finally, the relative proportion of subjects showing asymmetries is comparable between human and nonhuman primate species. However, Heilbroner and Holloway (1988) conclude that the variety of different procedures employed to study brain anatomy is considerable, and thus conclusions are limited.

Handedness in Great Apes

In addition to neuroanatomical asymmetries, numerous functional asymmetries are also extant in humans, with the most notable being related to handedness. Typically, research shows that individuals possess a strong hand preference for various activities, or show greater manual skills with one particular hand, although the preferred hand and skilled hand may not always be congruent. At the population level, the proportion of right-, left-, and ambidextrous handed individuals is relatively constant, with the majority being right-handed (Healey, Liederman, & Geschwind, 1986). The frequency of left-handedness (or non-right handedness) is greater in populations suffering from various developmental, learning, and medical disorders, suggesting numerous influences on handedness (Geschwind & Galaburda, 1985).

Most studies of monkeys and apes have failed to yield results suggestive of population-level handedness asymmetries (Warren, 1980). However, such null results may be due to inadequate assessment procedures which are not comparable to those measures used to assess human handedness. MacNeilage, Studdert-Kennedy and Lindblom (1987) have suggested that studies of handedness in nonhuman primates should emphasize the dual processing that each hand may control in relation to lateralized hemispheric functions, as has been suggested in the human literature (Bever, 1975; Bradshaw & Nettleton, 1981). Thus, the left hand, due to its control by the spatially-dominant right hemisphere, would be the more preferred hand in visually-guided tasks, while sequential tasks requiring repeated fine-motor operations on an object would show a greater right hand usage. Although there is some support for the model outlined by MacNeilage, et al. (1987), data from great apes are extremely limited.

Reach tasks involving unskilled unimanual activity, are the most common form of assessment of hand preference in nonhuman primates. The earliest reports of handedness in great apes came from Finch (1941) who assessed hand preferences in 30 chimpanzees using four different reach tasks. Roughly equal numbers of subjects showed right and left hand preferences. Marchant and Steklis (1986), studying five chimpanzees, reported similar data, with two subjects showing left hand preferences, two showing right hand preferences, and one mixed or ambidextrous hand preference. Fischer, Meunier, and White (1982) reported that their 4 gorillas all showed significant right hand preferences in a reach task. Similarly, 10 of 12 gorillas demonstrated a significant right hand preference on one reach task, although no significant preferences were found on two additional tasks (Olson, Ellis, & Nadler, 1990). In this same study, population levels of handedness were not found in 13 orangutans, but a left hand population bias was found in 6 gibbons. Interestingly, significant hand preferences were found only on the task for which the subject's posture was bipedal, supporting the suggestion that posture may be an important factor contributing to the expression of hand preferences (Sanford, Guin, & Ward, 1984). Fagot and Vauclair (1988) reported that 7 of 8 gorillas showed significant left hand preferences in the opening of a compartment box containing food. It is possible that a left hand preference was obtained due to the nature of the task, which did not contain an independent reach or manipulation component.

Finally, Shafer (1987, 1988) reported hand preferences for a variety of behavioral contexts in 47 gorillas. In short, the findings suggested that hand preference was biased to the right in her population, although the bias was overall less (67% right) than other reports of handedness in gorillas. Moreover, fine motor acts were the most sensitive measures of hand preference. Finally, age and sex appeared to be related to hand preference, with younger animals showing stronger preferences and males showing a greater degree of asymmetry than females. Of note were observations of hand use by the two signing gorillas, Koko and Michael. Koko appeared to be left-handed while Michael showed no preference, although his data showed a tendency in the direction of a left preference.

When the tasks used to assess hand preference become more than a simple reach test, many other factors appear to influence hand usage, since hand preference then interacts with hand skill and, in some cases, bimanual coordination. In addition, the use of a particular hand may not be due to a preference for that hand, but rather due to a preference for the hand used for support, thus biasing any reach or manipulation required. When no such postural support or clinging occurs, both hands are equally free to reach and/or to manipulate, and thus a preference, if it exists, may be expressed. For example, it has been shown that gorillas, chimpanzees, orang-utans, and gibbons all show a right hand population bias in the hand used in leading of locomotor behavior (Heestand, 1986).

As mentioned earlier, Marchant and Steklis (1986) reported that five chimpanzees failed to show a consistent bias on a variety of tests aimed at assessing hand preference. The tasks included unimanual reach tasks, unimanual sequencing tasks such as bar pressing, and bimanual tasks. Although they concluded that hand preference was mixed for their subjects, on the bar press task the subjects as a whole performed better (i.e., were more skilled) with the right hand when contrasted with the left. Thus, because a subject prefers to use one hand over another does *not* necessarily imply that that particular hand is better skilled on that task, unless this is experimentally manipulated and assessed, which is often not the case.

In our laboratory, we have examined hand preferences using a variety of procedures in order to address multiple methodological issues and to investigate the multivariate nature of handedness assessment. In one study, we analyzed videotapes of two adult male chimpanzees, Austin and Sherman, during daily activities to assess hand usage when they were between 5 and 10 years of age (Bolser & Morris, 1987). In a second study,

Table 9.1. Coded Activities and Actions from Video Tapes and Handedness Tasks Employed

Observational Study Coding Categories

Activities:	*Actions*:
1) KYB - Keyboard Use	1) MAN - Manipulation
2) TOL - Tool Use	2) RCH - Reach to Keyboard
3) GES - Gesture Made	3) PLC - Reach and Place
4) OBJ - Object Manipulation	4) OTH - Other Hand Actions
5) PLY - Playing	5) PUP - Reach and Pick Up
6) OTH - Other Hand Activities	6) REC - Receive
7) EAT - Eating	7) CRY - Carry
8) OET - Eating with Object	8) SUP - Support

Experimental Measures

LRC Handedness Battery:

I. Gross Motor & Ballistic Tasks	II. Nonvisual Reaching
a) Free Field Food Reaching	a) Blind Bag Reaching for Food
b) Coconut Breaking	
c) Ball Toss	
III. Fine Motor Tasks, Unimanual	**IV. Fine Motor Tasks, Bimanual**
a) Opening Latch Box	a) Raisin Board
b) Opening Stationary Lock	b) Opening Padlocked Box
	c) Opening Suspended Lock

U.C.L.A. Handedness Battery:

a) Tossing Ball, b) Hitting with Hammer, c) Picking up an M&M, d) Picking Up a Dime, e) Pointing, f) Drinking Out of a Glass, g) Eat with Spoon, h) Use a Toothbrush, i) Drawing with Crayon

specific experimental assessment of hand preference was made using 18 different measures. Some measures were taken from previous handedness studies on both human and nonhuman primates, and all varied in the degree of fine motor control needed to solve each task (Bolser, Runfeldt, & Morris, 1988). Shown in Table 9.1 are descriptions of the tasks used and the behavioral categories measured during daily activities coded from the video tapes.

In the videotaped observational study, it was found that at the gross level of activity type (eating, tool use, play, etc.), only Sherman showed a pronounced right hand preference across all activities, while Austin showed no overall preference. The overall hand preferences were stable across the

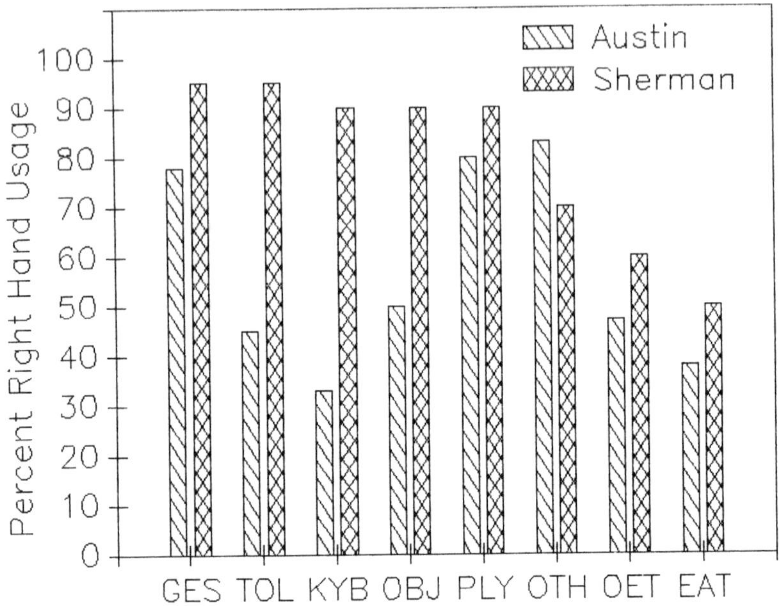

Figure 9.1. Percentage right hand use for each unimanual daily activity measured for Sherman and Austin.

5-year sampling period, in which Sherman continued to prefer using his right hand while Austin showed no significant overall preference. These activity data can be seen in Figure 9.1. However, when hand preference at the level of specific types of hand actions was explored, both chimpanzees were significantly more right-handed for some actions,(e.g. object manipulations) but not others (e.g. holding). These data can be seen in Figure 9.2. Many of the actions for which both Sherman and Austin were right-handed required fine motor skills, while the other actions, observed to be predominantly executed with the left hand or by either hand, required more gross motor acts. This dichotomy of hand use lends some support to MacNeilage et al.'s (1987) viewpoint, and is consistent with data from human subjects (Healey et al., 1986).

In the series of experimental studies Sherman and Austin were evaluated on 18 different measures, many replicating those used in previous handedness studies, which varied from simple reach tasks to fine motor sequencing tasks and bimanual tasks (Bolser et al., 1988). A subset of these tasks was taken from a study used to evaluate handedness in low functioning human subjects (Soper, Satz, Orsini, Van Gorp, & Green, 1987). Shown in Figure 9.3 is the percent right hand usage for each subject on each task

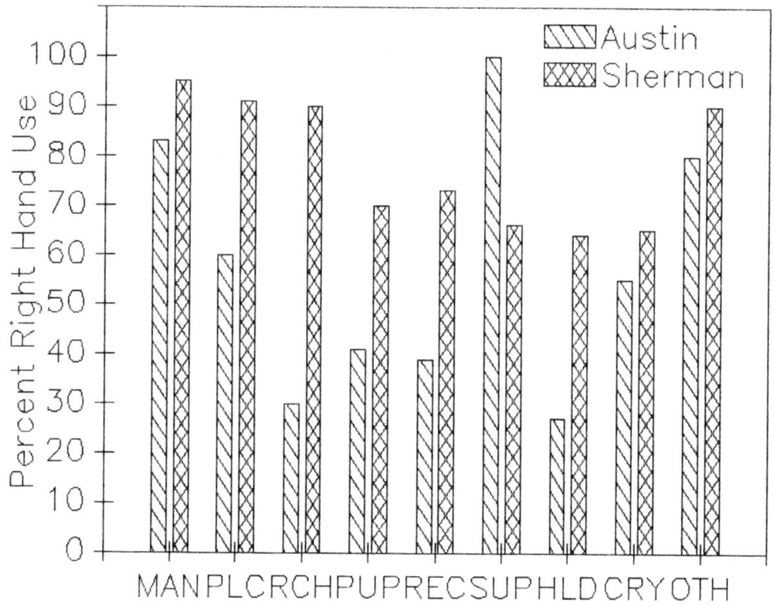

Figure 9.2. Mean right hand use for unimanual actions observed for Sherman and Austin.

classification. Included are data from another study by Hopkins, Washburn, and Rumbaugh (1989) on the manipulation of joysticks in these same subjects.

As can be seen, the tasks requiring fine motor movement clearly produced stronger right hand preferences in Austin and Sherman. On the joystick manipulation task, where the number of trials was equal for both subjects and their experiences comparable, both showed significant right hand preferences. In contrast, reach data (food retrieval), collected from these same subjects while performing the joystick manipulation tasks, yielded no significant preferences.

In summary, the strongest data for population hand preferences among great apes comes from studies with gorillas (see Fischer et al., 1982; Olson et al., 1990; Shafer, 1987). Data from chimpanzees and orang-utans are mixed; some report stronger right hand biases (Bolser et al., 1988; Heestand, 1986; Hopkins et al., 1989) while others indicate no hand biases (Bresard & Bresson, 1983; Finch, 1941; Marchant & Steklis, 1986; Olson et al., 1990). It is noteworthy that none of the studies indicate a left hand preference, with

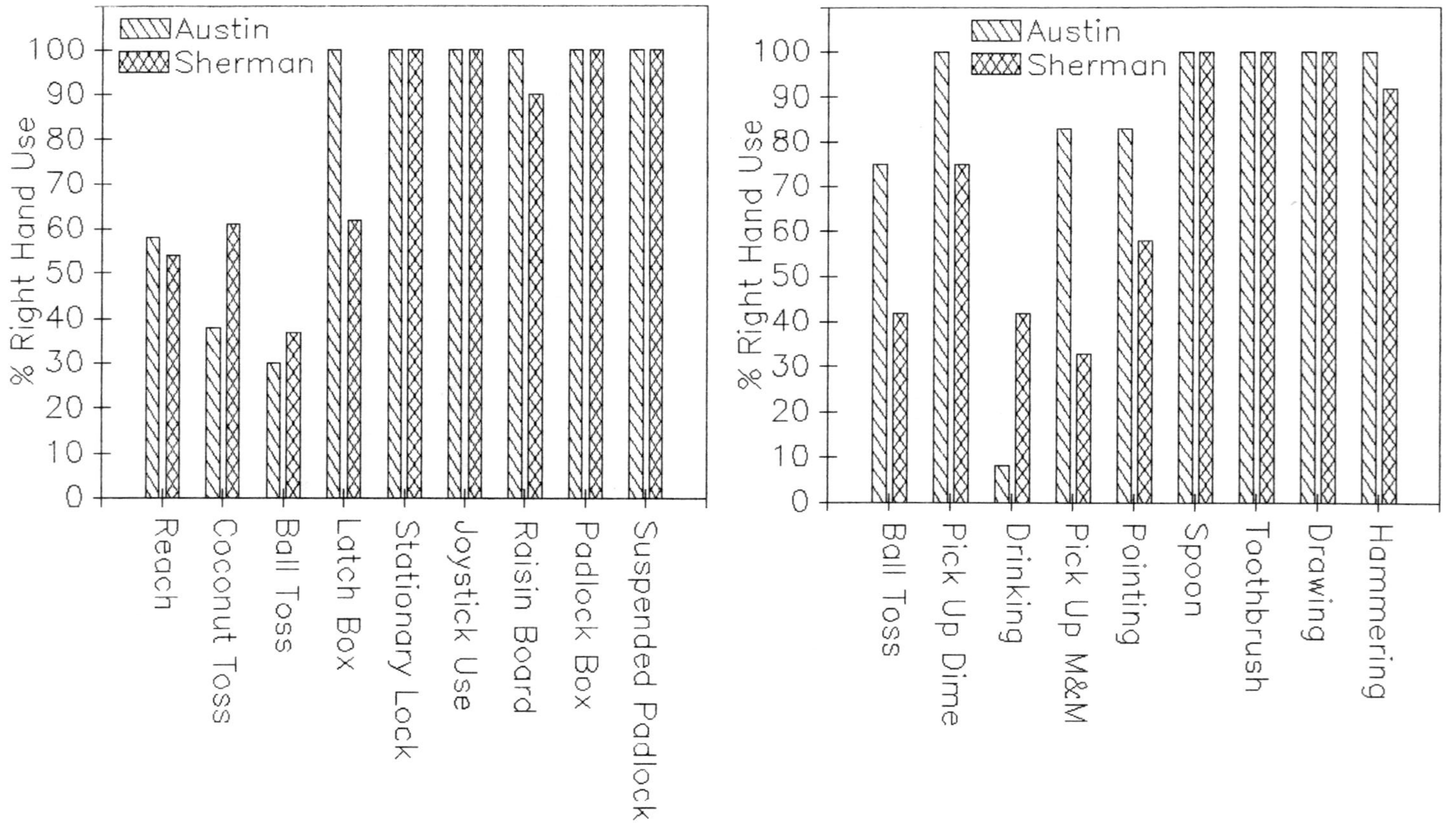

Figure 9.3 Percent right hand use for the L.R.C. Handedness measures (Left panel) and U.C.L.A. Handedness measures (Right panel) in Sherman and Austin.

the exception of Fagot and Vauclair (1988). However, the diversity of tasks employed in nonhuman primate studies does not allow for a clear evaluation of handedness in great apes, as different aspects of handedness are being measured in different studies, and the measures utilized are not always equitable with measures used to assess human handedness.

If handedness is to be studied comprehensively in nonhuman primate populations, we believe that tasks must be devised which assess both gross and fine motor abilities across varying test situations using various species. In other words, we believe that a multi-task, multivariate approach to handedness assessment in nonhuman primates is required before any conclusions regarding population attributes can be reached. By doing so, the tasks should not only measure hand preference, but also hand skill, since it is primarily hand skill which is measured in humans, such as the hand used for writing or sewing. Neither feature is mutually exclusive in the expression and assessment of overall handedness. Furthermore, bimanual tasks which control for positional biases and require varying hand processes simultaneously could provide additional information regarding handedness. Bimanual tasks should capitalize on skills already extant in ape species (e.g. tool use, drawing, rock pounding) to provide tasks which are more equitable with handedness tasks in human studies. Observational studies should record naturally-occurring behaviors requiring gross or fine motor abilities such as grooming and infant carrying or foraging behaviors and tool use, as these are most likely to have been most affected by lateralization.

We further suggest that factors such as posture, manual specialization, and the cognitive demands of the task be limited within a task so that intra-individual variability might be minimized. For example, when reaching for a piece of food, a nonhuman primate species, unlike humans, may either be in a quadrupedal, bipedal or tripedal posture. Moreover, location of the food relative to the subject's body position might be in the left, center, or right attentional hemispace. Reaching by a particular hand may be confounded by whether the subject is reaching with the intent to carry, throw, manipulate, or haptically inspect the food item. In each of these cases, hand use may be different due to interactions of any of these variables. Elimination of any one of these confounding factors, or covarying them via systematic task design, provides a clearer level of analysis in determining a subject's underlying handedness. Similar factors have recently been addressed in human handedess research (Healey et al., 1986).

Finally, defining hand preferences in nonhuman primates presents a good example of the difficulty faced by comparative psychology as a whole. What are the defining features of "left-handedness" or "right-handedness" in a given primate? In humans, we simply provide the subject with a 52-item checklist on hand preference for various actions, then correlate their responses to their verbal report of overall hand preference (Healey et al., 1986). As comparative psychologists, we have tested handedness simply by placing the subject in a cage and presenting food bits for them to pick up and we record what hand is used. It should be obvious to most observers that this is not what we do when we look for handedness in humans. In fact, it would be unlikely that any hand preference would be found in humans if we tested them with the same procedures and methods employed with nonhuman primates, although there has never been such a truly "comparative" study (however, see Seltzer, Forsythe, & Ward, 1990). The point we wish to make is that hand preference, as measured by traditional reach tasks (arm use), are not terribly sensitive, at least among great apes, and clearly do not follow the assessment tradition in the human literature.

We believe investigators need to look beyond the simple act of reaching, to improve the theoretical basis for task designs, to utilize a multivariate approach to handedness assessment, and to record the context or function in which the task occurs in order to more validly assess handedness in nonhuman primates. For example, the subjects left hand reaching behavior may be guided more by their preference to locomote tripedally with the right hand (Heestand, 1986) than to an overall "left-handedness". In this case, it could be argued that the subject is "right-handed". In short, handedness needs to be defined by the nature of the task requirements and by the context in which the animal is behaving rather than solely by a single type of task (i.e., reaching) or by the contextual similarity in which we typically (and somewhat arbitrarily) define human handedness.

Perceptual Asymmetries in Apes

An emerging perspective concerning laterality in nonhuman primates is the apparent lack of relationship between handedness and lateralized cognitive functions (Hamilton & Vermeire, 1988b), with the exception of studies showing an association between hemispheric learning and hand preference (Hamilton, 1983; Hamilton & Vermeire, 1982; Horester & Ettlinger, 1985). From an evolutionary perspective, the absence of observed, specific handedness in nonhuman primate populations does not necessarily imply that nonhuman primate brains are bilaterally symmetrical for all functioning. Instead, specialized functioning for any given hemisphere, as seen in humans, may operate independently of handedness, or the relationship between lateralized functions and specific hand use for nonhuman primate species may differ from that found in humans. Because of these possibilities, investigators have begun to focus on lateralization of functions where cognitive commonalities exists between human and nonhuman primates, particularly in areas of auditory processing (Dewson, 1977; Pohl, 1983), vocal communication (Petersen, Beecher, Zoloth, Moody, & Stebbins, 1978), spatial learning (Doty & Yamaga, 1973; Hamilton, 1983; Jason, Cowey, & Weiskratz, 1984), and facial expression (Hamilton, 1976; Hamilton &

Vermeire, 1988a; Overman & Doty, 1982).

Several investigators have studied the differences between the hemispheres in the processing of various auditory stimuli (species-specific calls or auditory pure tones) in two species of macaques. Processing of these auditory stimuli has been found to occur primarily in the left hemisphere (Dewson, 1977; Petersen et al., 1978). For example, unilateral left hemisphere lesions or bilateral lesions in the middle third of the superior temporal gyrus of rhesus monkeys resulted in a significantly impaired performance in discrimination between a tone or a burst of noise (Dewson, 1977). Similarly, Petersen et al. (1978) have reported a left hemisphere preference in the processing of meaningful species-specific vocalizations in Japanese macaques, and left hemisphere lesions in this species have been reported to decrease performance in the discrimination of these vocalizations (Heffner & Heffner, 1984). Only one study found evidence of right hemispheric processing of auditory stimuli; Pohl (1983) tested ear advantages of baboons, using four classes of auditory stimuli (pure tones, musical cords, vowels, and consonant-vowels). Significant left ear (right hemisphere) advantages, as determined by accuracy of responding, were found for the pure tone, consonant-vowel, and musical stimuli. Similar results for pure tone and musical stimuli have been reported in human studies (Bryden, 1982). It is important to note that Dewson (1977) and Pohl (1983) used similar types of auditory stimuli, but their findings differed with regards to ear preference. The two studies differed procedurally, one assessing the effects of specific lesions on discriminations, while the other used noninvasive accuracy methods. In addition, the studies differed in the species studied. Given these differences, it is difficult to conclude whether only specific aspects of auditory processing are lateralized, or whether such findings are robust phenomenon in various nonhuman primates.

With the exception of the results obtained by Pohl (1983), who used noninvasive methods and multiple types of auditory stimuli with baboons, the combined results do suggest that the processing of many types of acoustic information, especially those related to species-specific vocalizations, is performed better in the left hemisphere in at least one species of nonhuman primate (Japanese macaque). Such a finding is compatible with results found in human studies using related auditory speech/language stimuli (Beaton, 1986; Kimura, 1967).

Another area of investigation has been that of spatial processing through the use of visual discrimination paradigms (Jason et al., 1984) or maze performance (Doty & Yamaga, 1973). The results reported in these two studies are conflicting, again possibly due to significant methodological differences. In Doty and Yamaga's study (1973), macaques were taught to negotiate to certain endpoints within a maze. After reaching criterion, the optic chiasm was severed and a retest was performed for each subject using varying eye-hemisphere combinations. It was found that for one monkey, right eye-right hemisphere access to the maze yielded the highest recall performance. The remaining four subjects did not demonstrate these high performances. However, it was later revealed that the optic chiasm was not completely severed for these subjects, thus possibly confounding their results.

Jason et al. (1984) initially taught their subjects a visual discrimination based on the location of a dot within a central display. After reaching a pretraining criterion, the monkeys were subjected to unilateral lesions of the left or right occipital lobe. The findings indicated that left hemisphere lesions resulted in a greater decrement in performance than did right hemisphere lesions. The authors concluded that this species of macaque displayed a left hemisphere advantage for visual-spatial processing. These data supported an earlier report of a left hemisphere asymmetry for line orientation discrimination in split-brain monkeys (Hamilton, 1983).

Recently, we performed a study using a visual half-field paradigm in which two chimpanzees, Sherman and Austin, were required to make a visual-spatial discrimination based on the location of a short line contained within a geometric form (Hopkins & Morris, 1989). The subjects had been trained to fixate at midline by placing their cursor, using a joystick, on a midline stimulus. Once fixation was completed, a stimulus display occurred for less than 200 msec. in either visual field, and the subject was required to identify the location of the short line within the stimulus array by moving their joystick in either an up or down direction, depending on the location of the line. Both accuracy and reaction time measures were acquired using this paradigm.

The mean reaction times for each hand and subject are depicted in Figure 9.4. As can be seen, Sherman demonstrated a strong left visual field (LVF) bias regardless of which hand he used to perform the task. Austin also showed a LVF advantage, but it was in relation to his left hand responses. We concluded that both subjects had a right hemisphere processing bias for this task and these stimuli.

More recently, we examined the recognition of geometric forms presented unilaterally in Sherman, Austin, and a third chimpanzee, Lana. The subjects were trained to move a cursor to a fixation point where upon a stimulus was presented to either the right or left visual field. They were then required to move the cursor to select the presented stimulus from a two choice array. Stimulus presentations in each visual field were 100 msec. Subjects received 5 blocks of 20 trials using the same stimuli across test blocks. Both accuracy and reaction time served as dependent measures. Mean accuracy within each visual half-field as a function of test block are depicted in Figure 9.5. A significant interaction was found between block and visual half-field. In blocks 1 and 3 accuracy was significantly higher within the LVF contrasted with the RVF. Additional results from this study can be found in Hopkins, Washburn, and Rumbaugh (1990).

At this point, the limited data suggests that monkeys may evidence a left

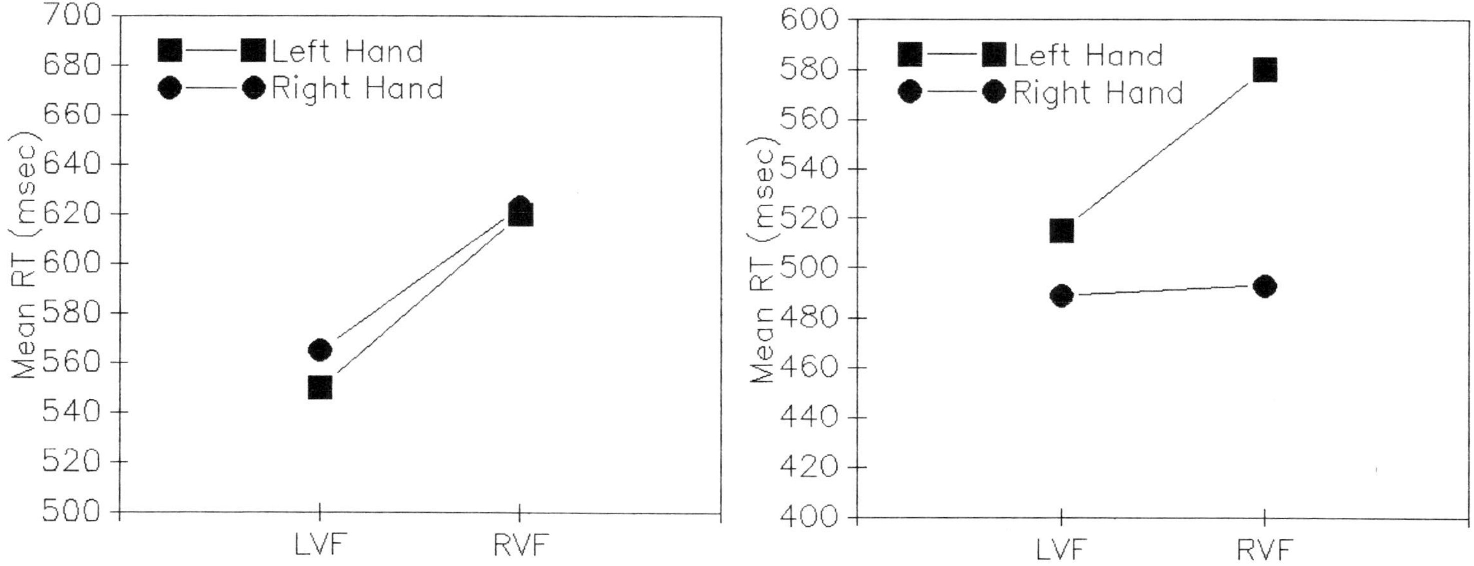

Figure 9.4. Mean reaction times for Sherman (Left panel) and Austin (Right Panel) hand as a function of hand and visual field of presentation. (Copyright the American Psychological Association).

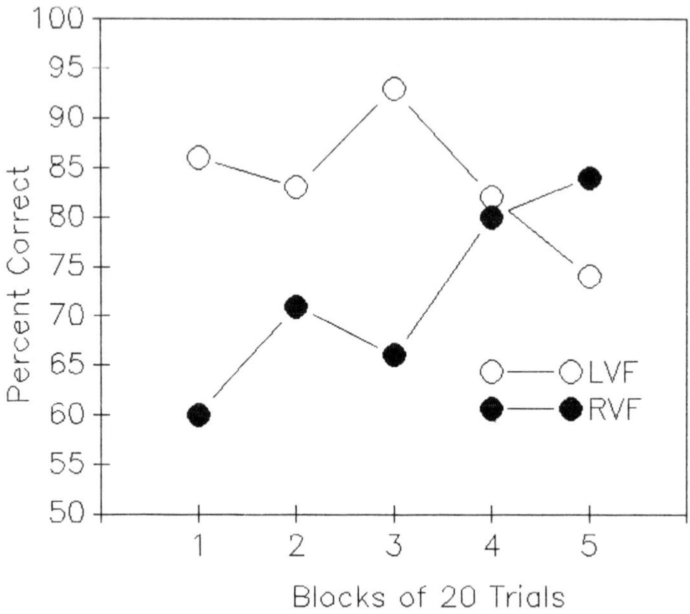

Figure 9.5. Mean accuracy in form recognition within each visual half-field and block. LVF = Left Visual Half-Field, RVF = Right Visual Half-Field. (Copyright the American Psychological Association).

hemisphere advantage for processing certain types of visual-spatial (i.e., geometric) information (Hamilton, 1983; Hamilton & Vermeire, 1988a, b; Jason et al., 1984), while apes with language-training have shown right hemisphere advantages for visual-spatial discriminations (Hopkins & Morris, 1989) and form recognition (Hopkins et al., 1990). Such findings may be due to a species difference but clearly this suggestion requires further exploration using similar procedures and stimuli with both species in order to warrant any conclusions.

The processing of facial expressions requires both spatial and affective perception and has been shown to be lateralized in many human studies (Heilman & Satz, 1983) although methodological factors can clearly affect one's findings. When a lateralized selection paradigm was used by Overman and Doty (1982), such human findings were easily replicated, although similar findings were not found in some monkey species (e.g. *Macaca nemistrina*). The human subjects selected the photos depicting faces which fell primarily in the left visual field (right hemisphere) while the monkeys did not. Moreover, the response was species-specific. That is, humans demonstrated a left visual field preference for human faces but not for monkey faces. Monkeys showed no overall preference.

It has been reported that split-brain macaques demonstrated significant right hemisphere advantages for learning individual discriminations of monkey faces and emotional discriminations of conspecifics (Hamilton & Vermeire, 1988a). Perret, Rolls, and Caan (1982), using single cell recording, have found that rhesus monkeys display as much as 10 times greater neuronal responsivity to slides of human and monkey faces in contrast to slides of nonfacial characteristics. Moreover, this responsivity of neurons is greater in the right hemisphere than the left. Split-brain monkeys have also been reported to display a greater a number of emotional responses to slides (faces or scenery) shown to the right hemisphere than to the left hemisphere (Ifune, Vermeire, & Hamilton, 1984).

Beyond the basic reports of individual recognition of conspecifics (Bauer & Phillips, 1982; Boysen & Bernston, 1986), there are no reports of hemispheric processing for faces in great apes. Recently, we examined visual field biases in the processing of human smiling faces presented as chimeric stimuli in a free vision task (Morris & Hopkins, in press). The task was identical to that used by Levy, Heller, Banich, and Burton (1983a) with human subjects. In this paradigm, neutral and smiling faces were split down the center, then disparate halves (e.g. left smiling with right neutral) were joined. These chimeric faces were presented in pairs as a vertical array. When human subjects were asked to judge which face was "happier", a significant LVF bias was found for both right- and left-handed subjects. In other words, subjects judged the chimeric face "happier" when the smiling half of the face fell on the LVF.

In our study, Sherman, Austin and Lana again served as subjects. Since it was not feasible to verbally ask the chimpanzee subjects "which face is happier?", they were initially taught a two choice discrimination between a human smiling face and human neutral face. Once they evidenced generalization to new exemplars, demonstrating they could discriminate between any smiling or neutral faces, they were subsequently tested for the ability to discriminate chimeric faces. The testing was identical to their training, except that chimeric faces were alternated with the normal faces over five test sessions. The intervening days between the test sessions varied from 1 to 14 days. Austin and Sherman showed significant LVF asymmetries while Lana's preference failed to reach statistical significance. Interestingly, we predicted and found that on 2 of the 5 test sessions no asymmetry was observed. However, these test sessions occurred 1 day after a previous chimeric test session compared to at least 14 days which intervened in the remaining test sessions. We believe this suggests that continual testing for laterality using faces may cloak asymmetries due to familiarity effects of the stimuli (Sergent & Bindra, 1981). Indeed, a similar decrement in accuracy occurs under conditions of familiarity in humans as well. The role of familiarity and related learning factors in laterality research on nonhuman primates is of significant concern given the extensive training and typical testing schedules which are required. How such factors may have affected the findings from other lateralization research on nonhuman primates has not been previously considered nor addressed (but see Vauclair & Fagot, this volume).

Hemispheric Activation in Apes

It has been proposed that conjugate lateral eye movement (CLEM) is an accurate method for assessing lateralized cognitive functions (Kinsbourne, 1972; Penfield & Roberts, 1959) although others have questioned its validity (Bryden, 1982; Ehrlichman & Weinberger, 1978). Presumably, direction of eye movement is indicative of the activation of a particular hemisphere during a specific task. It has been suggested that while solving linguistic problems human subjects tend to look to the right, but while solving visual-spatials they tend to look to the left or in no direction (Gur, 1975). Two studies have examined conjugate lateral eye movements in chimpanzees (Muncer, 1982; O'Neil, Stratton, Ingersoll, & Fouts, 1978).

Two chimpanzees (*Pan troglodytes*) exposed to language-training using American Sign Language (ASL) were presented with either linguistically or spatially-based problems (Muncer, 1982). Following each question, given in ASL, eye movement was coded manually. It was found that, for all types of questions, one subject demonstrated significant right movements, while the other showed no asymmetric eye movements. Asymmetry of eye movements due to the type of questions being posed as found in humans was not found in these chimpanzees.

O'Neil et al. (1978) studied four juvenile and four adolescent chimpanzees (*Pan troglodytes*) for conjugate lateral eye movements, all of which had received some language-training. In their study, specific types of questions were not asked of the subjects but, rather, direct eye contact was made. When eye contact was broken (presumably by the chimpanzee), the direction of the eye movement was recorded. Results showed a significant left movement bias for two of the 8 subjects. Because the two subjects which demonstrated significant results were the oldest, the authors concluded that a developmental effect may have accounted for their lack of findings in the younger chimpanzees.

Although the previous use of conjugate lateral eye movemments in nonhuman primates may not have utilized the strongest methodology (Bryden, 1982), the underlying theoretical concepts behind their use have potentially much to offer in the study of lateralized processes in nonhuman primates. The specific activation of one hemisphere in the processing of different types of stimuli has received significant attention in previous years as an alternative model and paradigm for predicting laterality findings in humans (Allen, 1983). Such theoretical and methodological options for the study of lateralized cognitive processes in nonhuman primates have not been applied until recently. We have reported that our two chimpanzee subjects, Sherman and Austin, demonstrated significant hemisphere activation using familiar and nonspecific color stimuli (Hopkins, Morris, Bolser & Washburn, 1987). The task was identical to that used with human subjects by Heilman and Van Den Abell (1979). In this task, subjects fixate on a cross presented in the center of a computer monitor, then depress a button. The cross remains on the screen for a variable foreperiod of time (one or two seconds). At the end of this foreperiod, a warning stimulus was presented for 40 msec. to the right or left visual field. For contrast purposes, on some trials no warning stimulus (NWS) was presented. After a variable delay (one or two seconds), a response cue in the form of a green square is presented in the center of the stimulus display. Subjects are to remove their finger from the button as quickly as possible upon seeing the response cue. The initial findings from this study indicated that both subjects showed a significant right hemisphere activation pattern for the nonspecific color stimuli.

In a recent series of studies, three categories of warning stimuli were employed. The categories of stimuli were nonmeaningful lexigrams, familiar lexigrams, and meaningful lexigrams. Lexigrams are visual-graphic symbols employed by these chimpanzees for interspecies and inter-individual communication (Savage-Rumbaugh, 1986). Both subjects appear to have demonstrated significant hand by visual field interactions. For the nonmeaningful symbols, the right hand/RVF combination was slower than all other hand/visual field interactions. For the meaningful symbols, the subjects showed a shift in asymmetry. For these stimuli, the right hand/RVF combination was faster than the right hand/LVF presentations. These preliminary data suggest that meaningful symbols may be processed more quickly by the left hemisphere while the nonmeaningful symbols were processed more quickly by the right hemisphere, a finding consistent with the human literature (Bowers & Heilman, 1980) using similar paradigms. These results of this series of experiments are described in detail elsewhere (Hopkins, Morris & Savage-Rumbaugh, 1991).

In summary, the conjugate lateral eye movement studies have yielded limited and conflicting results. Of the 10 total subjects used in both studies (Muncer, 1982; O'Neil, et al., 1978), only three demonstrated significant lateral eye movements. Our own hemispheric priming studies appear more promising in their ability to identify asymmetric processing in nonhuman primates, although these also have methodological limitations. In all of these studies, subjects who had not received language-training were not used as controls, and thus the activation findings may or may not have been related to their early language-training. The other side of this problem, of course, is the very limited population of language-trained apes that can be studied in a systematic manner. Studies of only a couple of subjects can provide interesting ideas and direction for research, and clearly can provide data to refute a null hypothesis, but are limited in their ability to make population statements regarding any lateralized functioning. Regardless, the underlying theoretical rationale for activation studies of asymmetric functioning has potential for numerous areas of investigation related to the lateralization of cognitive processes in nonhuman primates.

Summary and Theoretical Considerations

The data reviewed indicate that nonhuman primates do evidence anatomical asymmetries similar to humans. Handedness does not appear to be a robust phenomenon, but clearly more methodologically sophisticated research is needed before any definitive conclusions can be made. Positive, although very limited, evidence for functional or cognitive asymmetries has been reported in nonhuman primates, even though there is little correspondence between handedness and the documented perceptual asymmetries in nonhuman primates.

Even though the number of subjects is small in our own research, we believe that there is considerable internal validity to these findings. This belief is manifest by the consistency of findings across the tasks employed which are examining the same theoretical processes through different avenues of investigation. For example, the asymmetries in Austin have consistently been less than the indices of asymmetry in Sherman. Recall that in our handedness studies, Austin's overall right hand use was less than Sherman's. Also, in the visual half-field tasks, Austin's effects were significant only for the left hand while Sherman's were significantly different for both hands. Finally, the studies of hemispheric activation revealed greater hemispheric involvement for Sherman compared to Austin. We view these differences as quantitative, not qualitative. In other words, both chimpanzees show asymmetries which are similar in pattern but greater in degree for Sherman compared to Austin. It is important to note that Levy, Heller, Banich, and Burton (1983b) have shown that there are individual characteristic differences in the hemispheric asymmetries found in right-handers and these differences are indicative of arousal differences between the hemispheres, particularily the right hemisphere. It seems reasonable to suggests that similar patterns may emerge in nonhuman primates.

As mentioned, we have also collected some laterality data on a third subject, Lana, an adult female chimpanzee reared in a language environment (Rumbaugh, 1977). The extent to which we have examined her laterality is significantly less than Austin and Sherman, but her data resemble and differ from Austin's and Sherman's in several respects. First, the measures of handedness revealed mixed results for Lana, with strong left and right hand preferences emerging as a function of task requirements. Unlike Austin and Sherman, within a given task requirement (e.g. fine motor movement) her data were much more variable. For the chimeric faces task, her data revealed a LVF bias but this bias failed to reach significance while it did for Austin and Sherman. We have also tested Lana on the perception of geometric forms in the divided visual-field task described previously. Her data indicated a left visual field advantage. With regards to the activation tasks and visual-spatial discrimination task, no systematic evaluations have been made.

Why Lana's data differ from Austin's and Sherman's is difficult to answer at this point for several reasons. First, the differences may be due to her sex, as has been suggested in the human literature (McClone, 1980). Secondly, it may be that Lana's early rearing experiences, which were different from Austin and Sherman's, may have altered her laterality pattern. For example, her experience with tools was much more limited than Austin and Sherman's. Hence, if experiences with tools and object manipulation can influence later hand preferences, than this may account for some of the observed differences. Alternatively, her language-training was very different from Austin and Sherman's (Savage-Rumbaugh, 1986). If the development of language is fundamentally related to brain dominance and hand preference (Lennenberg, 1967), it is feasible that this could have affected the observed asymmetries. Notwithstanding, future studies with all three subjects should provide interesting comparisons as they pertain to possible factors which can affect the expression of lateralized functioning.

Unfortunately, nearly all the studies of cognitive and perceptual asymmetries have used great ape subjects who have been reared in enriched social and communicative environments and the number of subjects in all studies has been small. How such environmental factors might have influenced the observed asymmetries is an extremely important question and warrants further investigation (Denenberg, 1981). However, the data from language-trained apes have suggestions of functional asymmetries which indicate that their asymmetries closely model those of human subjects if one uses more recent theoretical models of human laterality (Allen, 1983).

The findings of hemispheric asymmetries irrespective of hand preference has prompted some researchers to reconceptualize aspects of laterality as they relate to the evolution of language (Falk, 1980; Hamilton & Vermeire, 1988b; Steklis, 1985). These researchers question whether the evolution of the cortical nervous system supported a manual or gestural communication system or "language" prior to the emergence of spoken language, or vice versa. From one perspective, gestural languages preceded spoken languages (Hewes, 1973, 1977), while from another, spoken languages preceded manual or gestural languages (Steklis, 1985). If we accept the notion that spoken language evolved first, then one must consider whether comprehension of spoken language preceded spoken language production evolutionarily, a question not easily answerable (Liberman, Cooper, Shankweiler, & Studdert-Kennedy, 1967; Lieberman, 1975). Neurologically, speech and fine motor movement (i.e., gestures) systems share different but adjacent neuroanatomical areas within the frontal and premotor cortex of the left hemisphere in humans. There is reason to believe that these two systems did not evolve as separate neural systems but developed from, or shared proximal neurological resources in the left hemisphere. Unfortunately, the tendency has been to view language as a discrete, independent phenomenon — either an organism has it or not. Laterality has been seen in this same light — as either present or absent.

Behaviorally in humans, there is at least a receptive communication sub-system (comprehension) and a productive sub-system (spoken language, written language, signs, etc.). Neurologically in humans, there is a concordant sensory system responsible for comprehension (e.g., Wernicke's area) and motor systems responsible for productive language (e.g., Broca's area). It may be that nonhuman primates share with humans similar neurobehavioral asymmetries in components of the comprehension sub-system but not in the productive sub-system. In fact, there is considerable evidence indicating that nonhuman primates perceive speech sounds categorically much in the manner that humans do (Kuhl, 1987), in addition to evidence for advanced skills in their speech and language comprehension (Herman, 1987; Savage-Rumbaugh, 1987). The fact that components of language comprehension have been found in nonhuman primates should not be taken as prima facie evidence for left hemisphere dominance for such processes. There is considerable evidence that humans can process components of human speech quite adequately with the right hemisphere when the response is gestural (pointing) rather than spoken (LeDoux, Wilson, & Gazzaniga, 1977; Zaidel, 1977).

In the evolution of primate cognition, a concordance between brain development and learning abilities is emerging (Fobes & King, 1982). Correlations between learning set performance and the amount of neocortex in a given primate species are about .50. Even better correlations (about .70) are found when criterional learning and subsequent reversal performance (the Transfer Index, see Rumbaugh & Pate, 1984), are correlated with the amount of neocortex in different species. At a more global level, greater brain development correlates with a general "intelligence" across species.

Levy (1977) has proposed that the development of cerebral lateralization nearly doubled the cognitive capacities of the human species. With this duality of processing capacity, there is less need for redundancy in function between the two hemispheres. At the neurological level, the auditory system developed rapidly, coincident with selection pressures for greater cognitive processes, suggesting an adaptive feature guiding the emergence of related skills. We would suggest that one adaptive feature which would have selected for duality of cerebral function might have been the ability to derive two types of information from a single stimulus event. Using speech as an example, it has been shown that human subjects with right hemisphere parietal lesions are capable of comprehending the content of the sentences spoken to them but are unable to identify the affective tone of the speakers voice (Heilman & Watson, 1977). In contrast, persons with left hemisphere lesions are unable to identify the content of the sentence but are capable of identifying the speakers affect (Bryden & Ley, 1983). Thus, depending on the cue subjects are required to attend to, a left or right hemisphere advantage can be found from the same single stimulus.

In conclusion, there is increasing evidence of greater theoretical and methodological sophistication in research devoted toward the understanding of cerebral asymmetries, or the lack thereof, in nonhuman primates. Only through the more systematic application of these models and methods to a wide variety of species will the relationship between basic cognitive processes or behavioral abilities and their supporting neurological systems become understood. Only then will their evolutionary or biobehavioral relevance become clear.

Acknowledgements

Funding for this research was supported from NIH grant RR-00165 from the Division of Research Resources and NICHD grant HD-06016 to the Yerkes Regional Primate Research Center, Emory University. The Yerkes Regional Primate Research Center is fully accredited by the American Association for Accreditation of Laboratory Animal Care. We gratefully acknowledge the extensive language-training work by Dr. Duane Rumbaugh and Dr. Sue Savage-Rumbaugh that forms the foundation for all of the studies reported herein with Lana, Sherman and Austin. Without their strong historical and current support, this unique group of subjects would not be available for such study. We also appreciate their helpful comments during the preparation of this work. We finally acknowledge the electrical assistance of Mr. Ray Taylor.

References

Allen, M. (1983). Models of hemispheric specialization. *Psychological Bulletin, 93*, 73-104.

Annett, M. (1985). *Left, right hand and brain: The right shift theory*. New York: Academic Press.

Bauer, H., & Phillips, K. (1982). Facial and vocal individual recognition in the common chimpanzee. *The Psychological Record, 33*, 161-170.

Beaton, A. (1986). *Left side, right side: A review of laterality research.* New Haven: Yale University Press.

Bever, T. G. (1975). Cerebral asymmetries in humans are due to the differentiation of two incompatible processes: Analytic and holistic. *Annals of the New York Academy of Sciences, 263*, 251-262.

Bolser, L., & Morris, R. D. (March, 1987). *A longitudinal study of handedness during daily activities in two language-trained chimpanzees.* Paper presented at the 54th annual meeting of the Southeastern Psychological Association, Atlanta, Georgia.

Bolser, L., Runfeldt, S., & Morris, R. (1988). Handedness in language-trained chimpanzees (*Pan troglodytes*) in daily activities and assessment tasks. *Journal of Experimental and Clinical Neuropsychology, 10*, 40.

Bowers, D., & Heilman, K. M. (1980). Material specific hemispheric activation. *Neuropsychologia, 18*, 309-320.

Boysen, S. T., & Bernston, G. G. (1986). Cardiac correlates of individual recognition in the chimpanzee (*Pan troglodytes*). *Journal of Comparative Psychology, 100*, 321-324.

Bradshaw, J. L., & Nettleton, N. C. (1981). The nature of hemispheric specialization in man. *Behavioral and Brain Sciences, 4*, 51-91.

Bresard, B., & Bresson, F. (1983). Handedness in *Pongo pygmaeus* and *Pan troglodytes*. *Journal of Human Evolution, 12*, 659-666.

Bryden, M. P. (1982). *Laterality: Functional asymmetry in the intact brain.* New York: Academic Press.

Bryden, M. P., & Ley, R. G. (1983). Right hemispheric involvement in the perception and expression of emotion in normal humans. In K. M. Heilman & P. Satz (Eds.), *Neuropsychology of human emotion* (pp.6-44). New York: The Guilford Press.

Cain, D. P., & Wada, J. A. (1979). An anatomical asymmetry in the baboon brain. *Brain, Behavior and Evolution, 16*, 222-226.

Denenberg, V. H. (1981). Hemispheric laterality in animals and the effects of early experience. *Behavioral and Brain Sciences, 4*, 1-50.

Dewson, J. H. (1977). Preliminary evidence of hemispheric asymmetry of auditory function in monkeys. In S. Harnad, R. W. Doty, L. Goldstein, J. Jaynes, & G. Krauthamer (Eds.), *Lateralization in the nervous system* (pp.63-74). New York: Academic Press.

Doty, R. W., & Yamaga, K. (1973). Maze behavior in macaques. *American Journal of Physical Anthropology, 35*, 404-405.

Ehrlichman, H., & Weinberger, A. (1978). Lateral eye movements and hemispheric asymmetries: A critical review. *Psychological Bulletin, 85*, 1080-1101.

Ettlinger, G. (1984). Humans, apes and monkeys: The changing neuropsychological viewpoint. *Neuropsychologia, 22*, 685-696.

Fagot, J., & Vauclair, J. (1988). Handness and bimanual coordination in the lowland gorilla. *Brain, Behavior and Evolution, 32*, 89-95.

Falk, D. (1978). Cerebral asymmetry in Old World Monkeys. *Acta Anatomica, 101*, 334-339.

Falk, D. (1980). Language, handedness and primate brains: Did the australiopithecines sign? *American Anthropologist, 82*, 72-78.

Falk, D., Cheverud, J., Vannier, M. H., & Conroy, J. C. (1986). Advanced computer graphics technology reveals cortical asymmetry in endocasts of rhesus monkeys. *Folia Primatologica, 46*, 98-103.

Finch, G. (1941). Chimpanzee handedness. *Science, 94*, 117- 118.

Fischer, R. B., Meunier, G. F., & White, P. J. (1982). Evidence of laterality in the lowland gorilla. *Perceptual and Motor Skills, 54*, 1093-1094.

Fobes, J. L., & King, J. L (1982). *Primate behavior.* New York: Academic Press.

Gardner, R. A., & Gardner, B. A. (1968). Teaching sign language to a chimpanzee. *Science, 165*, 664-672.

Geschwind, N., & Galaburda, A. M. (1985). Cerebral lateralization: Biological mechanisms, associations and pathologies: I. A hypothesis and a program for research. *Archives of Neurology, 42*, 428-459.

Geschwind, N., & Levitsky, W. (1968). Human brain: left-right asymmetries in temporal speech region. *Science, 161*, 186-187.

Gur, R. E. (1975). Conjugate lateral eye movements as an index of hemispheric activation. *Journal of Personality and Social Psychology, 31*, 416-420.

Hamilton, C. R. (1976). An assessment of hemispheric specialization in monkeys. *Annals of the New York Academy of Sciences, 299*, 222-232.

Hamilton, C. R. (1983). Lateralization for orientation in split-brain monkeys. *Behavioural Brain Research, 10*, 399-403.

Hamilton, C. R., & Vermeire, B. A. (1982). Hemispheric differences in split-brain monkeys learning sequential comparisons. *Neuropsychologia, 20*, 691-698.

Hamilton, C. R., & Vermeire, B. A. (1988a). Complementary hemispheric specialization in monkeys. *Science, 242*, 1694-1696.

Hamilton, C. R., & Vermeire, B. A. (1988b). Cognition, not handedness, is lateralized in monkeys. *Behavioral and Brain Sciences, 11*, 723-725.

Healey, J. M., Liederman, J., & Geschwind, N. (1986). Handedness is not a unidimensional trait. *Cortex, 22*, 33-53.

Heestand, J. (1986). *Behavioral lateralization in four species of ape.* Unpublished doctoral dissertation, University of Washington, Seattle, WA.

Heffner, H. E., & Heffner, R. S. (1984). Temporal lobe lesions and perception of species specific vocalizations by macaques. *Science, 226*, 75-76.

Heilbroner, P. L., & Holloway, R. L. (1988). Anatomical brain asymmetries in new world and old world monkeys: Stages of temporal lobe development in primate evolution. *American Journal of Physical Anthropology, 76*, 39-48.

Heilman, K. M., & Satz, P. (1983). *Neuropsychology of human emotion.* New York: Guilford Press.

Heilman, K. M., & Van Den Abell, T. (1979). Right hemisphere dominance for mediating cerebral activation. *Neuropsychologia, 17*, 315-321.

Heilman, K. M., & Watson, J. T. (1977). The neglect syndrome-A unilateral defect of the orienting response. In S. R. Harnad, R. W. Doty, L. Goldstein, J. Jaynes, & G. Krauthamer (Eds.), *Lateralization in the nervous system* (pp. 310-321). New York: Academic Press.

Herman, L. (1987). Receptive competencies in language-trained animals. *Advances in the study of behavior, 17*, 1-60.

Hewes, G. W. (1973). Primate communication and the gestural origin of language. *Current Anthropology, 14*, 5-24.

Hewes, G. W. (1977). Language origin theories. In D. M. Rumbaugh (Ed.), *Language learning by a chimpanzee: The LANA Project* (pp.3-53). New York: Academic Press.

Holloway, R. L., & de la Coste-lareymondie, M. C. (1982). Brain endocast asymmetry in pongoids and homonoids: Some preliminary findings on the paleontology of cerebral dominance. *American Journal of Physical Anthropology, 58*, 101-110.

Hopkins, W. D., & Morris, R. D. (1989). Laterality for visual-spatial processing in two language-trained chimpanzess. *Behavioral Neuroscience, 104*, 227-234.

Hopkins, W. D., Morris, R. D., Bolser, L., & Washburn, D. A. (March, 1987). *Hemispheric activation for color stimuli in chimpanzees*. Paper presented at the 98th annual meeting of the Southeastern Psychological Society, New Orleans, Louisiana.

Hopkins, W. D., Morris, R. D., & Savage-Rumbaugh, S. (1991). Evidence of asymmetrical hemispheric priming for known and unknwon warning stimuli in two language-trained chimpanzees (*Pan troglodytes*). *Journal of Experimental Psychology: General, 120*, 46-56.

Hopkins, W. D., Washburn, D. A., & Rumbaugh, D. M. (1989). Note on hand use in the manipulation of joysticks by rhesus monkeys (*Macaca mulatta*) and chimpanzees (*Pan troglodytes*). *Journal of Comparative Psychology, 104*, 91-94.

Hopkins, W. D., Washburn, D. A., & Rumbaugh, D. M. (1990). Processing of form stimuli presented unilaterally in humans, chimpanzees (*Pan troglodytes*), and monkeys (*Macaca mulatta*). *Behavioral Neuroscience, 104*, 577-582.

Horester, W., & Ettlinger, G. (1985). An association between hand preference and tactile discrimination performance in the rhesus monkey. *Neuropsychologia, 23*, 411-413.

Ifune, C. K., Vermeire, B. A., & Hamilton, C. R. (1984). Hemispheric differences in split-brain monkeys viewing and responding to videotape recordings. *Behavioral and Neural Biology, 41*, 231-235.

Jason, G. W., Cowey, A., & Weiskrantz, L. (1984). Hemispheric asymmetry for a visual-spatial task in monkeys. *Neuropsychologia, 22*, 77-784.

Kimura, D. (1967). Functional asymmetry of the brain in dichotic listening. *Cortex, 3*, 163-178.

Kimura, D. (1979). Neuromotor mechanisms in the evolution of human communication. In H. D. Steklis & M. J. Raleigh (Eds.), *Neurobiology of social communication in primates: An evolutionary perspective* (pp.197-220). New York: Academic Press.

Kinsbourne, M. (1972). Eye and head turning indicates cerebral lateralization. *Science, 176*, 539-541.

Kuhl, P. (1987). The special-mechanisms debate in speech: Contributions of tests on animals and infants. In S. Harnad (Ed.), *Categorical perception* (pp.355-386). Cambridge, MA: Cambridge University Press.

Laidler, K. (1978). Language in the orang-utan. In A. Lock (Ed.),*Action, gesture and symbol: The acquisition of language* (pp.133-155). New York: Academic Press.

LeDoux, J. E., Wilson, D. H., & Gazzaniga, M. S. (1977). Manipulospatial aspects of cerebral lateralization: Clues to the origin of lateralization. *Neuropsychologia, 15*, 743-750.

LeMay, M. (1985). Asymmetries of the brains and skulls of nonhuman primates. In S. Glick (Ed.), *Cerebral lateralization in nonhuman species* (pp.233-245). New York: Academic Press.

LeMay, M., Billig, M. S., & Geschwind, N. (1982). Asymmetries of the brains and skulls of nonhuman primates: In E. Armstrong & D. Falk (Eds.), *Primate brain evolution: Methods and concepts* (pp.68-81). New York: Plenum Press.

LeMay, M., & Geschwind, N. (1975). Hemipheric differences in the brains of great apes. *Brain, Behavior and Evolution, 11*, 48-52.

Lennenberg, E.H. (1967). *Biological foundations of language*. New York: Wiley.

Levy, J. (1977). The mammalian brain and the adaptive advantage of cerebral asymmetry. *Annals of the New York Academy of Sciences, 299*, 264-272.

Levy, J., Heller, W., Banich, M. T., & Burton, L. A. (1983a). Are variations among right handers in perceptual asymmetries caused by characteristic arousal differences between hemispheres. *Journal of Experimental Psychology: Human Perception and Performance, 9*, 329-358.

Levy, J., Heller, W., Banich, M. T., & Burton, L. A. (1983b). Asymmetry of perception in free viewing of chimeric faces. *Brain and Cognition, 2*, 404-419.

Liberman, A., Cooper, F. S., Shankweiler, D. S., & Studdert-Kennedy, M. (1967). Perception of the speech code. *Psychological Review, 74*, 431-461.

Lieberman, P. (1975). *On the origins of language: An introduction to the evolution of human speech*. New York: Wiley.

MacNeilage, P. F., Studdert-Kennedy, M. G., & Lindblom, B. (1987). Primate handedness reconsidered. *Behavioral and Behavior Sciences, 10*, 247-301.

Marchant, L. F., & Steklis, H. D. (1986). Hand preference among captive island groups of chimpanzees (*Pan troglodytes*). *American Journal of Primatology, 10*, 300-313.

McClone, J. (1980). Sex differences in human asymmetry: A critical survey. *Behavioral and Brain Sciences, 3*, 215-263.

Menzel, E. W. (1974). A group of young chimpanzees in a one-acre field. In A. M. Schrier & F. Stollnitz (Eds.), *Behavior of nonhuman primates* (pp.83-153). New York: Academic Press.

Morris, R. D., & Hopkins, W. D. (in press). Perception of human chimeric faces by chimpanzees: Evidence for a right hemisphere asymmetry. *Brain and Cognition.*

Muncer, S. J. (1982). Functional asymmetry in the chimpanzee. *Perceptual and Motor Skills, 54*, 147-152.

Olson, D., Ellis, J. E., & Nadler, R. D. (1990). Hand preferences in captive gorillas, orang-utans, and gibbons. *American Journal of Primatology, 20*, 83-94.

O'Neil, C. R., Stratton, H. T. R., Ingersoll, R. H., & Fouts, R. S. (1978). Conjugate lateral eye movements in *Pan troglodytes*. *Neuropsychologia, 16*, 759-762.

Overman, W. H., & Doty, R. W. (1982). Hemispheric specialization displayed by man but not macaques in the analysis of faces. *Neuropsychologia, 20*, 113-128.

Patterson, F. G. (1978). The gestures of a gorilla: language acquisition in another pongoid. *Brain and Language, 5*, 72-97.

Penfield, W., & Roberts, L. (1959). *Speech and brain mechanisms*. New Haven: Princeton University Press.

Perret, D. I., Rolls, E. T., & Caan, W. (1982). Visual neurons responsive to faces in the monkey termporal cortex. *Experimental Brain Research, 47*, 329-342.

Petersen, M. R., Beecher, M. D., Zoloth, S. R., Moody, D. B., & Stebbins, W. C. (1978). Neural lateralization of species specific vocalizations in Japanese macaques (*Macaca fuscata*). *Science, 202*, 324-327.

Pohl, P. (1983). Central auditory processing: Ear advantages for acoustic stimuli in baboons. *Brain and Language, 20*, 44-53.

Premack, D. (1970). Language in a chimpanzee? *Science, 172*, 808- 822.

Rumbaugh, D. M. (1977). *Language learning by a chimpanzee: The LANA Project*. New York: Academic Press.

Rumbaugh, D. M., & Pate, J. L. (1984). The evolution of cognition in primates: A comparative perspective. In H. L. Roitblat, T. G. Bever, & H. S. Terrace (Eds.), *Animal cognition* (pp.403-420). Hillsdale, NJ: Lawrence Erlbaum and Associates.

Rumbaugh, D. M., Savage-Rumbaugh, E. S., & Scanlon, J. L. (1982). The relationship between language in apes and human beings. In J. L. Fobes & J. E. King (Eds.), *Primate behavior* (pp.363-383). New York: Academic Press.

Sanford, C., Guin, K., & Ward, J. P. (1984). Posture and laterality in the bush baby (*Galago senegalensis*). *Brain, Behavior and Evolution, 25*, 217-224.

Satz, P., Baymur, L., & Van Der Vlugt, H. (1979). Pathological left handedness: Cross cultural tests of a model. *Neuropsychologia, 7*, 101-103.

Savage-Rumbaugh, E. S. (1986). *Ape language: From conditioned response to symbols*. New York: Columbia University Press.

Savage-Rumbaugh, E. S. (1987). A new look at ape language: Comprehension of vocal speech and syntax. *Nebraska Symposium on Motivation, 45*, 201-255.

Savage-Rumbaugh, E. S., McDonald, K., Sevcik, R. A., Hopkins, W. D., & Rubert, E. (1986). Spontaneous symbol acquisition and communicative use by pygmy chimpanzees (*Pan paniscus*). *Journal of Experimental Psychology: General, 115*, 211-235.

Seltzer, C., Forsythe, C., & Ward, J. P. (1990). Multiple measures of motor lateralization in human primates. *Journal of Comparative Psychology, 104*, 159-166.

Sergent, J., & Bindra, D. (1981). Differential hemispheric processing of faces: Methodological considerations and reinterpretation. *Psychological Bulletin, 89*, 541-554.

Shafer, D. D. (1987). *Patterns of hand preference among captive gorillas*. Unpublished masters thesis, San Fransisco State University, San Francisco, CA.

Shafer, D. D. (1988). Handedness in gorillas. *The Gorilla Foundation, 8*, 2-5.

Soper, H., Satz, P., Orsini, D. L., Van Gorp, W. G., & Green, M. F. (1987). Handedness distribution in a residential population with severe or profound mental retardation. *Journal of Mental Deficiency, 92*, 94-102.

Steklis, H. D. (1985). Primate communication, comparative neurology, and the origin of language re-examined. *Journal of Human Evolution, 14*, 157-173.

Wada, J., Clarke, R., & Hamm, A. (1975). Cerebral hemispheric asymmetry in humans. *Archives of Neurology, 32*, 239-246.

Warren, J. M. (1977). Handedness and cerebral dominance in monkeys. In S. Harnad, R. W. Doty, L. Goldstein, J. Jaynes, & G. Krauthamer (Eds.), *Lateralization in the nervous system* (pp.151-172). New York: Academic Press.

Warren, J. M. (1980). Handedness and laterality in humans and other animals. *Physiological Psychology, 8*, 351-359.

Witelson, S. F. (1977). Anatomic asymmetry in the temporal lobes: Its documentation, phylogenesis, and relationship to functional asymmetry. *Annals of the New York Academy of Sciences, 299,* 328-354.

Witelson, S. F. (1987). Neurobiological aspects of language in children. *Child Development, 58,* 653-688.

Witelson, S. F., & Pallie, W. (1973). Left hemisphere specialization for language in the newborn: neuroanatomical evidence of laterality. *Brain, 96,* 641-646.

Yeni-Komshian, G., & Benson, D. (1976). Anatomical study of cerebral asymmetry in the temporal lobe of humans, chimpanzees and monkeys. *Science, 192,* 387-389.

Zaidel, E. (1977). Auditory vocabulary of the right hemisphere following brain bisection or hemispherectomy. *Cortex, 12,* 191-211.

10
Ontogeny of Object Manipulation and Manual Lateralization in the Guinea Baboon: Preliminary Observations

Joël Fagot

The main focus of this paper is the early development of manual lateralization in a primate species (*Papio papio*). There are several reasons for studying the ontogeny of asymmetrical hand usage in nonhuman primates. First, developmental studies might provide an explanation about the forms of manual laterality in adults. Second, an ontogenetic approach helps address features of the underlying brain maturation. Third, such studies also raise the question of the resemblance, and thus of the dissimilarities, between the development of manual lateralization in man and forms of laterality during ontogeny in other primates. Although numerous studies have been carried out on manual lateralization in nonhuman primates (see references in Fagot & Vauclair, 1991; MacNeilage, Studdert-Kennedy, & Lindblom, 1987; Warren, 1980), there are many gaps in the literature with respect to the question of its development. Cunningham, Forsythe, and Ward (1989) have reported an unstable right preference for food reaches, but a stable preference for nonfood reaches in an orang-utan observed from 43 to 70 weeks of age. Chorazyna (1976) and Brésard and Bresson (1983) have found in the chimpanzee an initial instability in lateral preference with age. Similar fluctuations had been found in human newborns (e.g., Liederman, 1983). Bard, Hopkins, and Fort (1990) have reported that 8 of 10 infant chimpanzees preferred to suck their right thumb. Other data come from cross-sectional studies conducted with monkeys of different ages. It was shown that age variables can greatly influence either the strength of hand lateralization (Brooker, Lehman, Heinbuch, & Kidd, 1981; Lehman, 1980; Vauclair & Fagot, 1987a) or the direction of hand bias (Forsythe & Ward, 1988; Vauclair & Fagot, 1987a). In brief, the few available studies suggest that lateralization, which seems to characterize hand usage in nonhuman primate infants, does not necessarily develop monotonically. The observed fluctuations may reflect the intertwining effect of maturational and experiential processes.

In a recent paper, Fagot and Vauclair (1991, and see also Ch. 9 in this volume) have tentatively applied to nonhuman primates the distinction between "handedness" and "manual specialization" that had initially been developed in the human context (Young, Segalowitz, Corter, & Trehub, 1983). "Manual specialization" describes hand preference in novel tasks (as compared to routine activities) that are precise with respect to their spatio-temporal dimensions. By contrast, "handedness" characterizes preferences for routine, familiar, and gross motor actions (e.g., simple reaching for food). This distinction was applied to a large body of data collected in nonhuman primates, and it was found that manual behaviors revealing handedness demonstrated a symmetrically shaped distribution of hand preferences for the population. Conversely, tasks that required manual specialization presented overwhelmingly asymmetrical distributions of hand usage. This theoretical framework predicts a consistency within individual hand biases for the first trials of a given task, but not necessarily for further trials. Consequently, within a developmental perspective, this framework predicts (1) an inter-individual consistency in hand lateralization at the onset of a given behavior, but (2) this consistency is not expected to continue when, with practice, the behavior becomes part of routine daily activities.

Quantification of manual activities in the baboon has shown that a high percentage of manual behaviors are directed towards edible or unedible inanimate objects (branches, stones, etc.: Altmann, 1980; Joubert & Vauclair, 1986; Post, 1981). With the group under study in this chapter, our previous observations of spontaneous manual activities (unpublished data) have shown that 59% of adults' manual actions were directed to inanimate objects. That percentage was markedly stronger for the infants and juveniles (89.8%). Another characteristic of object manipulation in the baboon is its ontogenic precociousness compared to other taxa (e.g., chimpanzees and humans: Spinozzi & Natale, 1986). For example, one week old baboons touch and occasionally grasp objects (Chalmers, 1980; Flament, 1965). Given the precocity of object manipulation in infant baboons and its quantitative importance in the juveniles, we have chosen to study the development of hand lateralization in this species through analysis of this kind of behavior. In short, the main objectives of this chapter are threefold: 1) to provide a description of the ontogeny of object manipulation; 2) to search for the first appearance of manual preference; 3) to determine which kinds of activities primarily lead to manual preference.

Methods

Subjects and Setting Conditions

The subjects were two female (F-1 and F-2) and two male (M-1 and M-2) full-term infant baboons (*Papio papio*) living with their mothers in a social group of 30 monkeys. Both females were born in the winter of 1986 whereas both males were born in the spring of 1987. These subjects were the only monkeys born in the troop during the period dedicated to the study. Spontaneous hand usage of the four infant baboons was recorded from birth to 14 weeks (F-1), 24 weeks (M-1 and M-2), and 36 weeks (F-2). All observations had to be terminated in the fall of 1987, when the group was moved to a new compound. The observations of subject F-1 had to be prematurely interrupted because this monkey died of a sudden illness (enteritis) at 14 weeks of age.

Manual preferences of the mothers had been assessed at two different periods: one year and two years before the present investigation (Vauclair & Fagot, 1987a, b); F-1 and F2's mothers expressed a consistent right hand preference, whereas M-2's mother had a consistent left hand preference within the two studies. M-1's mother demonstrated a left hand preference in the second study (1987b) but no preference in the first one (1987a).

The social group lived in a large outdoor enclosure of 650 m^2 linked by a tunnel to a monkey-house. The enclosure contained two wooden constructions (each 4.5 m in height) and branches and stones of different sizes which were scattered on the ground. The monkeys were free to move from the compound to the monkey-house. However, for the purpose of the study, the group's movements were limited to the enclosure during the observation periods. Special monkey chow was regularly given both in the enclosure and in the monkey-house. Additional food (apples and bananas) was provided every day at 8:30 a.m. in the enclosure.

Procedure

Daily spontaneous activities were recorded using the focal animal sampling method (Altmann, 1974). Use of a JVC VHS video system with a focal lens of 300 mm made it possible to see in detail and record all manual activities. Behaviors were collected at 11 a.m. from an observation tower close to the compound. Table 10.1 presents the sampling schedule, which varied as a function the monkey's age.

Coding System

We have employed an original coding system which was devised to provide a typology of hand kinetics in relation to the target object. It was considered that a manual behavior involved one or more of the following four acts: 1) "directed extension" of the forelimb(s) toward the object; 2) "contact" of the object by the hand, either without any grip (simple touch), or by a palmar or by a precision grip; 3) "bimanual manipulation" of the object (e.g., bimanual shaking of the object); 4) "bring the object to mouth". Patterns of the hand's movements have been defined by several combinations of these

Table 10.1. Sampling Session Schedule

Age (in weeks)	Duration and frequency of the recording session
0 to 2	Two hours twice a week
3 to 12	One hour twice a week
13 to 24	Half an hour twice a week
25 to 36	Half an hour per week

four types of act (see Table 10.2).

An APPROACH pattern implied an "extension" which was not followed by a "contact". A HANDLE implied at least an "extension" and a "contact", or an "extension" and a "contact" followed by a "bimanual manipulation", but the goal object was never brought to the mouth. A DIRECT TO MOUTH pattern involved an "extension," a "contact," and a "bring to mouth" action without any "bimanual manipulation". INDIRECT TO MOUTH involved the same three acts but a "bimanual manipulation" occurred between the "contact" and the "bring to mouth" actions. The SECOND TO MOUTH pattern corresponded to a sequence of movements when the object was extracted from the mouth and was brought to mouth for a second time after an intermediate "bimanual manipulation." Spontaneous manual behaviors of infant baboons have been analyzed with respect to this coding system. In order to obtain data about hand laterality, we have also noted: 1) the hand(s) used (the left, the right, or both hands) for the "extension," the "contact," and for the "bring to mouth" acts; 2) the type of grip (simple touch, palmar grip, or precision thumb-index grip; see Napier, 1961) for the "contact"; 3) transfer of the object from the right to left or left to right hand was recorded during the phase of "bimanual manipulation". Only the behaviors emitted when the goal was in front of the subject or when it was contralateral to the hand used (cross-reaching) were scored. Occurrences of manual behaviors were recorded regardless of their duration.

Inter- and Intra-Observer Agreements

Inter- and intra-coder agreements were assessed using a Kappa analysis (Cohen, 1960). Although this statistic provides no standard threshold of reliability, one can assume that a Kappa superior to .70 indicates an acceptable reliability level. Four videotaped sessions of films (two hours) have been coded in parallel by the usual coder (coder A) and by two independent coders (B and C). Moreover, intra-observer agreement was also assessed two months after the initial coding. It has been empirically observed that coders B and C have had difficulties in categorizing the kind of grip

Table 10.2. Definition of the Five Patterns of Hand Movement

Patterns	Extension	Contact	Bimanual Manipulation	Bring to Mouth
APPROACH	yes	no	no	no
HANDLE !	yes	yes	no	no
!	yes	yes	yes	no
TO MOUTH:				
DIRECT	yes	yes	no	yes
INDIRECT	yes	yes	yes	yes
SECOND	no	no	yes	yes

(simple touch, palmar, or precision grip) exhibited by the monkeys. We thus proposed a Kappa analysis that dissociated the coding of prehension types from the coding of other categories. With respect to the coding of prehension, Kappas were as follows: A vs B, Kappa = .78; A vs C, Kappa = .61; B vs C, Kappa = .70; A vs A (intra-coder), Kappa = .88. For the remaining categories when excluding prehension types: A vs B, Kappa=.92; A vs C, Kappa=.85; B vs C, Kappa=.84; A vs A, Kappa=.95.

Results

Overall Distribution of Hand Usage

Individual scores were pooled by age intervals. Intervals were 2 weeks long, 4 weeks long, and 8 weeks long from W1 (Week 1) to W8, from W9 to W16 and from W17 to W36 respectively. A total of 6,540 infant manual actions were recorded. The analysis shown in Table 10.3 with respect to the five patterns of hand movement (as defined in Table 10.2) provides a good illustration of the ontogeny of the infant baboons' sensorimotor manual competencies.

The APPROACH and HANDLE patterns constituted almost all manual actions performed from birth to week 4, but their peak values progressively declined with age in favor of other gestural structures. DIRECT TO MOUTH was first observed at weeks 4-6 for all subjects except M-2 (W2-4; only one instance) and attained a maximum of 54% later in infancy. INDIRECT TO MOUTH appeared at weeks 4-6 and remained the less frequent behavioral pattern. The "SECOND TO MOUTH" category emerged during the second month and, with age, progressively reached a maximum of 24.8% of all manual actions. Other data are related to the kind of grip. Simple touches were observed during the first three days of life.

Table 10.3. Percentages of Each Behavioral Pattern Defined in Table 10.2 and Total Number of Actions Observed (N) as a Function of Age in Weeks for Each Animal

Subject	Age in Weeks	(N)	Approach	Handle	To the Mouth		
					Direct	Indirect	Second
F-1	0-2	(73)	33.0	67.0			
	2-4	(116)	34.5	65.5			
	4-6	(199)	15.6	58.8	23.1	2.5	
	6-8	(246)	4.1	49.6	36.2	7.7	2.4
	8-12	(416)	2.4	46.9	32.0	4.5	14.1
	12-14	(249)	.8	46.2	30.5	4.4	18.1
F-2	0-2	(129)	56.6	43.4			
	2-4	(101)	24.8	76.2			
	4-6	(138)	13.0	76.8	8.0	.7	1.5
	6-8	(183)	8.2	70.5	16.4	1.6	3.3
	8-12	(486)	1.0	51.7	29.8	6.1	11.1
	12-16	(366)	.8	36.1	34.4	3.9	24.8
	16-24	(621)		40.9	34.5	5.3	19.3
	24-32	(364)		32.7	40.9	7.1	19.3
	32-36	(174)		39.7	33.3	7.5	19.5
M-1	0-2	(52)	40.4	59.6			
	2-4	(59)	16.9	83.1			
	4-6	(128)	4.7	74.2	18.0	3.2	
	6-8	(145)	2.1	36.6	52.4	6.2	2.7
	8-12	(154)	1.3	35.7	48.1	2.6	12.3
	12-16	(176)		29.9	54.0	.6	15.5
	16-24	(511)		29.0	39.7	6.8	24.5
M-2	0-2	(22)	68.2	31.8			
	2-4	(87)	21.8	77.0	1.2		
	4-6	(171)	5.8	77.2	14.0	2.4	.6
	6-8	(206)		52.9	44.2	1.0	1.9
	8-12	(408)		42.9	44.6	4.2	8.3
	12-16	(146)		24.7	52.7	6.8	16.4
	16-24	(414)		32.8	40.6	7.3	19.3

Palmar grips occurred first at 10, 13, 9, and 11 days respectively for M-1, M-2, F-1, and F-2. Precision grips were recorded for the first time at the age of 45 (M-1), 48 (M-2 and F-2), and 55 days (F-1). Thus, changes in gestural structures were so rapid that each subject was able to express all patterns of hand movement before W8.

```
SUBJECTS

   xxxxXXXX    XXXX
   --------++++----++++++++++++
F-1

   xxxxXXXXxxxx    xxxxxxxx                        XXXXXXXXXXXXXXXXXXXX
   ------------++++--------++++++++++++++++++++++++--------------------++++
F-2

   XXXXxxxxxxxx
   ------------++++++++++++++++++++++++++++++++++++
M-1

       xxxx
   ++++----++++++++++++--------++++++++++++++++++++
M-2                    XXXXXXXX

   I---I---I---I---I-------I---I---I-------I---I---I---I-------I---I------I
   0   2   4   6   8       12  14  16      20  22  24  26      30  32     36

   WEEKS OF AGE
```

Figure 10.1. Overall hand preferences for the "extension" and "contact" components of manual activities. Signs above each axis indicate a right hand preference. Signs below each axis indicate a left hand preference.
x = $p<.1$; X = $p<.05$; + = no preference.

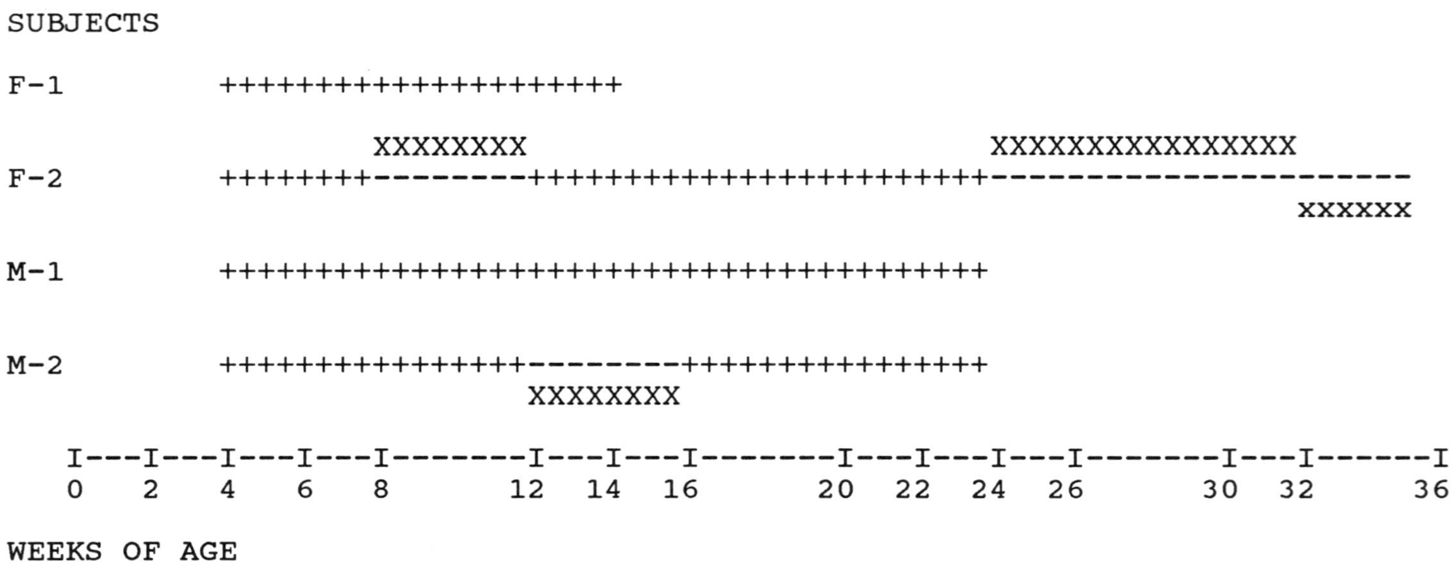

Figure 10.2. Overall hand preferences for the bringing to mouth component of manual activities. Signs above each axis indicate a right hand preference. Signs below each axis indicate a left hand preference.
x = p<.1 ; X = p<.05 ; + = no preference.

Subject	Pattern	Grip	0–2	2–4	4–6	6–8	8–12	12–16	16–24	24–32	32–36
F-2	APPROACH		R+	---	---	---					
	HANDLE	touch	---	R+	---	---	---	---	---	---	---
	HANDLE	palmar	R*	R*	R*	---	---	R+	---	---	---
	HANDLE	precision				---	---	---	---	---	---
	Direct To Mouth	palmar				R+	---	---	---	---	---
	Direct To Mouth	precision					---	---	---	L+	---
M-1	APPROACH		R+								
	HANDLE	touch	R*	R*	R*			L+	---		
	HANDLE	palmar		---	---	---	---	---	---		
	HANDLE	precision					---	---	---		
	Direct To Mouth	palmar			---	---	---	---	---		
	Direct To Mouth	precision				---	---	---	---		
M-2	APPROACH		---	---							
	HANDLE	touch		R+			---	---	---		
	HANDLE	palmar		---	---	---	---	L+	---		
	HANDLE	precision					---	---	---		
	Direct To Mouth	palmar			---	---	---	---	---		
	Direct To Mouth	precision					L+	---	---		
F-1	APPROACH		---	---	---						
	HANDLE	touch	R*	R*	---	R*	---	---			
	HANDLE	palmar		---	---	---	---	---			
	HANDLE	precision					---	R+			
	Direct To Mouth	palmar			---	---	---	---			
	Direct To Mouth	precision				R+	R+				

AGE IN WEEKS: 0 2 4 6 8 12 14 16 24 32 36

Figure 10.3. Hand preference per subject and per age class as a function of the three main patterns of hand movement and of the distinction between simple touch, palmar and precision grips. Hand preferences were calculated using a two-tailed *Chi square* (left versus right). R = right; L = left; + = $p<.1$; * = $p<.05$; --- indicates the absence of preference. *Chi square* was not computed if $N<10$.

Hand Preferences

Lateralizationswere determined for each type of hand movement as defined in Table 10.2. They were calculated using a two-tailed Chi-Square test (number left vs. number right hand usage) per subject and per age interval.

Extensions and Contacts

The hand used for the "contacts" is strictly dependent on the hand used for the "extensions". An overall lateralization, depicted on Figure 10.1, was thus determined for these two actions.

Manual lateralization has appeared precociously because three monkeys showed a hand preference (or at least a tendency) as early as the first two weeks of life. For the fourth monkey (M-2), preferences appeared for the first time at weeks 2 - 4. Initial preferences were all in favor of the right hand, so that a right hand inter-subjects consistency emerged at weeks 2-4. Further, hand lateralization consisted either of a stable state of ambidextry (e.g., M-1) or of a transient period of left (i.e., M-2) or of right hand preference (e.g., F-2) inserted within periods of no preference.

Bimanual Manipulations

Information on hand laterality for the "bimanual manipulation" could be derived from the number of transfers of the object from one hand to the other. First transfer actions were observed either at weeks 6-8 (F-1) or at weeks 8-12 (F-2, M-1, M-2). A total of 44 instances of transfer were observed. Unfortunately, the individual number of transfers for each subject were too small to conduct statistical analyses of lateral preferences.

Bring to Mouth

The search for manual preference was extended to the "bringing to mouth" component of manual behaviors regardless of the kind of grip and of the hand used in previous steps. The lack of significant findings concerning this step appear in Figure 10.2. Only two subjects (F-2 and M-2) demonstrated a transient lateral preference, but these preferences appeared at different ages and they also diverged with respect to their bias (left or right).

The APPROACH, HANDLE and DIRECT TO MOUTH patterns represented the largest proportion of manual behaviors throughout every observational period. In Figure 10.3, these data are shown in terms of the kinds of grips (e.g., touch, palmar prehension, and precision thumb-index prehension) employed in the execution of each category of behavior.

Only a minority of *Chi squares* revealed a significant preference for one side. Nevertheless, it appears that the right hand was generally preferred to the left during the first attempts in performing certain patterns. For example, F-2 and M-1 had a tendency to use their right hand preferentially

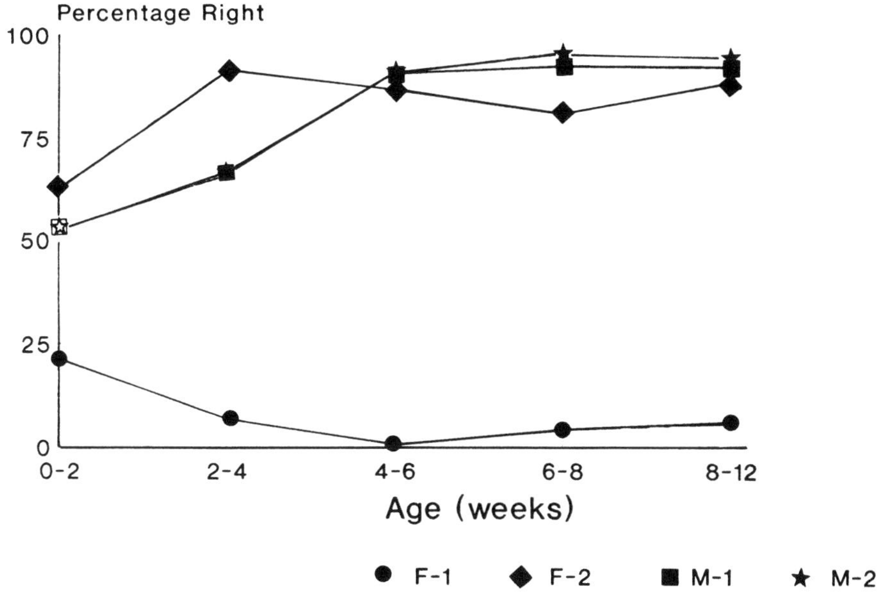

Figure 10.4. Infant's percentage of right breast sucking as a function of age. Black symbols indicate a significant preference (two tailed *Chi Square* number right vs. number left, $p < .05$). White symbols indicate non-significance.

for the APPROACHES observed from birth to week 2. A hand preference was observed for the first HANDLE patterns involving either a simple touch (M-1, M-2 et F-1) or a palmar prehension (F-2). Moreover, F-2 expressed a tendency to use the right hand for the first palmar DIRECT TO MOUTH patterns. Two subjects demonstrated a left (M-2) or a right (F-1) hand preference for the first DIRECT TO MOUTH patterns employing a precision grip (opposition between thumb and index).

Complementary Observations

Neonate baboons spend long periods of time in physical contact with their mothers and most of their manipulations are spatially constrained by the mother's body (e.g., Altmann, 1980). One can wonder whether such an interrelationship may influence individual manual preferences. This question was examined by studying two asymmetrical behaviors which might be considered as indicators of infants' asymmetrical positioning in the mothers'

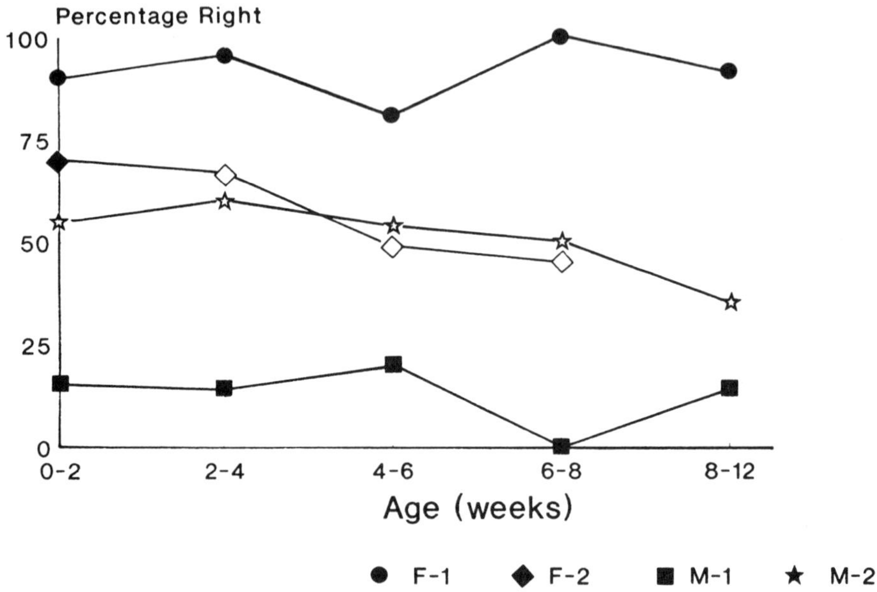

Figure 10.5. Mother's percentage of right hand uses to carry its infant as a function of infant's age. Black symbols indicate a significant preference (two tailed *Chi Square* number left vs. number right hand usage, $p<.05$). White symbols show non-significance.

arms. The first behavior examined was breast feeding by infants. The side of nursing (left or right) was recorded each time one breast was taken in mouth when the mother was in a sitting position. The second asymmetrical behavior recorded was the mothers' infant carrying, namely when the mother changed her posture from a sitting stance to a three-legged position. The hand used (left or right) by the mother to carry the infant was noted each time this behavior appeared.

Lateral preferences both for the nipple sucking and for the infant carrying were computed using a two tailed *Chi Square* goodness-of-fit (number left vs. number right; $p<.05$). Figure 10.4 indicates the percentage of right nipple sucking for each baboon as a function of age. Preferences for one breast were significant at weeks 0-2 for subjects F-1 and F-2. After week 2, all subjects exhibited a significant lateral preference either in favor of the left nipple (F-1) or in favor of the right nipple (F-2, M-1 and M-2).

With respect to the carrying behavior, the percentage of right and left preferences can be seen in Figure 10.5. Only the mothers of F-1 and M-1 exhibited significant and consistent hand preferences. F-1's mother displayed a right hand preference while M-1's mother displayed a left hand preference. These preferences of the two mothers were consistent with those previously recorded for them during spontaneous reaching (Vauclair & Fagot, 1987b). Thus, mother preferences for both side of nursing and for infant carrying were stable (i.e., did not shift with the development of the infant) by contrast to the development of infants' manual preferences.

Discussion

The present study investigated the early development of object manipulation in infant baboons. For that purpose, spontaneous manual behaviors of four mother-reared neonates have been videotaped from birth to three, six, and nine months. The coding scheme was initially designed to describe the kinetics of hand movements in relation to object manipulation. Data obtained by this analysis demonstrated a rapid improvement in manual sequences and a high level of achievement in sensorimotor skills as early as the end of the second month. For example, the first arm "extensions" and manual "contacts" employing a simple touch or a palmar prehension of the goal object occurred before 15 days of age. At weeks six to eight, young baboons were able to bring an object to their mouth either directly or indirectly after an intermediate bimanual manipulation. This chronology is in agreement with longitudinal data previously gathered on this species (Flament, 1965).

Another objective of this research was to study manual lateralization during early development. Several remarks are in order here. First, manual preferences were observed very early after birth because three baboons expressed manual lateralization (or at least a lateral tendency) before the end of the second week of age. The fourth baboon demonstrated a hand preference between the second and fourth weeks of age. Second, hand preference occurred mainly at the onset of certain behavioral patterns such as the HANDLE employing a simple touch (e.g., M-1) or DIRECT TO MOUTH movements employing a precision grip (e.g., F-1). Third, as it has been observed in a chimpanzee (Chorazyna, 1976), initial preferences were not necessarily consistent with later biases. Fourth, initial hand preferences were mostly right sided. It is obviously impossible to reject the hypothesis of a consistent subtle environmental artifact shaping an initial right hand preference. However, several arguments are against such an hypothesis. For example, the infant baboons lived in a naturalistic environment and never had direct physical contact with humans (e.g., to receive food directly by hand). A second argument is that the mothers' hand lateralizations were dissimilar (right sided for the mothers of F-1 and F-2 and left sided for the mothers of M-1 and M-2) to those observed in the infants. The examination of lateralization for infant carrying also demonstrated inconsistent hand biases within mothers. Finally, F-1 preferred the left breast whereas others displayed a preference for the right. It appears that our four baboons have been reared in very different environments. Consequently, their consistency in initial hand biases could be due to intrinsic factors. However, given that this investigation is based on a sample of only four subjects, the descriptive results must be viewed as preliminary and caution is advised in generalizing from these data.

The data presented above raise several questions regarding the developmental characteristics of manual lateralization in nonhuman primates. A working hypothesis to consider here is that novelty factors are fundamental in the expression of hand laterality. The novelty factor has rarely been included in the analysis of data on hand preferences, even thought its importance has been demonstrated from a cognitive perspective (Goldberg & Costa, 1981) and with regard to the neurological basis of motoric voluntary behaviors (Paillard, 1986). Fagot and Vauclair (1991) have proposed that manual gestures which are distant from daily routine activities (i.e., novel) may show an asymmetric distribution of hand preference for the group. The present findings of inter-subject consistency in hand preference in early "contacts" (weeks 0-4) provide some support of this hypothesis.

From a human perspective, the question of early manual lateralization has led to an abundant literature but with some contradictory results. Whereas some contributors (Gesell & Ames, 1947) have found a left hand preference for the first reaching attempts, other researchers have obtained data in favor of a right hand preference (Von Hofsten, 1982). These contradictory results were interpreted in a review of this literature by Young, Segalowitz, Misek, Alp, and Boulet (1983). According to these authors, the left hand is preferred when manual activity is not clearly directed (e.g., passive holding). By contrast, the right hand is involved for more directed, target-related activities. Since we have only recorded instances of clearly goal-directed behaviors, our findings of an initial right hand lateralization in neonate baboons appear to fit with Young et al.'s interpretation. Further investigations with a larger number of subjects may provided further evidence of the initial right hand bias we have observed and thus establish its generality.

Connolly and Elliott (1972) have already noted that very few data were available concerning the development of hand function in nonhuman primates. This observation is still true today. Besides the preliminary data we have provided in this chapter, the methods we propose for coding the hand's kinetics and hand coordination may be of use in further investigations of these ontogenetic questions.

Acknowledgements

I would like to thank M. P. Nesa, M. A. Juin and N. Benghezal for their assistance in the data collection. I am also grateful to Dr. B. Pallaud for providing facilities to conduct the observations. Special thanks to Dr. J. Vauclair for his help in all phases of this work.

References

Altmann, J. (1974). Observational study of behavior: Sampling methods. *Behaviour, 49*, 227-267.

Altmann, J. (1980). *Baboons mother and infant*. Cambridge: Harvard University Press.

Bard, K. A., Hopkins, W. D., & Fort, C. L. (1990). Lateral bias in infant chimpanzees *(Pan troglodytes)*. *Journal of Comparative Psychology, 104*, 309-321.

Brésard, B., & Bresson, F. (1983). Handedness in *Pongo pygmaeus* and *Pan troglodytes*. *Journal of Human Evolution, 12*, 659-666.

Brooker, R. J., Lehman, R. A. W., Heinbuch, R. C., & Kidd, K. K. (1981). Hand usage in a colony of bonnett monkeys *(Macaca radiata)*. *Behavior Genetics, 11*, 49-56.

Chalmers, N. R. (1980). Developmental relationships among social manipulatory, postural and locomotor behaviours in olive baboons *(Papio anubis)*. *Behaviour, 74*, 21-37.

Chorazyna, H. (1976). Shifts in laterality in a baby chimpanzee. *Neuropsychologia, 14*, 381-384.

Cohen, J. (1960). A coefficient of agreement for nominal scales. *Educational and Psychological Measures, 20*, 37-46.

Connolly, K., & Elliott, J. (1972). The evolution and ontogeny of hand function. In Blurton Jones, N. (Ed.) *Ethological studies of child behavior* (pp. 329-383). Cambridge, MA: Cambridge University Press.

Cunningham, D., Forsythe, C., & Ward, J. P. (1989). A report of behavioral lateralization in an infant orang-utan *(Pongo pygmaeus)*. *Primates, 30*, 249-253.

Fagot, J., & Vauclair, J. (1991). Manual laterality in nonhuman primates: A distinction between handedness and manual specialization. *Psychological Bulletin, 109*, 76-89.

Flament, F. (1965). Quelques aspects du développement psychomoteur chez le babouin. *Cahiers de Psychologie, 8*, 75-78.

Forsythe, C., & Ward, J. P. (1988). Black lemur *(Lemur macaco)* hand preference in food reaching. *Primates, 29*, 369-374.

Gesell, A., & Ames, L. B. (1947). The development of handedness. *The Journal of Genetic Psychology, 70*, 155-175.

Goldberg, E., & Costa, L. (1981). Hemisphere differences in the acquisition and use of descriptive systems. *Brain and Language, 14*, 144-173.

Joubert, A., & Vauclair, J. (1986). Reaction to novel objects in a troop of Guinea baboons: Approach and manipulation. *Behaviour, 96*, 92-104.

Lehman, R. A. W. (1980). Distribution and changes in strength of hand preference of cynomolgus monkeys. *Brain, Behavior and Evolution, 17*, 209-217.

Liederman, J. (1983). Mechanisms underlying instability in the development of hand preference. In Young, G., Segalowitz, S. J., Corter, C. M., & Trehub, S. E. (Eds.). *Manual specialization and the developing brain* (pp. 71-92). New York: Academic Press.

MacNeilage, P. F., Studdert-Kennedy, M. G., & Lindblom, B. (1987). Primate handedness reconsidered. *Behavioral and Brain Sciences, 10*, 247-303.

Napier, J. R. (1961). Prehensibility and opposability in the hand of primates. *Symposia of the Zoological Society, London, 5*, 115-132.

Paillard, J. (1986). Development and acquisition of motor skills: A challenging prospect for neurosciences. In Whiting, H. T. A. & Wade, M. G. (Eds.), *Motor development in children: Aspects of coordination and control* (pp. 416-441). La Haye: M. Nijhoff.

Post, D. G. (1981). Activity patterns of yellow baboons (*Papio cynocephalus*) in the Amboseli National Park, Kenya. *Animal Behavior, 29*, 357-374.

Spinozzi, G., & Natale, F. (1986). The interaction between prehension and locomotion in macaque, gorilla and child cognitive development. In Else, J. G. & Lee, P. C. (Eds.). *Primate ontogeny, cognition and social behavior*, (pp. 155-160). Cambridge, MA: Cambridge University Press.

Vauclair, J., & Fagot, J. (1987a). Visually guided reaching in baboons. *Behavioral and Brain Sciences, 10*, 287.

Vauclair, J., & Fagot, J. (1987b). Spontaneous hand usage in a troop of baboons. *Cortex, 23*, 265-274.

Von Hofsten, C. (1982). Eye-hand coordination in newborns. *Developmental Psychology, 18*, 450-461.

Warren, J. M. (1980). Handedness and laterality in humans and other animals. *Physiological Psychology, 8*, 351-359.

Young, G., Segalowitz, S. J., Corter, C. M., & Trehub, S. E. (1983). Manual specialization and the developing brain: An overview. In Young, G., Segalowitz, S. J., Corter, C. M., & Trehub, S.E. (Eds.), *Manual specialization and the developing brain* (pp. 3-12). New York: Academic Press.

Young, G., Segalowitz, S. J., Misek, P., Alp, I. E., & Boulet, R. (1983). Review of manual specialization research. In Young, G., Segalowitz, S. J., Corter, C. M., & Trehub, S. E. (Eds.), *Manual specialization and the developing brain* (pp. 13-32). New York: Academic Press.

11
The Ontogeny of Lateralized Behavior in Nonhuman Primates with Special Reference to Chimpanzees (*Pan troglodytes*)

William D. Hopkins and Kim A. Bard

The development of hand preference and other lateralized behaviors has received a considerable amount of attention by a number of investigators (Hahn, 1987; Molfese & Segalowitz, 1988). Understanding the ontogeny of hand preference is important from a both a developmental and a comparative perspective because of the differing theoretical perspectives utilized in explaining "handedness" in its adult form. For example, in the review articles on nonhuman primate handedness by MacNeilage, Studdert-Kennedy, and Lindblom (1987), as well as Fagot and Vauclair (1991), the testing of immature or younger subjects are cited as possible factors influencing the expression of hand preference in nonhuman primates. This is particularly important because recent data on reaching in human infants indicates that preferences can switch from an initial left hand preference to a right hand preference with increasing age (Harris & Carlson, 1988; Young, Segalowitz, Corter, & Trehub, 1983). Thus, comparing hand preference distributions of infant and adult subjects of different species are important for understanding how asymmetries emerge in the course of development.

The most typical explanations for changes or stability in hand preference or other lateralized behaviors in humans are either: a) genetic, b) maturational, c) in a general sense environmental, if culture is included as an environmental stimulus which can serve to shape or modify behavioral development (see Witelson, 1987 for review). The genetic models (see Annett, 1985; Levy & Nagylaki, 1972 for reviews) of laterality suggest that hand preference of the offspring is influenced by the genetic makeup of the father and mother. There is some debate over the strength of each genetic model and how the phenotype is affected by the environmental input, but both can account for a variety of data which currently exists in the literature. The maturational explanation argues that the two hemispheres mature at different rates, even in utero, and that this different rate of development influences the initial motoric responses of neonates (Lenneberg, 1967; Corballis & Morgan, 1978). Finally, the environmental model argues that cultural or experiential factors influence preference. In this case, the two hemispheres start equipotentially where upon stimulation or reinforcement for certain preference patterns serves to enhance one pattern over another.

Until recently, hand preferences in nonhuman primates have largely been explained by use of the environmental model. That is to say, hand preferences were characterized as being idiosyncratic, inconsistent and modifiable due to training (Warren, 1980). Even if one assumes that population asymmetries are not present in nonhuman primates or that they are explained by experience rather than etiology or genetics, it is still important to understand how those preferences develop and how they interact with the biology of different organisms. However, as the many chapters of this volume attest, it is clear that hand preference can be found in a variety of nonhuman primate species using a variety of different measures. But how these lateralized behaviors developed across the lifetime of the organism has seldom been studied. Few studies have specifically examined hand preferences in nonhuman primates from different age groups, particularly among great apes. This is not to suggest that younger individuals have never been tested, but often these subjects are subsumed under the entire group of scores, thus age is disregarded as an important factor. This is unfortunate, because in human children the data clearly suggest that strength of preference, as well as direction of preference, can change from birth to childhood (Young et al., 1983).

In addition, data examining genetic aspects of hand (or paw) preference in nonhuman animals have yielded conflicting results. For example, Collins (1985) has found no evidence indicating that direction of paw preference in mice is influenced by the paw preferences in either the mother or father, but the strength of paw preference does appear to be influenced by heritability factors. In nonhuman primates, the data are mixed. Some researchers have shown that offspring will often show the same hand preference as their mothers (see Brinkman, 1984; Shafer, 1987) whereas others have failed to find significant relationships (Brooker, Lehman, Heinbuch, & Kidd, 1981; Vauclair & Fagot, 1987). However, even in those cases where significant hand preference relationships are found between mothers and offspring, it does not necessarily mean that genetic factors are the sole explanation. It is possible, and equally important, that hand preferences are influenced by observational learning or other environmental factors.

The focus of this chapter is to summarize those studies which have examined the effects of age on the expression of hand preference. In addition, recent data examining the ontogeny of lateralized behaviors in infant chimpanzees will be presented. The relationship between the expression of lateralized behaviors and other behavioral measures will also be discussed.

Developmental Studies

Depicted in Table 11.1 is a listing of studies which have reported significant findings for age or have examined age as a possible factor affecting the expression of hand preference. Fagot and Vauclair (1991) have reported similar findings in their recent review of nonhuman primate handedness studies. The authors, genus studied, and general results are provided. The Table is further broken down into those studies which have been developmental contrasted with those studies which have accounted for age by use of cross-sectional analyses. Developmental studies are those where the same individuals have been studied for a specified period of time. Cross-sectional studies are those in which age served as an independent variable or grouping variable distinguishing different subjects within the population under investigation.

As can be seen, there are very few studies which have examined the ontogeny of hand preference in nonhuman primates. Moreover, the number of subjects is extremely limited within each of these studies. In general, the findings suggest a right hand preference emerging from birth through infancy.

The findings from cross-sectional analyses are somewhat mixed. Although most authors believe that strength of preference becomes stronger with age, there is some disagreement as to whether direction of preferences emerges across time. For example, Ward, Milliken, Dodson, Stafford, and Wallace (1990) reported stronger directional preferences (i.e., left-handed) in younger subjects, whereas Lehman (1978, 1980) failed to find any change in direction of hand preference as a function of age. The data from great apes are equally mixed. Heestand (1986) found no effect for age in the expression of lateralized leading limb preferences. In contrast, Shafer (1987) reported stronger overall right hand preferences in adolescent gorillas contrasted with adults and infants.

A possible explanation for these diverse findings may be the different types of methods employed, as well as the different behaviors being measured. For example, in the study by Vauclair and Fagot (1987) on baboons, the age analysis indicated a trend toward left-handedness in younger subjects and right-handedness for older subjects. However, younger subjects were more active than older subjects. Thus, an interaction may exists between the frequency of responses and age of subjects. An obvious solution to this problem would be to simply score the same number of behaviors for each subjects, independent of age. However, since time is continuous, a sufficiently long time may elapse before comparable data are collected between subjects of different ages.

Other possible explanations may be the species under investigation, as well as the type of measure and/or task being assessed. Of those studies from Table 11.1, very few have recorded identical behaviors between species.

Table 11.1. Cross-Sectional and Developmental Studies of Left Hand (LH) and Right Hand (RH) Preference in Nonhuman Primate Subject Samples (N)

Author(s)	Genus	N	Findings
Developmental Studies			
Bard, Hopkins, & Fort (1990)	*Pan*	12	RH preference in hand-to-mouth behaviors in infant chimpanzees. Stronger effect by 3 months of age.
Bresard & Bresson (1983)	*Pan* *Pongo*	1 1	Both subjects exhibit RH preference in manipulation tasks.
Chorazyna (1976)	*Pan*	1	Infant chimpanzees show generalized RH preference by 37 weeks of age.
Cunningham, Forsythe, & Ward (1989)	*Pongo*	1	RH preference for reaching and manipulation; LH preference for body and head touching.
Kellogg & Kellogg (1933)	*Pan*	1	RH preference for skilled actions. No significant preference in reaching.
Robinson (1979)	*Pan*	3	Subjects of 3 different ages tested for hand preference. 1 was LH, 1 ambidextrous and 1 RH.
Cross Sectional Studies			
Boesch (1991)	*Pan*	67	A significant RH preference for wadge dipping. Also, younger subjects showed greater prevalence of RH preferences than adults.
Brooker, Lehman, Heinbuch, & Kidd (1981)	*Macaca*	69	No population preference found. Age of subjects unrelated to direction or strength of hand preference.
Fagot & Vauclair (1988)	*Gorilla*	8	7 of 8 subjects exhibit LH preference for visual-motor alignment task. No effects for age reported.
Heestand (1986)	*Gorilla* *Pan* *Pongo*	29 20 8	Overall, population RH preference in leading quadrupedal locomotion. No effect for age reported, younger animals had stronger leading limb asymmetries contrasted with older animals.
Lehman (1978)	*Macaca*	171	No population hand preference found but strength of hand preference higher in older subjects.
Lehman (1980)	*Macaca*	58	No population hand preference found. Correlation between strength of preference and age reported.
Shafer (1987)	*Gorilla*	47	Averaged across behaviors, population asymmetry 24 subjects exhibiting RH preferences, 10 LH preferences. No age effects reported. Among adults, 12 were RH, 9 LH. For juveniles, 9 were RH, 1 LH. For infants, 3 were RH and none LH.
Stafford, Forsythe, & Ward (1990)	*Hylobates*	19	RH preference in older females compared to younger females.
Vauclair & Fagot (1987)	*Papio*	18	No age effects for unimanual activities. Greater percentage of LH subjects found among younger subjects compared to older for bimanual activities.
Ward, Milliken, Dodson, Stafford, & Wallace (1990)	*Lemur*	194	Stronger LH preferences in younger subjects, as well as in males.

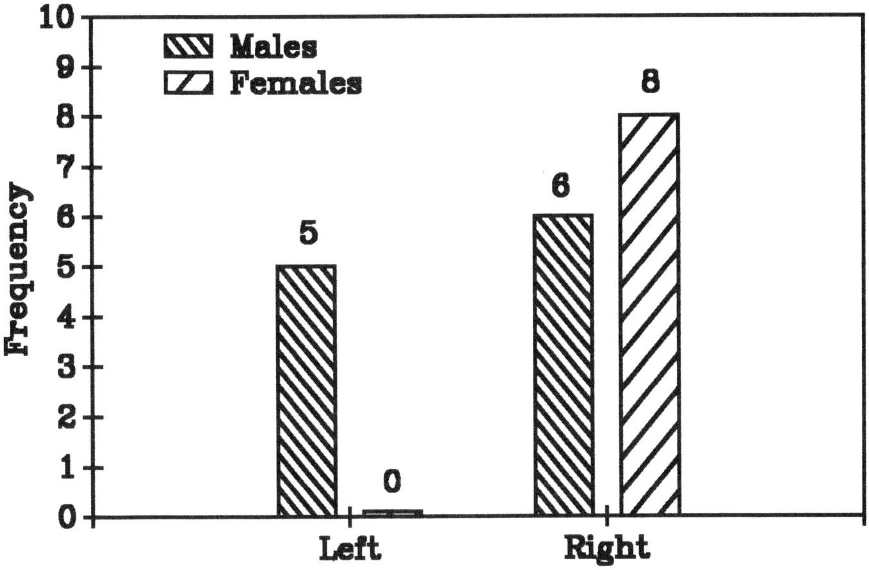

Figure 11.1. Distribution of right and left handed subjects in hand-to-mouth behavior as a function of sex.

Notable exceptions are the studies by Heestand (1986) and the three studies by Ward and her colleagues (Cunningham, Forsythe, & Ward, 1990; Stafford, Milliken, & Ward, 1990; Ward et al., 1990). Thus, comparing the effects of age across studies is very difficult at this point in time.

We would also point out that, from the standpoint of comparative data and developmental effects, different pathways to the expression of handedness may exist. In one case, different species may exhibit different directions of population hand preference using identical tasks. Alternatively, different species may exhibit the same direction of population hand preference but the developmental time course in the expression of hand preferences may differ among species. Finally, these two factors may interact, with some species showing a different direction and developmental time course of hand preference than other species. In short, the manner in which two species arrive at the same behavioral outcome may be derived from different developmental pathways.

The Ontogeny of Laterality in Infant Chimpanzees

Recently we reported a study on the presence, strength, and direction of

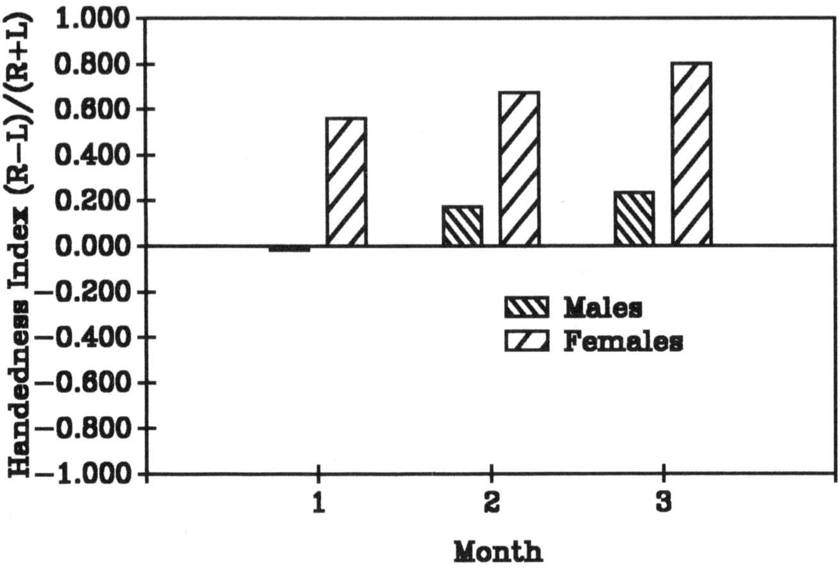

Figure 11.2. Average handedness index for months 1, 2, and 3 for males and females.

lateralized behaviors in infant nursery-reared chimpanzees (Bard, Hopkins, & Fort, 1990). Twelve nursery-reared chimpanzees, 7 males and 5 females, were assessed using reflexes and behaviors from the Neonatal Behavioral Assessment Scale (NBAS) from birth through 12 weeks of age (Brazelton, 1984). The lateral bias of two self-calming behaviors and two other behaviors were measured: hand-to-mouth, hand-to-hand, defensive grasps, and stepping reflex. Hand-to-mouth behavior was significantly lateralized for the entire sample. Eight out of the 10 chimpanzees who showed hand-to-mouth behaviors used their right hand. The degree of laterality in hand-to-mouth movements was inversely related to the degree of laterality in hand-to-hand movements. The degree of laterality for defensive grasps was positively related to the degree of laterality of the first step. The strength of lateralized behavior increased as the age of the chimpanzees increased. The results suggested that females demonstrated a greater degree and strength of laterality than males but a significant sex difference was not found.

With the collection of data from nine additional chimpanzee subjects, further evidence has accumulated supporting our early suggestions of a sex and age effect in the expression of hand-to-mouth behavior (Hopkins & Bard, 1991). Of the total 21 subjects tested, 14 display a right preference for hand-to-mouth behaviors. Of the 8 females tested, all of them showed a right hand preference. Of these 8, 6 of the subjects exhibited a significant lateral bias. For the males, of the total 12 subjects tested, 6 showed a right hand preference whereas 5 showed a left hand preference. Examining only those subjects with significant lateral biases, 3 males used their right hand whereas 2 used their left. The distribution of subjects with either a left or right hand preference are depicted in Figure 11.1.

Regarding developmental effects, a significant relation exists between the age of subjects and the direction and strength of lateralized hand-to-mouth behaviors. To determine this effect, a handedness index was calculated for each subject by subtracting the number of right hand responses minus the number of left hand responses and dividing by the total number of responses (R-L/R+L). The index results in values ranging from -1.0 to 1.0 with positive values representing right hand preferences and negative values representing left hand preferences. The handedness index was calculated for each subject at months 1, 2, and 3 of life. NBAS testing terminated after three months of age. Depicted in Figure 11.2 are the average handedness index values for males and females within months 1, 2, and 3. An analysis of variance revealed significant main effects for sex $F(1,17)=7.08, p<.02$ and month $F(2,34)=3.76$, $p<.03$. Irrespective of month, females evidenced significantly stronger right-handedness index values than males. Also, handedness index values increased linearly from months 1 to 3 with the strongest effects occurring in month 3.

Mechanisms of Laterality in Hand-to-Mouth Behavior

Our attention has recently focused on determining the mechanisms which are related to the expression of hand-to-mouth behaviors. Specifically, we have been interested in the relations between different behavioral cluster scores of the NBAS and lateralized hand-to-mouth behaviors. From the 28 behavioral items scored on the NBAS, 5 behavioral clusters scores are derived and are categorized as autonomic central nervous system stability (AUTO), motor performance (MOTOR), state regulation (REGULATE), orientation (ORIENT), and range in state (RANGE) (Lester, 1984). Our previous analyses have revealed several preliminary findings (Hopkins & Bard, 1991). First, with the exception of one of the behavioral items (the number of smiles) no sex differences were found in the remaining 27 behavioral items. In addition, there are no sex differences in the cluster scores derived from these behavioral items. Second, significant correlations exist between the z-score measures of laterality in hand-to-mouth behaviors and the cluster scores for range in state measured on day 2 (r = -.46) and day 30 (r = -.51). Finally, a significant negative correlation was found between the z-score measures of the two self-calming behaviors, hand-to-mouth and hand-to-hand (r = -.67).

These preliminary analyses indicated that one possible mechanism related to lateralized hand-to-mouth behavior is range in state, which is characterized as a general measure of arousal. Interestingly, there are no sex differences in the behavioral items comprising the cluster RANGE nor are there any significant sex differences in the RANGE cluster score. Thus, the correlation between lateralized hand-to-mouth behavior and range in state is not due to the overriding sex difference found in the direction of lateralized hand-to-mouth behaviors.

Extending our previous findings and to further examine the relation between the NBAS behavioral items and hand-to-mouth behaviors, we report here results from multiple analysis of variance performed on each behavioral item of the NBAS. The data were analyzed two ways. In one condition, the subjects were classified as having either a significant right hand preference (right) or a nonright hand preference (nonright). The nonright-handed subjects included those with nonsignificant hand preferences as assessed by binomial z-scores and subjects with significant left hand preferences. Right-handed subjects were those with significant right hand preferences in hand-to-mouth. In condition two, the subjects were dichotomized as either being left- or right-handed as determined by the sign of their binomial z-score. Left-handed subjects were those with negative values whereas right-handed subjects were those with positive z-scores regardless of their strength of laterality. These two different analyses were done to determine how hand preference classification affected the statistical findings. Although it could also be argued that three groups were needed, a left, right, and nonpreference group, but we had so few subjects with significant left-handed preferences that such a comparison would have been statistically impossible.

Of the 21 subjects tested on the NBAS, two subjects failed to exhibit hand-to-mouth behavior and therefore were excluded from the analysis. Furthermore, a complete set of NBAS data were absent in three additional subjects. Thus, for the analyses to be described, a total of sixteen subjects were included, 8 males and 8 females. For the analyses performed, handedness served as a grouping factor and day (Day 2 vs Day 30) served as repeated measure. A t-test was also performed on birthweight as a function of handedness classification to determine whether this factor could account for differences in hand-to-mouth behaviors.

The results of the first analysis comparing right-handed subjects with nonright-handed subjects revealed a borderline significant main effect for smiling $F(1,14)=4.42, p<.06$, a borderline interaction between day and group for irritability $F(1,14)=4.48$, $p<.06$, and a significant interaction between group and day for lability of state $F(1,14)=4.95, p<.05$. These data can be seen in Table 11.2. For irritability, nonright-handed subjects became less irritable from Day 2 to Day 30 while right-handed subjects were more irritable from Day 2 to Day 30. Regarding lability of state, a similar finding emerged. Across time, right-handed subjects showed more changes in state whereas nonright-handed subjects showed fewer state changes. No significant

Table 11.2. Average Score for Each Behavioral Item on the NBAS for Subjects Classified With Right and Non-Right Hand Preferences in Hand-to-Mouth

Behavior	Nonright Day 2	Nonright Day 30	Right Day 2	Right Day 30
Smiles	1.25	3.13	0.63	4.50 +
Autonomic				
Tremor	3.38	3.50	2.13	3.75
Startle	3.00	2.75	1.75	2.13
Regulation				
Cuddliness	8.88	8.88	8.25	8.38
Self-Quiet	7.13	7.75	7.86	6.29
Hand-to-Mouth	5.25	8.50	7.13	7.75
Motor				
Tone	6.13	6.13	5.50	6.25
Maturity	7.25	8.75	6.88	8.38
Pull-to-Sit	8.00	8.75	8.25	9.00
Defensive	6.50	7.75	5.75	7.63
Activity	4.50	4.88	4.00	5.13
Range				
Peak	4.75	4.25	4.75	5.13
Build-up	2.25	1.38	2.50	1.89
Irritability	3.50	3.00	2.63	4.13 +
Lability of State	2.75	2.13	2.38	3.13 **

** $p<.03$, * $p<.05$, + $p<.06$

difference was found for birthweight.

The second analysis revealed similar findings. These data are shown in Table 11.3. Right-handed subjects smiled more than left-handed subjects $F(1,14)=4.82$, $p<.05$. There were significant interactions between day and group for tone $F(1,14)=10.50$, $p<.007$ and irritability $F(1,14)=4.60$, $p<.05$. Right-handed subjects exhibited increased tone from Day 2 to Day 30 whereas left-handed subjects show decreased tone from Day 2 to Day 30. A similar pattern was found in irritability. Right-handed subjects exhibited slightly increased irritability from Day 2 to Day 30 whereas left-handed subjects exhibited considerable irritability at Day 2 with decreased irritability

Table 11.3. Average Score for Each Behavioral Item on the NBAS for Subjects Classified With Left- and Right-Hand Preferences in Hand-to-Mouth

	Left		Right	
Behavior	Day 2	Day 30	Day 2	Day 30
Smiles	0.25	2.00	1.17	5.75 *
Autonomic				
Tremor	3.25	2.58	3.75	3.58
Startle	2.75	2.13	2.00	2.33
Regulation				
Cuddliness	9.00	8.42	9.00	8.50
Self-Quiet	6.00	8.00	7.25	7.00
Hand-to-Mouth	6.50	6.08	8.00	8.17
Motor				
Tone	6.75	6.00	5.50	6.25 **
Maturity	7.25	7.00	9.00	6.42
Pull-to-Sit	8.25	8.08	8.75	8.92
Defensive	6.50	6.00	8.00	7.58
Activity	5.00	4.00	4.75	5.48
Range				
Peak	5.00	4.67	4.50	4.75
Build-up	3.00	2.17	1.75	1.58
Irritability	4.50	2.58	3.25	3.67 *
State Change	3.00	2.42	2.25	2.75

** $p<.03$, * $p<.05$, + $p<.06$

by Day 30. Finally, no significant difference was found in birthweight.

Discussion

Developmental aspects of asymmetric behaviors were documented in chimpanzees as evidenced by consistent relationships between neonatal behaviors and laterality in hand-to-mouth self calming behavior at 3 months of age. Whether the subjects laterality is categorized by a more rigid criterion (all significant right vs nonright) or a more intuitive criterion (positive versus negative z-scores), the results are similar. The results of the two analyses, taken together, indicate that smiling and irritability were the two most consistent measures differentiating right- and left-handed subjects. In our previous analyses, a sex difference was found in smiling, with females showing more smiles than males. Given the higher distribution of right-handed females (see Figure 11.1), the findings of different frequencies in smiles was not surprising. The differences in irritability cannot be attributed to sex and therefore reflect a different pattern among right-handed subjects and nonright-handed or left-handed subjects. Most likely, since right-handed subjects are more irritable at day 30 but not day 2, they exhibit this behavior in greater frequency and may develop a more preferred pattern of hand-to-mouth behaviors. Why they choose the right hand remains unclear.

There are several plausible explanations for the right hand preferences but they remain speculative and clearly warrant further examination. One explanation may be maturation. Some evidence exists that the left hemisphere matures more slowly in males than females. Given that more males show nonright-handed preferences in hand-to-mouth, a possible explanation is that the left hemisphere is not mature enough to regulate or control this behavior. However, neither right- or left-handed subjects, nor males and females, significantly differ in birthweight, a rather crude measure of maturation. Thus, we do not feel that maturation is the most plausible explanation.

A second explanation may be development. The analyses indicated that left- and right-handed subjects differed in irritability and muscle tone. If right-handed subjects are more irritable and therefore display more hand-to-mouth behavior, then there may be greater stimulation of the right hand (i.e, left hemisphere). However, we failed to a find a significant correlation between irritability and hand-to-mouth behavior, a finding that would be predicted if the two variables were interrelated. Notwithstanding, it appears that the ability and developmental time course of hand-to-mouth behaviors are different for males and females. It should also be pointed out that no differences were found in subjects' ability to perform the hand-to-mouth behavior as a function of either sex or handedness classification. Thus, when observed, all subjects appeared equally facile at performing this behavior.

A third explanation may be heritability. Among the chimpanzees studied thus far, several of the infants are either full or half siblings. For example, two pairs of females are full siblings. For one pair, the binomial z-scores were 3.74 and 3.71, respectively. For the other pair of full siblings, the binomial z-scores were 4.43 and 4.01. It should be noted that all four of these subjects have the same mother. Among the males, two are full siblings. One of these subjects failed to show the behavior and the other was nonsignificantly left-handed (z = -.65). Finally, one pair of half sibling exists between males and females. One subject had a significant right hand preference and one was a nonsignificant right-handed subject.

Previous analyses, focusing on the relationship between cluster scores of the NBAS and laterality in hand-to-mouth behavior, revealed that arousal at Day 2 and Day 30 was linked to laterality at 3 months of age (Hopkins & Bard, 1991). The current findings concerning the 28 items of NBAS pinpoint irritability, tonus, and smiles as behaviors whose developmental course differs between right- and left-handed subjects. Interestingly, Korner (1974), discussing sex differences in the human neonate, suggests that males and females "discharge neural energy potential" via different mechanisms. Females exhibit reflex smiles and rhythmic mouthing. Males exhibit greater muscular strength, irritability, and whole body activation (e.g, startles). Korner suggests that the underlying mechanisms responsible for these sex differences are hormonal, which differentially sensitizes male and female central nervous systems.

There also appears to be a relationship between the time of separation of infants from mothers and the frequency of hand-to-mouth behaviors. Four of the 21 subjects tested were not separated from the mothers at birth. These subjects stayed with their mothers between 4 to 20 days prior to entering the nursery. Two of these 4 subjects failed to show the behavior whereas the remaining two exhibited the behavior a total of 4 times. Although it could be concluded that early separation is critical to the expression of hand-to-mouth behaviors, it is unclear whether these subjects are less stressed, aroused, or irritable than the remaining subjects. It is our belief that these subjects are so stressed and tense that they cannot exhibit the behavior. They often will hold their arms rigidly, fists tightly clenched, and simply not move. Thus, as indicative of a lack of hand-to-mouth behaviors, staying with less than adequate mothers appears to have adverse affects on the subjects ability to regulate their state.

Finally, it should be pointed out that these data reflect findings on chimpanzees reared in a nursery. Many chimpanzees stay with their mothers and hand-to-mouth behavior is not readily apparent. Thus, whether this asymmetry is specific to some aspect of nursery rearing remains unclear. That mother-reared infants do not exhibit the behavior should not be interpreted to mean that they are not lateralized. To the contrary, it is equally likely that this asymmetry may manifest itself in different behavioral patterns in mother-reared chimpanzees. It is also possible that the asymmetry we observe in the nursery reared chimpanzees may not be a motor asymmetry but rather an asymmetry in "emotion" or affective state. What we are observing in the hand-to-mouth behavior is a self calming behavioral response to that emotional or affective arousal. In the nursery setting, the infant must moderate the arousal since the mother is not available. In mother-reared infants, the mother provides moderation of the infant's arousal, and the behavior exhibited by the infant's hands is clinging to the mother. In mother-reared chimpanzees, the hands are not free as often to make the hand-to-mouth behavior observed in the nursery. Thus, the asymmetry in emotional or affective arousal may exist at a population level in both groups of chimpanzees but the individual responsible for its moderation differs, as does the expression of the behavior.

In conclusion, the findings indicate that a population right hand asymmetry in hand-to-mouth behaviors exists in nursery-reared chimpanzees. This effect is largely due to a sex difference, with females showing a higher incidence of right-handedness than males. In both males and females, the strength of hand-to-mouth preferences are strongest by three months of age but females, in months 1, 2, and 3, show stronger patterns of lateralized hand-to-mouth than males. The lateralized hand-to-mouth behaviors in chimpanzees appear to be linked to aspects of arousal or other temperamental factors related to range of behavioral state as measured by the Neonatal Behavioral Assessment Scale. It is important to note that range of behavioral state is unrelated to the sex of the chimpanzees or the chimpanzees ability to perform hand-to-mouth behaviors. Thus, it is the relation between arousal and laterality that accounts for the largest degree of variability in the data.

Why the right hand contrasted with the left hand remains unclear. A possible explanation may be that the right and left hemispheres differ in the degree of neurotransmitter present when subjects are in these aroused behavior states. This hyper- or hypo-activation of each cerebral hemisphere may then manifest itself in the hand-to-mouth behaviors we are observing in these chimpanzees. This activation model is similar to that which has been described for humans with a variety of different psychiatric as well as neuropsychological disorders (Tucker & Williamson, 1984). Future research should enable us to pinpoint those neurotransmitter or physiological factors which are related to lateralized self-calming behaviors.

Acknowledgements

Support for this research was provided by NICHD grant RR-00615 and HD-06016 to the Yerkes Regional Primate Research Center. Additional support was provided by grants RR-06158 to K. A. Bard, NICHD Research Fellowship HD-07105 to K. A. Bard, NICHD Intramural Research Program Funds to the Laboratory of Comparative Ethology, and RR-03591 to R. B. Swenson. We gratefully acknowledge the professional contributions of Carolyn L. Fort, Dr. Duane M. Rumbaugh, and Dr. Stephen J. Suomi during the course of this research.

References

Allen, M. (1983). Models of hemispheric specialization. *Psychological Bulletin*, *93*, 73-104.

Annett, M. (1985). *Left, right, hand and brain: The right shift theory*. New York: Academic Press.

Bard, K. A., Hopkins, W. D., & Fort, C. (1990). Lateral bias in infant chimpanzees (*Pan troglodytes*), *Journal of Comparative Psychology, 104*, 309-321.

Boesch, C. (1991). Handedness in wild chimpanzees (*Pan troglodytes*). *International Journal of Primatology, 6*, 541-558.

Brazelton, T. B. (1984). *Neonatal behavioral assessment scale*. Philadelphia: Spastics International Medical Publications.

Bresard, B., & Bresson, F. (1983). Handedness in *Pongo pygmaeus* and *Pan troglodytes*. *Journal of Human Evolution, 12*, 659-666.

Brinkman, C. (1984). Determinants of hand preference in *Macaca fascicularis*. *International Journal of Primatology, 5*, 325.

Brooker, R. J., Lehman, R. A. W., Heinbuch, R. C., & Kidd, K. K. (1981). Hand usage in a colony of bonnett monkeys (*Macaca radiata*). *Behavior Genetics, 11*, 49-56.

Chorazyna, H. (1976). Shifts in laterality in a baby chimpanzee. *Neuropsychologia, 14*, 381-384.

Collins, R. L. (1985). On the inheritance of direction and degree of asymmetry. In S. Glick (Ed.), *Cerebral lateralization in nonhuman species* (pp.41-71). Orlando: Academic Press.

Corballis, M. C., & Morgan, M. J. (1978). On the biological basis of human laterality: I. Evidence for a maturational left-right gradient. *The Behavioral and Brain Sciences, 2*, 261-336.

Cunningham, D., Forsythe, C., & Ward, J. P. (1989). A report on behavioral lateralization in an infant orang-utan (*Pongo pygmaeus*). *Primates, 30*, 249-253.

Denenberg, V. H. (1981). Hemispheric laterality in animals and the effects of early experience. *Behavioral and Brain Sciences, 4*, 1-50.

Fagot, J., & Vauclair, J. (1988). Handedness and bimanual coordination in the lowland gorilla. *Brain, Behavior and Evolution, 32*, 89-95.

Fagot, J., & Vauclair, J. (1991). Manual laterality in nonhuman primates: A distinction between handedness and manual specialization. *Psychological Bulletin, 109*, 76-89.

Geschwind, N., & Galaburda, A. M. (1985). Cerebral lateralization: Biological mechanisms, associations and pathologies: I. A hypothesis and a program for research. *Archives of Neurology, 42*, 428-459.

Hahn, W. K. (1987). Cerebral lateralization of function: From infancy through childhood. *Psychological Bulletin, 101*, 376-392.

Harris, L. J., & Carlson, D. F. (1988). Hand preference in visually guided reaching in human infants and adults. *Behavioral and Brain Sciences, 11*, 726-727.

Heestand, J. E. (1986). *Behavioral lateralization in four species of apes?* Unpublished Doctoral Dissertation, University of Washington, Seattle, WA.

Hopkins, W. D., & Bard, K. A. (1991). Hemispheric specialization in infant chimpanzees: Evidence for a relation between gender and arousal. Paper submitted for publication.

Kellogg, W. N., & Kellogg, L. N. (1933). *The ape and the child: A study of environmental influence upon early behavior*. New York: Whittlesey House.

Korner, A. F. (1974). The effect of infant's state, level of arousal, sex, and ontogenetic stage on the caregiver. In M. Lewis & L. A. Rosenblum (Eds.), *The effects of the infant on its caregiver* (pp.105-121). New York: Wiley & Sons.

Lehman, R. A. W. (1978). The handedness of rhesus monkeys. I: Distribution. *Neuropsychologia, 16*, 33-42.

Lehman, R. A. W. (1980). The handedness of rhesus monkeys - III. Consistency within and across activities. *Cortex, 16*, 197-204.

Lenneberg, E. H. (1967). *Biological foundations of language*. New York: Wiley.

Lester, B. M. (1984). Data analysis and prediction. In T. B. Brazelton (Ed.), *Neonatal behavioral assessment scale* (pp.85-96). Philadelphia: Spastics International Medical Publications.

Levy, J., & Nagylaki, T. (1972). A model for the genetics of handedness. *Genetics, 72*, 117-128.

MacNeilage, P. F., Studdert-Kennedy, M. G., & Lindblom, B. (1987). Primate handedness reconsidered. *Behavioral and Brain Sciences, 10*, 247-301.

Molfese, D. L., & Segalowitz, S. J. (1988). *Brain lateralization in children: Developmental implications*. New York: Guilford Press.

Robinson, C. E. (1979). *The development of hand preference in children and young chimpanzees*. Unpublished doctoral dissertation, University of Nevada, Reno, Nevada.

Shafer, D. D. (1987). *Patterns of hand preference among captive gorillas*. Unpublished Masters Thesis, San Francisco State University, San Francisco, CA.

Stafford, D. K., Milliken, G. W., & Ward, J. P. (1990). Lateral bias in feeding and brachiation in *Hylobates*. *Primates, 31*, 407-414.

Tucker, D. M., & Williamson, P. A. (1984). Asymmetric neural control systems in human self-regulation. *Psychological Review, 91*, 185-215.

Vauclair, J., & Fagot, J. (1987). Spontaneous hand usage and handedness in a troop of baboons. *Cortex, 23*, 265-274.

Ward, J. P., Milliken, G. W., Dodson, D. L., Stafford, D. K., & Wallace, M. (1990). Handedness as a function of sex and age in a large population of *Lemur*. *Journal of Comparative Psychology, 104*, 167-173.

Warren, J. M. (1980). Handedness and laterality in humans and other animals. *Physiological Psychology, 8*, 351-359.

Witelson, S. F. (1987). Neurobiological aspects of language in children. *Child Development, 58*, 653-688.

Young, G., Segalowitz, S. J., Corter, C. M., & Trehub, S. E. (1983). *Manual specialization and the developing brain*. New York: Academic Press.

12

Patterns of Handedness: Comparative Study of Nursery School Children and Captive Gorillas

Dulce D. Shafer

Humans seem drawn to compare themselves with other primates in a search for similarities and differences. It is accepted that primates have certain anatomical characteristics in common but also are unique in species-specific ways. A biased right hand preference has been considered a uniquely human trait. The laterality of nonhuman primates has been studied mostly by means of food rewards as a consistent way to elicit hand preference behaviors. Until recently, no bias has been found between right and left hand behaviors in a nonhuman primate species.

An excellent review of primate laterality studies by MacNeilage, Studdert-Kennedy, and Lindblom (1987) showed hand-use asymmetries for some tasks in some studies and no asymmetry in other studies. Okada (1980) suggested that hand preference in monkeys seemed to be task-specific and the strength of the preference was possibly influenced by experience as well as by the difficulty of the task. Beginning with bushbabies as subjects (Sanford, Guin, & Ward, 1984), Jeannette Ward and her graduate students have written a series of papers showing left hand preference in various species of prosimians. The most recent study is the 1990 work on lemurs by Ward, Milliken, Dodson, Stafford, and Wallace. These papers presented the first evidence of nonhuman primate laterality bias. In 1986, Heestand reported a significant right preference in the leading limb for locomotion in 19 of 29 gorillas, 13 of 20 chimpanzees, 7 of 8 orangutans, and 9 of 13 siamangs. In four other behavioral categories — sitting, climbing, and eating with hands or fingers, no significant limb preference was found.

Individual laterality patterns were observed by the author while participating in a study of gorilla play behavior at the San Francisco Zoo (Goerke, Fleming & Creel, 1987). Because of anatomical similarities to humans, it was felt that gorillas as a species might have approximately the same pattern of handedness behaviors as humans between the ages of three and five years, with laterality not as strongly manifested as in adult humans. A search of the literature revealed very little about hand preference of very young children. There was some controversy in the literature as to how laterality was to be defined and measured and at what age hand use laterality appeared in young children.

A child's hand preference has been positively correlated with the head orientation preference of an infant in the supine position during the first three months (Coryell & Michel, 1978). This in turn has been positively correlated with intrauterine fetal position (Michel & Goodwin, 1979). In the infant, both sides of the body operate as a single unit, with bilateral activities dominant until about six months of age, when general movements become reciprocal (Fiorentino, 1972). According to Caplan (1973) the 12 month old prefers the use of one hand, holding with one hand and maneuvering with the other. The capacity to use both sides of the body together, or one side at a time, operates throughout life. De Quiros and Schrager (1977) noted that during the third and fourth years of life the body neurologically expands, dividing into two more discrete halves in order to coordinate different motor activities on each side. Then by five or six years, body laterality tends to be fully established and standing for some length of time on one leg is possible. (Three year olds are able to stand on one leg for a few seconds.) There has been some agreement that human laterality, defined by Cheney and Kephart (1968) as "complete motor awareness of the two sides of the body," is generally established between six and seven years of age. Cognitive awareness of the two sides of the body and the ability to distinguish the right and left sides of the body, of objects, and of space, is probably not fully established until about nine or ten years of age. The preschool child draws and paints, can make circles at three, squares at four, and triangles at five (Rand, 1973), developing fine motor skills as part of lateralization. After reviewing the literature, it was apparent that more information was needed on the actual hand-use of preschool children in order to make comparisons with that of gorillas.

Two parallel studies were planned. It was decided to examine the general hand preference patterns of gorillas by observing hand behaviors and categorizing them, then beginning the study, noting hand behaviors by category as they occurred. The children could then be studied by asking them to do tasks modified from the gorilla categories.

Method

Gorillas

Subjects

Forty-seven gorillas from five zoos were observed for naturally occurring behaviors. The zoos were: The San Francisco Zoo; The Lincoln Park Zoo,

Chicago; The Denver Zoo; The Woodland Park Zoo, Seattle; and the San Diego Wild Animal Park. There were 19 male and 28 female gorillas. Twenty-six were adults (ten years and older), 15 were juveniles (four to nine years), and six were infants (birth to three years).

Scoring Categories

Ten categories were used to score behaviors. Right, left, and bimanual behaviors were scored separately into the following categories:

1. *Touch Self*. Any touch of self except arm-fold, clap, and chest beat. (These two-handed behaviors were listed in "Other" in order to keep them separate and accounted for. A purposeful arm-fold with right over left or vice versa was thought to be a significant measure of human handedness by some researchers, such as Sakano (1982), and not by others, such as Plato, Fox, and Garruto (1985). Scratching was counted by "session" rather than by individual scratch. When the hand doing the scratching was lifted away from the body, the behavior was counted once.)
2. *Touch Other Gorilla*. Any touch of another gorilla, except slaps.
3. *Hit/Slap*. Any hit or slap of another gorilla, glass, grass, rock, etc.
4. *Throw*. The throwing of any object - branch, grass, feces, etc. (This was kept separate because it has been a behavior category in many human studies.)
5. *Eat*. Any eating behavior. The hand listed was the hand brought to the mouth. The other hand was often used to hold (the branch, banana, etc.). Both hands were listed only when both hands were brought to the mouth.
6. *Manipulate Large Object*. The manipulation of any object with the whole arm, distinguished by elbow flexing.
7. *Manipulate Small Object*. The manipulation of any object with just the hands or fingers, distinguished by wrist and/or finger flexing.
8. *Digging/Sifting*. Digging with the hand or hands only. The hand noted for sifting was the one letting the dirt or grass fall through the fingers. Each dig and each sift were recorded separately.
9. *Gestures*. Any hand motions interpreted as signalling to another gorilla or as solitary gestures.
10. *Other*. Any hand behaviors not listed above. (Needed in order to document all hand use.)

Procedures

Available observation time was divided as equally as possible among the total gorilla population at each zoo. The observation of each gorilla was divided into thirty minute segments, listing the name and time of day. Each gorilla was observed at different times of day because activities and the number of behaviors per hour varied during the course of the day. The average time observed per gorilla was five hours, fifty-seven minutes. The range was from three hours to eleven hours, eighteen minutes. When a gorilla was not visible (in the night quarters, behind a rock, etc.) for more than one minute, it was noted. Only the time during which a subject was visible was used to calculate hours observed. In general, observations were made from the visitor's viewing area. Some observations were made in the night quarters and other areas. In all zoos some form of food was available during the day, such as tall grass, cut branches, hay, vegetables, etc., so it was possible to observe a variety of eating behaviors.

Children

Subjects

The subjects were 28 nursery school children ranging in age from two years nine months to five years one month. There were 9 males and 19 females. The tester taught a special program at the school, so the children knew her and were comfortable with her. (Forty-six children were evaluated altogether, but only 28 were considered for comparison as the rest did not meet observation criteria. See Procedures below.)

Scoring Categories

Each gorilla category was changed to a specific request for the children, with the category "Other" modified to "Can you touch the ground?" The requests were as follows:

1. *Touch Self*. "Can you touch your nose?"
2. *Touch Other*. "Can you touch my hand?"
3. *Hit/Slap*. "Can you spank the table?"
4. *Throw*. "Can you pick up the little nerf ball and throw it?"
5. *Eat*. "Here is a raisin. You may eat it"
6. *Manipulate Large Object*. "Can you move this little chair?"
7. *Manipulate Small Object*. "Can you spin this top?"
8. *Digging/Sifting*. "Can you dig a little hole in the cornmeal?"
9. *Gestures*. "Can you wave goodbye?"
10. *Other*. "Can you touch the ground?"

Procedures

Right, left, and bimanual hand-use were noted. It was intended that each child would be observed three times for hand preference in each of the ten tasks requested. The reason for three observations instead of just one was

Table 12.1a. Individual Gorillas: Twenty-one Infants and Juveniles

Name	Age	Sex	Zoo	Total Behaviors (R+L+Bi)	*z*-scores
Six Infants - Birth to 3 years					
Bebac	8 mo	M	LPZ	760	6.11*
Schroeder	8 mo	M	SDWAP	1207	0.80
Jumoke	1 yr	F	WPZ	1687	2.11*
Hope	2 yr	F	LPZ	737	1.95
Brooks	2 yr	M	LPZ	817	2.33*
Matadi	3 yr	F	LPZ	845	2.55*
Fifteen Juveniles - 4 to 9 years					
Zura	5 yr	F	SFZ	607	2.77*
Gino	5 yr	M	LPZ	925	5.25*
Jojo	5 yr	M	LPZ	718	2.26*
O.J.	5 yr	M	SFZ	1226	9.78*
Bawang	6 yr	F	SFZ	563	-2.79*
Kivu	6 yr	F	LPZ	616	0.00
Jitu	6 yr	M	SDWAP	1747	7.23*
Alberta	7 yr	F	SDWAP	2846	4.70*
Kowali	7 yr	F	LPZ	929	0.86
Bassa	8 yr	F	LPZ	486	0.54
Jo-Ray-K	8 yr	F	LPZ	498	0.64
Kamilah	9 yr	F	SDWAP	3203	13.97*
Kisuma	9 yr	F	LPZ	990	4.63*
Sunshine	9 yr	M	SFZ	1337	8.66*
Wanto	9 yr	M	WPZ	1437	3.62*

* $p<.05$

Note. Negative *z*-scores denote left hand preference; positive *z*-scores denote right hand preference.

Table 12.1b. Individual Gorillas: Twenty-six Adults

Name	Age	Sex	Zoo	Total Behaviors (R+L+BI)	*z*-scores
Mkubwa	11	M	SFZ	662	4.62*
Binti	11	F	WPZ	1757	0.79
Patty	13	F	LPZ	649	5.53*
Benga	14	F	LPZ	1052	-4.87*
Maguba #	14	F	DZ	1762	6.50*
Winston +#	14	M	SDWAP	1510	-7.20*
Kumba	15	F	LPZ	489	0.43
Bibi #	17	F	DZ	1921	-14.89*
Nina #	18	F	WPZ	1710	0.96
Pete #	18	M	WPZ	1268	4.57*
Debbie #	19	F	WPZ	1001	4.64*
Kiki +#	19	M	WPZ	1752	-10.62*
Alvilah	21	F	SDWAP	1437	2.47*
Frank +#	21	M	LPZ	744	3.03*
Lenore #	22	F	LPZ	651	6.58*
Otto +#	22	M	LPZ	841	6.26*
Dolly #	23	F	SDWAP	1662	11.74*
Pogo #	23	F	SFZ	1040	-5.17*
Mumbi #	24	F	LPZ	1052	-0.85
Kisoro +#	24	M	DZ	1857	-4.50*
Bwana +#	25	M	SFZ	835	-6.69*
Terra #	26	F	LPZ	600	2.37*
Fred +#	26	M	LPZ	613	6.06*
Helen #	27	F	LPZ	732	-4.09*
Missus #	27	F	SFZ	1573	-3.41*
Trib +#	27	M	SDWAP	399	-1.69

* = $p<.05$
+ = Dominant male
# = Wild born

Table 12.2a. Total Number of Behaviors by Category and Hand Use for All Gorillas (N=47).

BEHAVIORS								
		Total	Total	Total	Grand	Percent		
Category	Mean(Range)	Right	Left	Bilat.	Total	R	L	Bi
Touch Self	244 (39-590)	5,785	4,838	849	11,472	51	42	7
Touch Other	78 (0-392)	1,502	1,251	915	3,668	41	34	25
Hit/Slap	29 (0-213)	653	433	364	1,350	48	32	20
Throw	4 (0-44)	79	54	43	176	45	31	24
Eat	429 (51-1329)	9,329	9,741	1,086	20,156	46	48	6
Large Object	87 (0-425)	1,804	1,527	779	4,110	44	37	19
Small Object	173 (3-1297)	4,247	3,397	495	8,139	52	42	6
Dig/Sift	6 (0-182)	68	45	146	259	26	17	56
Gestures	14 (0-157)	246	192	225	663	37	29	34
Other	80 (12-225)	1,597	1,219	941	3,757	43	32	25
All Behaviors	1,144(399-3203)	25,310	22,697	5,743	53,750	47	42	11

Table 12.2b. Number of Subjects Displaying a Right Preference (R), Left Preference (L), Nonsignificant (NS) Preference, Insufficient Data (ID), or No Data (ND) for Each Behavioral Category.

INDIVIDUALS								
	By z-scores					Percentage Index		
Category	R	L	NS	ID	ND	R	L	NP
Touch Self	18	14	25	0	0	29	17	1
Touch Other	6	3	17	19	2	28	14	5
Hit/Slap	5	1	12	27	2	26	14	7
Throw	0	0	1	30	16	15	9	23
Eat	19	11	17	0	0	28	18	1
Large Object	7	4	24	11	1	30	16	1
Small Object	15	5	22	5	0	31	16	0
Dig/Sift	0	0	1	15	31	10	5	32
Gestures	1	1	1	36	8	22	10	15
Other	15	3	25	4	0	27	11	9

the possibility that a child would use the right hand to perform a task the first time and the left hand the next time. A third observation would show a bias for one side or the other. Twenty-eight children, out of 46, were evaluated three times, eight children twice, and ten children only once.

Observations were kept light and playful. If a child did not wish to do one of the tasks, the observer simply went on to the next task. At the beginning of each session, the child was seated in the center of one side of a small table, with the observer on the other side backed to a wall to preclude unseen distractions. Tasks were assigned one at a time, the necessary equipment was placed directly in front of the child, and then put away before starting on the next task. For the category "Touch Other," the observer placed one hand over the other, centered. This technique consistently elicited a one-hand response. Most of the children enjoyed the raisin offered for the category "Eat," and called the whole observation process "The Raisin Game." If a child did not like the raisin, a small marshmallow was offered on the centered paper napkin. The category "Gesture" was requested last as it was natural to finish with a farewell wave.

Results

Gorillas

A total of 53,750 hand-use behaviors were observed with 25,310 (or 47%) being right-handed behaviors, 22,697 (or 42%) left-handed behaviors, and 5,743 (or 11%) were bimanual. Out of 47 gorillas, 34 (or 72%) used their right hand more than their left, 12 (or 26%) used their left hand more than their right, and one (or 2%) used right and left equally. When tested by z-scores ($p<0.05$), 26 gorillas were evaluated as right-handers and 10 as left-handers. Eleven gorillas did not show statistically significant preference; however, eight of these were biased to the right. See Tables 12.1 and 12.2. Table 12.1 shows the total behaviors in all categories performed by each individual gorilla and the corresponding z-scores arranged by age. Handedness by age showed that of six infants, all were right-biased, four significantly. Of 15 juveniles, 13 were right biased, 10 significantly. Adult gorillas also showed predominant right-handedness. Of 26 adults, 15 were right-biased, 12 significantly. However, a large number of adults were left-handers. Eleven adults were left-biased, nine significantly. Table 12.2a combines the behaviors of all the gorillas for each category. Table 12.2b depicts the number of right- and left-handed individuals in each category as calculated by z-scores. Handedness was also calculated by percentages for each gorilla in each category. The smaller percentage of right- or left-handed behaviors was subtracted from the larger percentage to give a handedness

Table 12.3. Hand Use as Right (R), Left (L), and Bilateral (Bi) for 28 Children Each Evaluated with a 10-item Laterality Performance Test on Three Separate Occasions.

Child	Age	Sex	R	L	Bi	Total	*Z*-Scores
1	2.9	M	22	0	4	26	4.68*
2	3.0	F	21	2	7	30	3.96*
3	3.0	F	26	2	2	30	4.53*
4	3.1	M	24	1	5	30	4.60*
5	3.1	F	24	0	6	30	4.90*
6	3.2	F	26	0	4	30	5.10*
7	3.2	F	27	1	2	30	4.91*
8	3.2	M	26	0	4	30	5.10*
9	3.2	M	25	0	5	30	5.00*
10	3.4	F	16	9	4	29	1.40
11	3.5	F	13	13	2	28	0.00
12	3.5	F	19	1	2	22	4.02*
13	3.5	F	21	1	8	30	4.26*
14	3.6	M	17	6	5	28	2.29*
15	3.6	F	28	0	2	30	5.28*
16	3.6	F	28	0	2	30	5.28*
17	3.7	F	20	7	2	29	2.50*
18	3.7	M	27	0	3	30	5.19*
19	3.8	F	27	0	3	30	5.19*
20	3.8	F	28	0	2	30	5.28*
21	3.9	F	2	19	9	30	-3.71*
22	3.10	F	23	1	6	30	4.49*
23	3.10	F	29	1	0	30	5.11*
24	4.2	F	24	3	3	30	4.04*
25	4.9	M	29	1	0	30	5.11*
26	4.9	F	26	0	4	30	5.10*
27	4.11	M	27	0	3	30	5.19*
28	5.1	M	19	1	7	27	4.02*
Total Behaviors:			644	69	106	819	
Percentage:			79	8	13		

* $p<.05$
Note. Age given as Years.Months

Table 12.4. Analysis of Behaviors by Category for All Children by Frequency and Percentage of Responses as Right (R), Left (L), and Bilateral (Bi) Behaviors on the 10 Tasks of the Laterality Performance Test by 28 Children Each Tested on Three Separate Occasions.

Category	Frequency⁺				Percentage		
	Total	R	L	Bi	R	L	Bi
Touch Self	84	70	11	3	83	13	4
Touch Other	83	75	8	0	90	10	0
Hit/Slap	80	73	3	4	91	4	5
Throw	84	72	6	6	86	7	7
Eat	79	70	8	1	89	10	1
Large Object	82	21	5	56	26	6	68
Small Object	84	76	8	0	90	10	0
Dig/Sift	83	71	11	1	86	13	1
Gesture	80	74	4	2	91	5	4
Other	80	42	5	33	53	6	41
Total All	819	644	69	106	79	8	13

⁺ Frequency: 84 possible behaviors per category Total; 840 possible behaviors Total All categories.

Table 12.5. Frequency of Lateral Hand Bias in Gorillas as Compared to Human Children and Percentage (%) of Total Number (N) of Subjects in Each Group.

	N	RIGHT BIAS	LEFT BIAS	NO BIAS
Gorillas				
Adults	26	16 (62%)	10 (38%)	0
Juveniles & Infants	21	19 (90%)	1 (5%)	1 (5%)
Children	28	26 (93%)	1 (3.5%)	1 (3.5%)

Table 12.6. Frequency of Hand Preference in Right (R), Left (L), and Bilateral (Bi) Behaviors as a Function of Sex in Gorillas and Children; Also in Number of Individuals as Right, Left, and No Statistically Significant Preference (NS); with Percentage Bias (%) of Total Subject Sample (N) by Sex and Species.

INDIVIDUALS	N	R	L	NS	
GORILLAS					
Males	19	13 (68%)	4 (21%)	2 (11%)	
Females	28	13 (46%)	6 (21%)	9 (32%)	
CHILDREN					
Males	9	9 (100%)	0 (0%)	0 (0%)	
Females	19	16 (84%)	1 (5%)	2 (11%)	
BEHAVIORS	**N**	**R**	**L**	**Bi**	**TOTAL**
GORILLAS					
Males	19	9,596 (46%)	8,651 (42%)	2,408 (12%)	20,655
Females	28	15,714(48%)	14,046(42%)	3,335 (10%)	33,095
CHILDREN					
Males	9	216 (83%)	9 (3%)	36 (14%)	261
Females	19	428 (77%)	60 (11%)	70 (12%)	558

index for each gorilla by category. In some categories there was a lack of data. Certain behaviors were performed more than others, and laterality was stronger in some categories than in others. Though not defined by statistically significant hand preferences, "Manipulate Small Object," was the only category where all of the gorillas used the right hand more than the left (31 individuals) or the left hand more than the right (16 individuals). In all other categories at least one gorilla or more used the right and left hand equally or did not perform the behavior at all (Shafer, 1987).

Children

A total of 819 behaviors were observed: 644 (or 79%) were right-handed behaviors, 69 (or 8%) were left-handed, and 106 (or 13%) were bimanual. Of the 28 children, 26 used their right hand more than their left, one used left more than right, and one used right and left equally (see Table 12.3).

Table 12.3 shows the total behaviors of each individual child in all categories combined and the corresponding z-scores. Twenty-five children were significantly right-handed and one was significantly left-handed. Table 12.4 shows the frequency and percentage of responses with right, left, and both hands arranged by category. The category "Manipulate Small Object" received the most right-handed responses (76 right-handed responses out of 84). The greatest number of total responses were equally distributed among three categories: "Touch Self," "Throw," and "Manipulate Small Object."

When the 28 children who were evaluated on three occasions are compared with the gorillas, similarities in lateral biases of the two species are apparent, but in diminishing strength (see Table 12.5). Eighty-nine percent of the children and 67% of the juvenile and infant gorillas were right-biased. Forty-six percent of the adult gorillas were right-biased. More gorillas than children showed no significant hand preference, especially young gorillas. Nineteen percent of adult gorillas and 28% of juvenile and infant gorillas showed no significant preference, whereas only 7% of human children showed no preference.

The data were also examined for sexual bias. Male and female laterality patterns were calculated by the frequency of right and left behaviors of gorillas and children. Individual handedness of children and gorillas, determined by z-scores, was also used to calculate male and female laterality patterns (see Table 12.6). Male gorillas and children were more strongly right-biased than female gorillas and children. Sixty-eight percent of gorilla males and 46% of gorilla females were right-biased. One hundred percent of male children and 84% of female children were right-biased.

Discussion

Because the great apes (gorillas, chimpanzees, and orangutans) are similar to humans in bone, muscle, and nerve structure, among other characteristics, one would assume that muscle/nerve behaviors such as right- or left-handed movements would be similar. We know that laterality is more complex than reflex. The concept of right and left preference must include ideas about motor skills, strength and precision, efficiency of information storage overriding the redundancy of other systems, and the integration of various systems which create new and more complex abilities. Other factors involved in laterality include the tactile stimulation of the handling of infants, the motivation of reaching for an object, and most importantly, upright posture. When infants of a species are carried, as opposed to walking about on their own, not only is the tactile system stimulated, but the vestibular and visual system as well, in ways different from quadrupedal stimulation. As a human infant begins standing and walking, vestibular and visual changes occur to accomodate the new posture. Denenberg and Yutzey (1985) noted that the laboratory rat was lateralized for spatial and effective behaviors to some degree. Handling stimulation given in infancy exaggerated these lateral differences. Tactile stimulation affects laterality in primates as well by helping to develop body awareness (where the parts and sides of the body are) and spatial awareness (bumping into an obstacle, for example, teaches a sense of space and assists in visual development).

The integration of sensory input has a profound effect on motor output and behavior, as is known from many deprivation studies, beginning with Harlow (Harlow & Harlow, 1962). Motivation is a factor in many behaviors. A purposeful reach for an object such as food or a toy brings a reward for the behavior and encourages repetition, building skills in eye-hand coordination and many other areas.

Upright posture is the key to hand preference. MacNeilage (1991) suggests that left hemisphere specialization for vocal communication as well as right-handedness might have had a postural origin. The normal development of upright posture in humans begins in the prone and supine position with the raising of the head, then the upper body, sitting upright and quadrupedal crawling, then finally standing and walking upright — with visual and vestibular development, eye-hand and eye-foot coordination inextricably woven together at the same time. The author believes that gorilla development is very similar and that gorillas spend more time upright than the other great apes. Juvenile gorillas in particular have been observed spending a great deal of time in the upright position. Further study is needed to verify this.

In this study, hand preferences of gorillas and children were determined by z-scores. When examining the gorilla percentage index (Table 12.2) and comparing it with the z-score analysis, it becomes apparent that many more individuals had a hand preference biased to the right in the percentage index. As in humans, it appears that hand preference in gorillas is not an "either-or" preference, but rather a continuum of bias from left to right with individual variations in strength of bias, yet definitely skewed to the right.

Two kinds of behavior in this study were observed which merit discussion. The first is evidence that the most strongly lateralized behavior occurred when fine motor tasks were performed. The category "Manipulate Small Object" was the only category in the gorilla study in which every gorilla used one hand more than the other. In all other categories one or more gorillas had an equal percentage of right and left behaviors. This same category in the children study elicited the greatest number of right hand responses to the restricted number of requests. The author has found that spinning a top manually with thumb and forefinger is the most reliable measure of hand preference in young children. The top is usually spun outward from the midline of the child's body (clockwise for right-handers and counterclockwise for left-handers). If only one measure can be used, the hand preference of adult humans is generally decided by the hand used for writing. Using a pen or pencil to write or spinning a top are tasks which fit into the category of manipulating a small object.

The other behavior seems to validate the importance of upright posture in lateralized movement. The gorilla category "Other" was modified to the request "Can you touch the ground?" for the children. This category and "Manipulate Large Object" (child's chair) elicited the largest number of two-handed behaviors (33 and 56 respectively, with the next highest number only six). The children generally needed two hands to move the chair. The assumption would be that greater strength is needed for moving larger objects, therefore two hands were used instead of one. What caused such a large number of two-handed responses to "Can you touch the ground?" Changing the head position so drastically would cause a vigorous vestibular response and a postural adaptation. This "four-legged" position, though only momentary, may be eliciting the pre-lateralized behavior of an infant.

Male gorillas showed a greater percentage (68%) of right-handedness than females (46%). This was consistant with the child data. One hundred percent of male children in the study were right-handed, and the only clearly left-handed individual was female. Annett (1985) drew up a table of human hand preference studies showing that a greater number of males than females were left-handed. Gorilla males and females in this study had the same percentage of left-handers (21%). There were ten gorillas, four males and six females, whose left hand bias was statistically significant. All four of these males, as well as an additional male with a simple majority left bias, were adult dominant males. The group of six significantly left lateralized females was composed of five adults and one juvenile, and there was one additional adult female with a simple majority left bias. Of the five adult males with significant right-handedness, three were dominant males. All the dominant males were wild-born. The large number of wild-born gorillas in the adult group (20 of 26) reflects the fact that in the past, gorillas were acquired by zoos from the wild. The trend towards group displays in zoological parks and species preservation measures have provided the opportunity to create the large young captive-born population.

The child study was basically a preliminary one, gathering information about a small population. It is quite possible that another population of three to five year old children could include more left-handers, perhaps with a different sexual bias. It has been the author's experience that there is a dichotomy in the left-handed population of children, with some establishing laterality more slowly than right-handers, particularly males, whereas another group of left-handers establishes laterality as right-handers do. The question is, which are the "true" left-handers? Are there perhaps five laterality groups? The standard model right-hander makes up the greatest percentage of the population. Then there are mixed-handers: right-handed for some tasks and left-handed for others. This group can be divided into the skillful ambidextrous group and the "ambi-awkward": not skillful with either hand. Left-handers seem to be divided into the slow developers and the standard developers. Although there are many studies of left-handed humans, the very young pre-schooler needs much more study in all areas of developing laterality, including hand preference, eye preference, and foot preference.

Levy and Gur (1980) found human females more bilaterally organized, which is consistent with the gorilla data. Thirty-two percent of female gorillas had no significant hand preference, whereas only 11% of males were not significantly biased. The child study was also consistent, with 11% of females and no males without significant hand preference. The small population follows the pattern but needs to be reinforced with more data.

The finding that juvenile and infant gorillas in this study were more strongly right-biased as a group than adults is consistent with Heestand's (1986) data on infant limb preference for locomotion. In the search for primate similarities, it seems necessary to look for them in younger primates. The older the primate, the more species-specifically unique they seem to be. However, Gesell and Ames (1947) noted that human infants were less clearly lateralized than adults, and this is consistant with the gorilla data. Twenty-nine percent of juveniles and infants, but only 19% of adults, had no statistically significant hand preference. Observations of one gorilla as a juvenile in a preliminary study, then again as an adult in the present study, showed that one gorilla developed stronger laterality with age just as humans do.

While a direct comparison of children and gorillas cannot be made from this study, the measures to determine hand preference in both species enable indirect comparisons to be made. It is speculated by the author that gorillas, particularly young gorillas, spend more time upright and bipedal than any other primate except humans, with the more arboreal chimpanzees next on the scale. Siamangs and gibbons also spend a great deal of time upright and bipedal, but more time brachiating. The size and weight of humans and gorillas, as well as their terrestrial existence, causes a different gravitational effect on the vestibular and visual systems. This may account for the difference in general lateral preference of humans and the study gorillas as opposed to only the locomotor preference of the *Hylobatidae*. It is possible that the relative degree of bipedalism between adult and young gorillas may be responsible for the greater strength of hand use in the young.

The most important finding of the research reported here is that the majority of the study population of gorillas showed right-biased hand use. This is the first instance of a non-human primate population study showing the same lateral bias as humans.

References

Annett, M. (1985). *Left, right, hand and brain: The right shift theory*. London: Lawrence Erlbaum Associates.

Caplan, F. (1973). *The first twelve months of life*. New York: Grosset & Dunlap.

Cheney, C. M., & Kephart, N. (1968). *Motoric aids to perceptual training*. Columbus, Ohio: Charles E. Merrill.

Collins, R.L. (1985). On the inheritance of direction and degree of asymmetry. In S. Glick (Ed.), *Cerebral lateralization in nonhuman species* (pp. 41-71). Orlando: Academic Press.

Coryell, J.F., & Michel, G.F. (1978). How supine postural preference of infants can contribute towards the development of handedness. *Infant Behavior and Development, 1*, 245.

Denenberg, V.H., & Yutzey, D.A. (1985). Hemispheric laterality, behavioral asymmetry, and the effect of early experience in rats. In S.Glick (Ed.), *Cerebral lateralization in nonhuman species* (pp. 110-133). Orlando: Academic Press.

de Quiros, J.B., & Schrager, O.L. (1977). *Fundamentos neuropsicologicos en las discapacidades de aprendizaje.* (Neuropsychological Fundamentals in Learning Disabilities). Buenos Aires: Panamericana.

Fiorentino, M. R. (1972). *Normal and abnormal development the influence of primitive reflexes on motor development.* Springfield, Illnois: Charles C. Thomas.

Gesell, A., & Ames, L. B. (1947). The development of handedness. *Journal of Genetic Psychology, 70*, 155-175.

Goerke, E., Fleming, L., & Creel, M. (1987). Behavioral changes of a juvenile gorilla after a transfer to a more naturalistic environment. *Zoo Biology, 6*, 283-295.

Harlow, H. F., & Harlow, M. K. (1962). The effect of rearing conditions on behavior. *Bulletin of the Menninger Clinic, 26*, 213-224.

Heestand, J. (1986). Behavioral Lateralization in Four Species of Apes? Doctoral dissertation, University of Washington, Seattle, WA.

Levy, J., & Gur, R. C. (1980). Individual differences in psychoneurological organization. In J.Herron (Ed.), *Neuropsychology of left-handedness* (pp.199-210). New York: Academic Press.

MacNeilage, P. F. (1991). The postural origins theory of neurobiological asymmetries in primates. In N. Krasnegor, D. Rumbaugh, M. Studdert-Kennedy, & R. Schiefelbush (Eds.), *The biological foundations of language development* (pp.165-188). Hillsdale, NJ: Lawrence Erlbaum.

MacNeilage, P. F., Studdert-Kennedy, M. G., & Lindblom, B. (1987). Primate handedness reconsidered. *Behavioral and Brain Sciences, 10*, 247-303.

Michel, G. F., & Goodwin, R. (1979). Intrauterine birth position predate newborn supine head position preferences. *Infant Behavior and Development, 2*, 29.

Okada, M. (1980). Lateral dominance in nonhuman primates with emphasis on hand preference. *Advances in Neurological Science, 24*, 529-542. (Japanese; abstract in English).

Plato, C. C., Fox, K. M., & Garruto, R. M. (1985). Measures of lateral function dominance: Foot preference, eye preference, digital interlocking, arm folding and foot overlapping. *Human Biology, 57*, 327-334.

Rand, C. W. (1973). Copying in drawing: the importance of adequate visual analysis versus the ability to utilize drawing rules. *Child Development, 44*, 47.

Sakano, N. (1982). *Latent left-handedness: Its relation to hemispheric and psychological functions*. Leipzig: Gustav Fischer Verlag.

Sanford, C., Guin, K., & Ward, J.P. (1984). Posture and laterality in the bushbaby (*Galago senegalensis*). *Brain Behavior and Evolution*, *25*, 217-224.

Shafer, D. (1987). *Patterns of hand preference among captive gorillas*. M.A. Thesis, San Francisco State University, San Francisco, CA.

Ward, J. P., Milliken, G. W., Dodson, D., Stafford, D., & Wallace, M. (1990). Handedness as a function of sex and age in a large population of *Lemur*. *Journal of Comparative Psychology*, *104*, 167-173.

13

Hand Preference for Visually-Guided Reaching in Human Infants and Adults

Lauren Julius Harris and Douglas F. Carlson

As the collection of new research reports in this volume shows, the question of handedness in non-human primates has become a subject of intense interest among psychologists and primatologists. The historical evidence also shows that this interest is of long standing (see Harris, Ch. 1, this volume). Still, the question remains controversial. There is a measure of agreement that individual animals show reliable hand preference but no consensus that this is a species characteristic.

Before going further, we want to address some issues about terminology. The literature on human handedness has seen the growth of an abundance of terms, not all of which are clearly defined or distinguished. One is "hand preference," by which we mean only that an individual consistently uses one hand rather than the other for a *given* task. By "handedness," we ordinarily mean that the hands are asymmetric in use and function so as to reliably favor one hand or the other across a *range* of skillful acts. Our reference point will be the modal form, right-handedness, with estimates of prevalence ranging from 85 to 92 percent of the adult population (Annett, 1985; Bryden, 1982).

Handedness is typically manifested for acts that call for complex, coordinated sequences of movements involving the distal musculature -- the hands and fingers. In the case of the right-hander, the right, or "dominant," hand is also more skilled and more often used in acts calling for finely modulated application of force, such as hammering or drawing. Because of the structure of the hand-brain system, this right-hand specialization is assumed to reflect the inherent specialization of the contralateral (left) cerebral hemisphere for analysis of sensory input (irrespective of modality) as discrete, finely tuned stimuli within a temporal frame and for the programming of serially organized discrete movements (Bradshaw & Nettleton, 1981; Kimura, 1982; Levy, 1974; Witelson, 1985).

The "non-dominant" left hand, by contrast, is typically used for actions that support the role of the right hand; for example, holding while the right hand manipulates. Certain of these supportive roles, however, require specialized skills in which the *left* hand may truly excel, for example, in analyzing the spatial contours or configurational properties of objects prior to manipulation by the dominant hand (e.g., Harris, 1980) and for controlling spatial positions of the effectors, as demonstrated by accuracy in open-loop ballistic aimed movements (Guiard, Diaz, & Beaubaton, 1983; but see Todor & Cisneros, 1985). These special functions may be seen as reflecting the inherent specialization of the contralateral (right) cerebral hemisphere for the synthesis of stimuli over the dimensions of time and space into holistic or gestalt-like configurations (Cohen, 1973; Levy, 1974).

Where the left-hander is concerned, the normative evidence is much less abundant, but what there is suggests that left-handers are not simply the mirror-image of right-handers in hand skill. For example, in regularity and speed of finger tapping, left- and right-handers do not differ in skill with the preferred hand, but left-handers are superior to right-handers in performance with the non-preferred hand, thus resulting in smaller between-hand differences (Peters & Durding, 1979).

To assess handedness in human adults, that is, in individuals whose handedness is presumed to be fully established, both questionnaires and tests of skill are commonly used. In questionnaires (e.g., Crovitz & Zener, 1962; Healey, Liederman, & Geschwind, 1986; Oldfield, 1971; Provins, Milner, & Kerr, 1982), the respondent names the preferred hand for such motorically demanding acts as writing, drawing, hammering, and using a toothbrush, all of them so-called "primary-action" items because factor analyses show them loading on a primary "handedness" factor (Bryden, 1977; see also Bryden, 1982).

Strictly speaking, questionnaires are measures only of "hand preference," where a demonstration of hand preference can usually, but not necessarily, be taken as an indication that the preferred hand is the more skillful, or specialized, hand for the task in question. In contrast to questionnaires, performance tests assess the actual skill of each hand. Because the tasks listed in questionnaires are presumed to be influenced by prior practice, performance tests assess less practiced skills, such as peg-moving (Annett, 1970), square-drawing (Bishop, 1980), dot-marking (Tapley & Bryden, 1985), and finger-tapping (Peters & Durding, 1979).

As psychologists whose primary research is on the development of human handedness, in the double sense of preference and specialization, one of the things that impress us about the studies of handedness in apes and monkeys is their reliance on measures that are in common use in research with human infants. One measure in particular — visually-directed reaching — is one that we have used in our own work. The use of visually-directed reaching in primate studies goes back many years and has yielded mixed results. For example, among the older studies, Hollis (1875) and Finch (1941), along with others, failed to find hand preference at the species level.

Among the positive reports, Ogle (1871) found that zoo monkeys showed no hand preference for food placed within easy reach but instead used the nearer, and so the more convenient, hand; however, when the bait was offered at a distance requiring the monkey to stretch out one of its arms as far as possible through the bars of the cage, nearly all the monkeys consistently reached with their right arms. Ogle likened the few exceptions to the way "a right-handed man may occasionally extend his left arm instead of his right to grasp an object" (1871, p. 287).

Ogle's remark about the exceptional cases implies that human adults would reach preferentially (although with occasional exceptions) for objects with their 'dominant' hand. (Whether Ogle might also have assumed that the preference would appear only for objects at a distance, we cannot say.) As already noted, visually-directed reaching is a common measure of handedness in research with human infants, but, so far as we know, it has never been part of any standard assessment battery in human adults. We expect that the reason is the tacit assumption by clinicians and researchers that, for adults, a simple reaching task makes insufficient demands on the kind of manual praxis, or motor skill, that is incorporated into the usual measures of handedness.

Although hand use in reaching has not been part of standard assessments of handedness in adults, on occasion it has been included in questionnaires. For example, in a study by Teng, Lee, Yang, and Chang (1979) over 4,000 Taiwanese Chinese 4th- and 5th-grade children and university students answered the question "which hand do you use for reaching into a jar?". Forty-nine percent answered "right hand," 48 percent answered "either hand" (meaning no preference), and 3 percent answered "left hand." By contrast, 97 percent of the subjects said that they used their right hand for writing, whereas only 2 percent answered "either hand," and 1 percent answered "left hand." (This last very low percentage was assumed to reflect a cultural bias against left-hand writing in this population.) By this self-report measure of reaching, then, less than 50 percent of the population would be judged right-handed.

If visually-directed reaching has never become part of the standard assessment of handedness in individuals whose handedness is known (human adults and older children) and for whom the existence of handedness at both the individual and population level is not in doubt, it is fair to ask, how valid is reaching as a measure of handedness in a species for which the existence of handedness is the very question at issue? The question is more pressing in the negative than in the positive case. That is, if, as Ogle (1871) found, most of the individual members of a non-human primate species show right-hand preference for reaching for an object, we perhaps can conclude, with Ogle, that handedness is a species characteristic, even if humans (contrary to Ogle's assumption) were not to show dominance for the same measure or if the degree of dominance were equivocal, as Teng et al's (1979) findings suggest. But if apes or monkeys failed to show hand preference, we could not validly conclude, as others have, that the animals are not handed if the implication is that they would prove not to be so on some other task. That would amount to accepting the null hypothesis.

In fact, those of us who study the development of handedness in human infants are in the same bind as those who study handedness in non-human primates because, in using hand preference for visually-directed reaching as a measure of handedness in human infants, we often seem to assume that this measure is as good an index of infant handedness at the age tested as are any of the standard measures used to assess adult handedness. This assumption also often carries a correlary that there is consistency and continuity across age, meaning that the hand preferred for reaching in infancy will be the same as the hand preferred for other skilled acts, regardless of whether the next assessment is 1 month, 4 years, or 10 years later. In fact, we know that in infancy, reliable hand use may vary according to the task used and even that, depending on the task and the child's age, the reaching hand may not be the same as the hand that grasps and manipulates the object (Cornwell, Harris, & Fitzgerald, 1991; Michel, Ovrut, & Harkins, 1986). Of course, the justification for using reaching tests to assess hand preference in infancy is that human infants are motorically incapable of performing the kinds of skilled tasks used in adult inventories. This also means that the reaching task places a much larger demand on skill for infants than for adults.

Because developmental psychologists have focused on human infants, we may also have been less sensitive to the problem of test validity than if we had been studying non-human primates, since we know that human beings show handedness at the population level. We might even have been more disposed to accept hand differences in reaching as an early index of handedness because we know that nearly every one of our infants eventually will be right- or left-handed. It is as though, because we are studying human beings, we (unconsciously) relax our standard of evidence[1].

Keeping these qualifications in mind, in this chapter we shall describe data from our studies on reaching in infants and in adults. Our original

---

[1] The same criterial difference has been said to be at work in the controversy over the existence of language skills in the great apes, namely, that the apes are being held to a stricter criterion of evidence than are human beings. That is, it has been suggested that in attacking the inference that great apes can perform proto-linguistic acts, the critics have been readier to credit human infants with linguistic speech acts than they have been ready to credit great apes with linguistic-signing acts of the same level of complexity because they *know* that human infants ultimately will speak, whereas the question is in dispute in the case of our primate cousins.

purpose in collecting the adult data was to find out how adults would perform on the same reaching task as we had given the infants. That is, having tracked the development of visually-directed reaching in infants, we were curious to know the mature form or, 'end point', of that development, something that is often taken for granted in developmental studies. At the time we were not thinking of the implications of the adult data for the animal literature, but now we see some potentially useful lessons.

Visually-Directed Reaching in Infants

Like the studies of non-human primates, the use of visually-directed reaching to assess handedness in human infants also goes back many years (Harris, 1983). The inspiration for our own work was possibly the first such study -- an investigation in 1890 by the American psychologist James Mark Baldwin of his daughter, Helen (see Harris, 1986).

From the beginning of Helen's 5th month to the end of her 10th, Baldwin administered over 2,000 trials at short distances (9 in., within easy reach) and always at midline. On these trials, Helen showed no trace of hand preference. For a 24-day-long period, during the 8th month, Baldwin added trials at longer distances (12-15 in.). This resulted in "very hard straining with all the signs of physical effort" (Baldwin, 1895, p. 61). Under these conditions, the right hand was preferred by a ratio of almost 15:1. Finally, during the last 10 days of this period, Baldwin presented objects from 1-5 in. to the right or left. The result was further enhancement of the right-hand preference. Baldwin therefore concluded that Helen's dextrality could not have derived from using the closer hand for such acts as reaching, grasping, and holding because her right hand "intruded regularly upon the domain of her left" (Baldwin, 1895, p. 63). By her 13th month, Helen was "a confirmed right-hander, so to speak" (Baldwin, 1890, p. 248). From his results, Baldwin concluded that hand movements were controlled by the "motor arm centers" in the contralateral cerebral hemisphere. Although today there is no consensus about the neuroanatomical locus of lateral asymmetries, we can accept the point of Baldwin's conclusion as meaning that handedness was an expression of *inherent* differences in cerebral functioning. Why, though, had the right hand preference appeared only on the far-distance trials? Ogle (1871) offered no explanation when he found a similar effect with monkeys. Baldwin, however, proposed that the crucial factor was the greater muscular effort required.

Familial Handedness and Sex

Since Baldwin's time, there have been many more studies of handedness in infants as measured by reaching (see Young, Segalowitz, Corter, & Trehub,

1983). Over this period, attention also has turned to the possible role of certain subject variables. One is the handedness of close family members on the assumption that handedness is a heritable trait. According to what is probably the most widely-known genetic model (Annett, 1981, 1985), most individuals are right-handed because they have inherited a "right-shift" factor, that is, a predisposition for left-hemisphere control of motor functions including speech and manual skill. The remaining minority lack this predisposition and become right- or left-handed with equal likelihood, although in practice a slight preponderance of right-handedness might result through environmental pressures.

To the extent that familial left-handedness (FLH) marks the absence of the right-shift factor, a familial left-handed infant should be more likely to become left-handed than a familial right-handed (FRH) infant. The model also predicts that individuals who lack the right-shift factor will be less clearly lateralized than individuals with the factor. This difference has been found for right- and left-handed adults, that is, left-handers typically show weaker laterality effects than right-handers (e.g., McKeever & Van Deventer, 1975), and a few cross-sectional studies suggest similar differences between FRH and FLH infants as well (e.g., Hawn & Harris, 1983; Saling, 1983).

Another possible contributor to the development of handedness is the sex of the individual. There are two reasons. First, girls show a faster rate of overall physical development than boys (Tanner, 1978). It has been suggested that a specific marker of this sex difference is the earlier maturation of the left hemisphere (Shucard, Shucard, Cummins, & Campos, 1981) and that this gives females the lead in the acquisition of certain left-hemisphere skills, speech in particular (cf. Moore, 1967; Wells, 1979). Second, in adults, evidence suggests that the neural mechanisms underlying praxis and speech are more focally organized in the left hemisphere of women than men (Kimura & Harshman, 1984). Kimura (1983) proposes that this may explain the slightly higher prevalence of right-handedness in women than in men (e.g., Bryden, 1982) and why, in right-handers, the margin of right-hand superiority in motor performance is stronger in females than in males (e.g., Annett, 1970). To use Tanner's (1978) term, the developmental pathways for handedness would appear to be more "canalized" in females than in males in the sense that handedness is a more clearly and directly specified outcome of developmental pressures. As Tanner remarked, "One thing does seem to be clear about canalization. Regulation is better in females than in males" (1978, p. 160). From the evidence cited here for both adults and children, we therefore might expect to find earlier emergence and more stable organization of right-hand preference in girls than in boys.

In one of our own studies, we examined the subject variables of sex and familial handedness by testing 32 infants — four equal-size groups (N = 8) of familial right- and familial left-handed boys and girls (Carlson & Harris,

1985). Each infant was tested every 3 weeks between 24 and 39 weeks of age, and once again at 52 weeks, for a total of 7 sessions per infant. For each session, the infant was placed upright in an infant seat against a testing table. The table was designed to fit closely against the infant's abdomen, allowing the infant's forearms to rest freely on the table. The experimenter sat opposite the infant.

On each trial, the experimenter placed a single object on the table before the infant. In order to examine the possible relation between different types of grasp and hand preference, different-size objects were used — a cereal piece, a 1 inch plastic cube, and a 2.5 inch ball. Would a small cereal piece, by requiring a pincer grasp that places greater demands on fine motor skill, elicit a more lateralized response than the ball, our largest test object, which required a palmar grasp? Following Baldwin (1890), the objects were also placed at different distances from the infant (4 inches, 8 inches, and 12 inches) and at different lateral positions (midline or 4 inches to the left or right). Finally, in order to help sustain the infant's interest in the task, we pushed the pace by immediately presenting a new object as soon as the infant completed a reach.

Test Reliability

The first question was whether individual patterns of responses were reliable across the two 27-trial test sequences within each testing session. To find out, the scores were compared, trial by trial, across test sequences. For this analysis the raw data were scored by grouping reaching responses into three categories: left (left hand used alone or predominantly), bimanual (both hands used equally), and right. The results showed that the percentage of agreement across test sequences rose from 11 percent at 24 weeks of age to 65 percent at 52 weeks (with a mean of 43 percent agreement across all ages combined).

Age and Sex Trends, All Trials

At 24 weeks, the reliability was low because the infants often seemed inattentive and failed to reach on many trials. Also, when they reached, they often used both hands equally. Over time, the infants showed increasingly sustained effort, as well as strong visual engagement in the task, and made an increasingly large number of scorable reaches. The number of bimanual reaches also declined. From 36 weeks on, infants reached on nearly all trials, and all reaches were either completely or predominantly unimanual (that is, one hand clearly led the other and contacted the target first).

To examine hand preference, the raw data were scored by grouping reaching responses into the same three categories that were used for the assessment of within-session reliability. In this chapter, we present the data in the form of a laterality index, using a single number to represent the predominant lateral bias across all reaching trials within a test session. Using the formula: $(R_{reaches} - L_{reaches}) / (R_{reaches} + L_{reaches})$, a separate laterality index was calculated for those trials in which the test objects were placed on the infant's right or left side or at midline. The numerator of the laterality index included only right and left unimanual reaches. The denominator included all trials for which there was a scorable reach, regardless of whether the reach was unimanual or bimanual. In addition, a grand mean laterality index was computed for all trials combined. This calculation yields a positive index to indicate a rightward bias, where +1.00 = 100% right hand responses, and a negative index to indicate a leftward bias, where −1.00 = 100% left hand responses.

Grand Mean Scores

The laterality index for the grand mean of all reaching trials combined disclosed marked differences among the four sex by familial handedness groups. For both FRH boys and girls, the trend over age was approximately linear. For FRH boys, it was from initial left bias toward no lateral bias. In other words, across all trials FRH boys, at 24 weeks, began by making more left-hand responses than right-hand responses (grand mean = −.25), but the strength of this bias fluctuated through 39 weeks, culminating in essentially no lateral bias at 52 weeks (grand mean = −.07). The scores for the FRH girls showed a modest right preference at all ages, the laterality indices varying from between +.10 at 24 weeks to +.23 at 52 weeks).

By contrast, the age trends for FLH infants were curvilinear, with no lateral preference at 24 weeks, a marked increase in right bias by 27 weeks, and then a gradual shift to a left bias by 52 weeks, with this shift occurring somewhat earlier in boys than girls.

Position Effects

We also used laterality index scores to assess the effects of object position on hand use. Because the analyses revealed clear and comparable effects for all infant groups at all ages, we have provided a summary of the results for only the 52-week test.

To begin with the midline position, for both FRH infants and FLH infants, the pattern for midline-trial reaches closely matched the grand means. For example, the laterality index for FRH boys showed a left-hand preference on these trials, whereas the FRH girls showed a preference for right hand reaches.

On the lateral position trials, the summary data show that the left hand was used more for objects placed to the infant's left side, resulting in negative laterality index scores, and the right hand was used more for objects on the right, resulting in positive laterality index scores. On these same trials, however, the hand biases shown in the grand mean scores were still evident; the FRH girls used their right hands for right-position objects more often than they used their left hands for left-position objects, meaning that, on a certain number of trials they reached for a laterally-positioned object with the contralateral hand, thus crossing the midline, and on those occasions, they reached from right to left more often than from left to right. Because there were equal numbers of right-position and left-position trials, these results mean that in FRH girls the right hand, as Baldwin said of his daughter, frequently "intruded into the domain" of the left. By contrast, the FRH boys showed more frequent left-hand use for the left position than right-hand use for the right position, meaning that when they crossed the midline, they reached from left to right more often than from right to left.

For FLH infants, object position effects similarly reflected the overall lateral bias as seen in the grand mean scores.

Distance and Object Size Effects

In contrast to what we would have predicted from Baldwin's (1890) report, object distance did not appear to affect hand preference. For example, the FRH girls, who showed overall right-hand preference, did not show the preference more strongly at the far distance than at the nearer distances. Nor was there any evidence that object size affected hand preference even though the infants, especially at later ages, were adjusting their grasp to the different objects in the way described earlier.

Individual Differences

The data indicate that both familial handedness and sex are related to the pattern of hand preference in this reaching task. The sub-group analyses, however, do not tell the whole story so far as individual differences are concerned because approximately a third to a half of the infants showed pronounced fluctuations in hand preference across testing sessions. This means that hand preference at any given time could not be predicted from preference on a test given only three weeks previously. This change presumably reflects some combination of test unreliability (as indexed by within-session agreement scores) and a genuine shift across age in lateral preference. However, inasmuch as infant reaching becomes more reliable with age, we can be more confident that between-session changes increasingly come to represent a genuine age change. What remains to be determined is whether fluctuating patterns are more common in FLH than in FRH infants, and more in boys than in girls.

In summary, the infant data suggest that by 52 weeks, handedness as indexed by reaching is still in flux, since the percentage of infants showing right-hand dominance falls short of the estimated percentage of right-handedness in the adult population. The trend toward right-handedness, however, is clearest in the FRH girls (as the sex-difference model predicts). The data also suggest that hand preference is modulated by familial handedness in this age period. Finally, we did not find that object size or target distance contributed significantly to performance.

Visually-Directed Reaching in Adults

As already noted, the *reliable* use of one hand for visually-guided reaching in a 1-year-old child does not mean that handedness is firmly established in any way comparable to what we know to be the case in the older child or adult. The next question is, what would adults — individuals with established handedness — do under similar circumstances? Given the low skill demand of the task, we might predict no preference at all, with object position perhaps being the only influence on choice. Alternatively, following the questionnaire data reported by Teng et al. (1979), we might simply predict weaker hand preference than for the primary action measures mentioned earlier. We also wondered whether the sex and handedness effects reported on fine-motor tasks would appear on a simple reaching test. To find out, we tested hand preference for reaching in a sample of young adults.

The subjects were 40 right-handed and 40 left-handed college students (20 men and 20 women in each group). Initially, subjects were selected on the basis of self-described handedness. Handedness was then confirmed with a 6-item questionnaire.

The reaching test was designed to correspond as closely as possible to that used in the infant study, but with all test objects and distances scaled to adult proportions. Each subject was tested once in a small room. During the test, the subject sat at a table with forearms resting comfortably on the table and with hands positioned approximately at midline. The experimenter sat on the opposite side.

On each trial the experimenter placed a single object on the table before the subject. Once again, to examine the possible effect of object size, we used three different-size objects: a pellet, a 2.25 inch cube, and a 5 inch diameter ball. Again, we wondered whether the pellet, which could be picked up only with a pincer-like grasp, would elicit a more lateralized (dominant-hand) response than either the ball, which could be picked up only with a palmer grasp, or the cube, which could be picked up with either kind of grasp. We also varied distance from the subject (8, 19, and 30 inches), and lateral position (midline, and 12 inches to the right and to the left).

The test sequence consisted of 27 trials presented in a counterbalanced combination of these three variables. The test sequence was presented twice for a total of 54 trials. The second test sequence thus provided a means to check for test reliability.

The subjects were instructed to reach for the object in any way that was most comfortable, to grasp the object and to hand it to the experimenter, and then to return their arm to the resting position. To discourage subjects from reflecting on the task, we pushed the pace as we had for the infants by presenting a new test object immediately after the previous object was picked up.

Test Reliability

Again, the first question was whether individual patterns of responses were reliable across the two 27-trial test sequences. To find out, the scores were compared, trial by trial, across test sequences as they had been for the infants. The analysis for adults, however, was simpler than the analysis for infants because, unlike the infants, none of the adults ever reached with both hands. Instead, their reaches were all strictly unimanual. The results showed an overall percentage of agreement for all four sex by handedness groups (overall mean = 87.8 percent; range: 82.8 - 90.7 percent), percentages that in all cases were higher than the highest percentage for the infants. Differences among the four groups were not significant.

Position Effects

Having found that the response patterns were consistent across sequences, all subsequent analyses were performed on the combined set. For the midline position (Figure 13.1), the dominant hand was used from 68% to 85% of the time across the four sex by handedness groups, with both the highest and lowest percentages being shown by the two left-handed groups. The percentages, however, were not significantly different from each other.

As was true for the infants, lateral position of target objects influenced hand choice; the adults reached more often with the same-side hand, regardless of whether it was the dominant or non-dominant hand. Fig.13.2 shows the proportions of dominant-and non-dominant-hand reaches when objects were positioned on the dominant-hand side (Fig.13.2A) and on the non-dominant-hand side (Fig.13.2B).

The results show that, like the infants, adult reaches frequently crossed the midline. But whereas for the infants only the FRH females showed more crossing by the hand that we would predict to be the eventual dominant hand (that is, the right hand), the adults in all four sex by handedness groups crossed more with the dominant hand than with the non-dominant hand. In each group, when the object was on the dominant-hand side (Fig.13.2A), the dominant hand was used on 90 percent or more of the trials, whereas when the object was on the non-dominant-hand side (Fig.13.2B), the non-dominant hand was used on only 60 to 75 percent of the trials. Positioning the object on the side of the non-dominant hand thus predisposed use of that hand but not to the same extent as when the object was on the side of the dominant hand. In other words, the dominant hand 'invaded the domain' of the non-dominant hand (Fig.13.2B) more often than the reverse occurred (Fig.13.2A). The difference was statistically significant (X^2 = 290.71, df = 176; p = .0001).

The comparison also suggests that the effect was stronger for the women than for the men. For objects placed on the dominant-hand side (Fig.13.2A), men and women showed equivalent proportions of dominant-hand reaches, whereas for targets placed on the non-dominant side (Fig.13.2B), women crossed over with the dominant hand more often than the men. The women therefore appeared to be more lateralized on this aspect of the task.

The comparison between men and women is revealed in a different way in Table 13.1, which tallies the percentage of reaches crossing the midline by the dominant and non-dominant hand. Right-handed women crossed right hand to left side (R to L) on 41% of the left-side trials, but left hand to

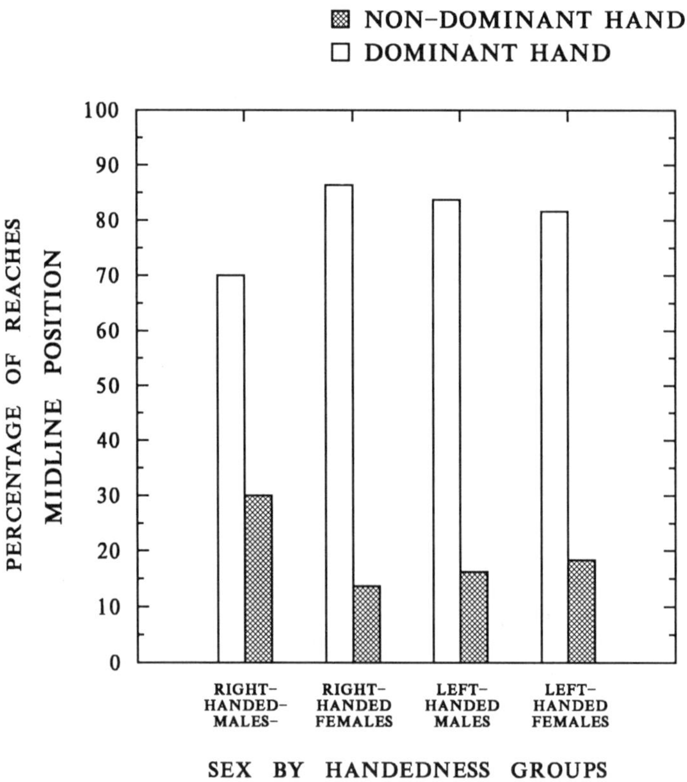

Figure 13.1. Percentage of dominant- and non-dominant-hand reaches for objects placed in the midline position.

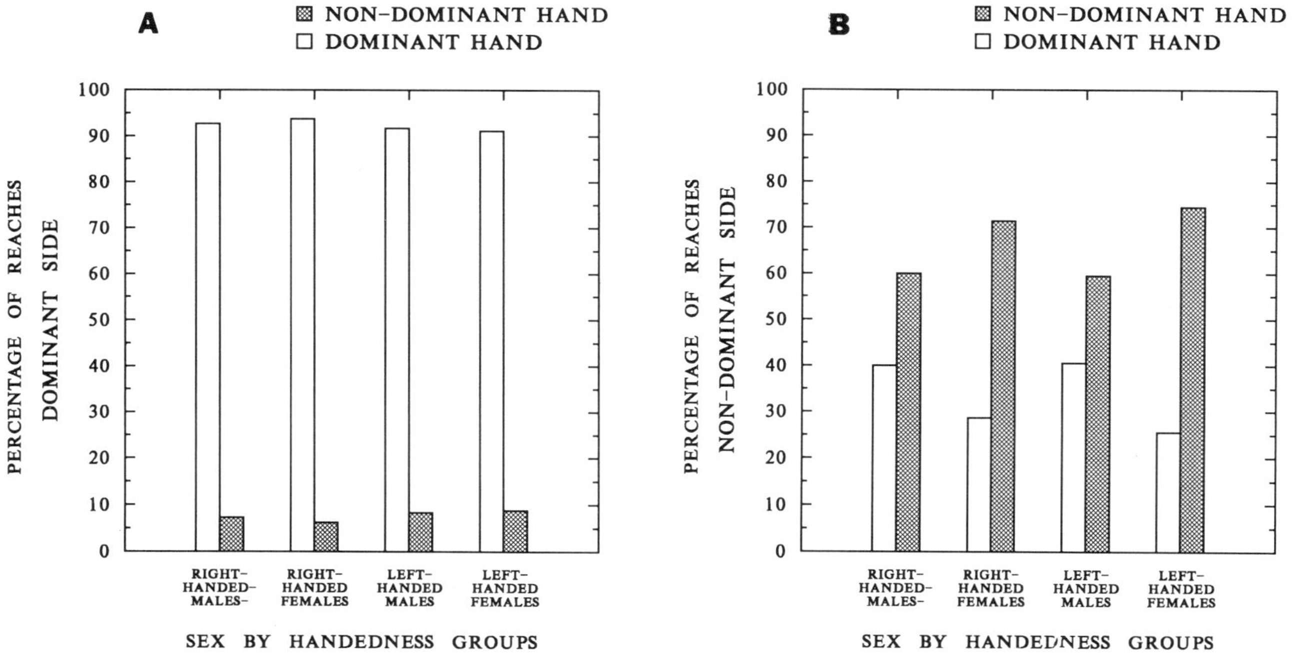

Figure 13.2 A and B. Percentage of dominant- and non-dominant-hand reaches for objects placed on the dominant-hand side (A) and the non-dominant-hand side (B).

Table 13.1. Percentage of reaches crossing the midline by sex and handedness.

		Dominant hand	Non-dominant hand
Right-handed women	(n=20)	41%	7%
Left-handed women	(n=20)	40%	7%
Right-handed men	(n=20)	26%	9%
Left-handed men	(n=20)	29%	7%

right side (L to R) on only 7% of the left-side trials. For right-handed men, the figures were 26% and 9%. Left-handed women crossed L to R on 40% of the right-side trials, but R to L on only 7% of the left-side trials. For left-handed men, the figures were 29% and 7%.

Despite these suggestive differences according to sex of subject, the sex difference for crossed reaches proved to be non-significant according to an ANOVA for the *percentage* of all reaches to that side (F = 2.66, df = 1,76; p = .11), although it at least approached significance with an analysis of the *frequency* of crossed reaches by individual subjects (Mann-Whitney U = .993, p = .06).

Distance and Object Size Effects

As in the infant study, analyses failed to disclose any effect for distance on hand use for reaches. However, there was some indication of an influence of object size, but only for objects placed in the midline, and then only for the ball (the largest object) versus the pellet and cube, which did not differ. The effect was reversed for right- and left-handers. For both handedness groups, the percentage of dominant-hand midline reaches ranged from 76% to 82% for the pellet and cube. For the ball, the percentage decreased to 58% for the right-handers, but increased to 88% for the left-handers. We found no sex differences here. Object size therefore was related to the extent of dominant hand use, but in different ways for right- and left-handers.

Conclusions

The data suggest several conclusions about the use of visually-directed reaching as a measure of handedness in human beings and, by extension, in non-human primates. To begin with our infant study, we would hardly want to assume that hand differences in infant reaching are the exact counterpart of hand differences in skilled performance in adults or older children. For one thing, we are measuring only the frequency of reaches, not the skill with which the reach is made. Our data, along with other data, at least suggest that we are on the right track in choosing this measure. They also show that the lateral placement of test objects strongly affects the hand used. This might be explained by the principle of "minimal effort," that is, that when an object is to one side, the individual is more likely to reach with the ipsilateral hand because that is less effortful (other things being equal) than a reach across midline by the other hand. But, like Baldwin's daughter, our infants also frequently crossed the midline to grasp objects on the opposite side. Therefore, one potentially sensitive index of hand preference may be the relative frequency with which one hand "intrudes into the domain" of the other in a visually-directed reaching task. Another reason to emphasize this measure is that the hand taking the lead on these trials also took the lead on the midline trials, where each hand was equally close to the object.

If we accept visually-directed reaching as a valid measure of handedness for infants, we must also conclude that, through 52 weeks of age, handedness is still undergoing considerable change. Other studies have disclosed similar changes in lateral bias in this early period (see Liederman, 1983). We also are impressed with the variability that Franz (1913) found in his study of hand preference in monkeys (see Harris, Cp. 1 this volume). Two of Franz's animals showed relatively long periods of exclusive use of one hand, followed by a period when the other hand was used. Franz assumed that, in a long series, these alternatives should balance if the preference was due to external influences that presumably would vary over time. By contrast, our assumption for our human subjects is that the alternatives will *not* balance because that preference represents an endogenous bias at some level. Although other evidence (for a sample of girls) suggests that stability will be achieved by about 20 months of age (Cornwell et al., 1991), only with our test of adults are we in a position to tell whether hand preference for reaching is also evident in someone whose handedness is unquestionably established.

What the adult data indicate is that, under appropriate testing conditions, adults do show hand preference for a reaching task that puts few demands on skill. The hand preference is highly reliable across test sequences (more so for adults than for infants), and it is in the direction consistent with their known handedness as indexed by self-report and by questionnaire for the so-called primary-action measures (writing, drawing, eating with utensils, etc.). The data also show, however, that the extent of dominance is weaker than that reported for primary-action measures, at least in part because, as in infants, it is susceptible to other influences, in particular, the lateral position of the target and target size. Nevertheless, it would seem that Ogle (1871) was not far wrong in supposing that a human adult, with "few exceptions," would reach preferentially with the dominant hand.

The question may be raised about how to explain the discrepancy between our data and those by Teng et al. (1979). Recall that whereas only 48 percent of their subjects (older children and adults) said that they would favor the right hand for "reaching into a jar," and about an equal percentage said that they would not favor either hand (no preference), both the right- and left-handed adults in our study reached with the dominant hand on the great majority of the trials. The discrepancy could be due to the difference between Teng et al.'s task (reaching into a jar) and ours (reaching for an object placed on a table), but we think the more likely difference has to do with how the data are obtained. Borod, Caron, and Koff (1984) have shown that hand preference assessed by questionnaire (the method used by Teng et al.) yields a different estimate of hand preference from that assessed by performance measures. Particularly for a relatively undemanding task like reaching, we have more confidence in a measure of actual performance than in self-report on a questionnaire. In other words, it is understandable that an adult, asked which hand he would use for an undemanding task, would underestimate his degree of preference relative to how he would actually perform. This is what our data suggest is the case.

We were surprised by the absence of a distance effect in our infant study in contrast to what Baldwin (1890) found for his daughter. Presumably, the far distance, even though it required a full-arm extension from our subjects, did not increase the task demands sufficiently to give any further advantage to the preferred hand, whichever hand it was at any given age. Alternatively, our group analyses of the data may have obscured these effects in individual subjects. For example, not all infants stretched hard and exerted real effort to reach the object on the far-distance trials, the element that Baldwin believed was crucial for eliciting the lateral bias. As for our adult subjects, despite our attempt to create testing conditions comparable to those used for the infants, none of the adults showed any real effort on the far-distance trials, so it is possible that the far distance was not far enough.

We are likewise uncertain how to account for the object size effect. In the infants, no such effect appeared even though grasp style was usually adapted to object size in the way we expected. On further reflection, the crucial factor may not have been object size per se but, instead, how the object could be manipulated once it was grasped. For example, in their study of 9- to 20-month-old infants, Cornwell et al. (1991) found little differentiation among different objects in the hand used for a simple reach. Objects used for bimanual manipulation, however, elicited different patterns of hand preference from objects used as a target for a simple reach.

Given that our simple reaching task makes fewer motoric demands on adults than on infants, it is perplexing that, where object size had any effect, it was more on the adults than on the infants, and then only for the right-handed adults. Whereas the right-handers reached more often with the right hand for the pellet and cube than for the large ball, the left-handers showed no such differentiation. In other words, for objects of different sizes, the left-handers used their dominant hand more consistently than did the right-handers. This result is contrary to the patterns shown in handedness questionnaires, where, as mentioned earlier, right-handers typically are more consistent across different test items. Given the difference mentioned earlier between questionnaire and performance assessments of handedness, we are uncertain how to interpret this result.

Validity of Reaching as a Measure of Handedness in Human Adults

Let us return to our original question: Is reaching a valid measure of handedness in non-human primates — that is, in species for which the existence of handedness is the very question at issue? First, insofar as human adults show hand preference on this task consistent with their known handedness, it suggests that the task, at least in principle, would be appropriate as a test of handedness in other primates. But as we said before, if other primates were *not* to show hand preference on this measure, it would not necessarily mean that they lacked handedness. It might mean, instead, that handedness is so well-established in human adults that it is manifest even in a task that makes minimal demands on skill. On the other hand, the fact that hand preference on a reaching task is weaker than on more conventional handedness measures of the kind mentioned earlier, and that it is sensitive to peripheral factors, such as object size and lateral position, suggests that these or similar peripheral factors would operate even more strongly in a non-human primate. At the very least, the findings underscore the necessity for controlling for lateral position, object size, and other variables (including probably distance even though it had no obvious effect in our data) when the reaching task is used for non-human primates. Finally, if hand preference for reaching is not yet clearly established in human infants, we should be all the more cautious about using this measure to test hand preference in the young of other primate species.

Two other possible differences between humans and other primates should be mentioned. One is that, as MacNeilage, Studdert-Kennedy, and Lindblom (1987) have pointed out, at least certain non-human primates (monkeys, in their example) normally favor reaches that do not cross the body midline. If so, a test in which target objects are laterally-positioned may obscure genuine hand preference (e.g., Cronholm, Grodsky, & Behar, 1963; Lehman, 1970). As our data show, human infants as well as adults show a similar bias, except that the midline is far from being an absolute topographic obstacle. Indeed, hand preference for crossing over the midline proved to be especially revealing.

Another possible difference between humans and other primates, also suggested by MacNeilage et al. (1987), is that in non-human primates the right hand is preferred for object manipulation, whereas the left hand is preferred for grasping. MacNeilage et al. also note that this differentiation is not found in human beings. This leads them to propose that non-human primates and human beings are "separated by an evolutionary progression in which the importance of the ability to operate on the environment (including the use of bimanual coordination and the consequent right-hand preference) has so increased that the right hand [in human beings] now normally preempts the left, even for visually guided movement" (1987, p. 260).

Our adult data, for both right- and left-handers, look consistent with this analysis. Furthermore, at least in the case of our FRH infant girls, the preemption of the left hand by the right seems to occur very early in life. The evidence from human infants and adults thus would seem to support MacNeilage et al.'s proposal. Or does it? One possible objection is that our test of visually-guided reaching is like what MacNeilage et al. call a "simple reach" test, in which the subject merely picks up a single item placed on a flat surface. In such a test, as Beck and Barton (1972) found, monkeys likewise made more right- than left-hand reaches. The reason, according to MacNeilage et al., is that with only a small number of food locations along with the small range of required movements, the animal can plan and execute its reaching with minimal visual information. In our test, demands for visual guidance likewise may appear to have been relatively reduced, except that on each trial, the location of the test object varied unpredictably across nine different positions in two dimensions. As already noted, our impression was that for the infants, at least after 24 weeks, the task elicited strong visual engagement and sustained effort. Although this was less true for the adults, even for them the pace of the task and the unpredictability of the target location and in the type of grasp required for the different objects at least meant that no reach could be initiated and completed without visual inspection. So perhaps our test was not a "simple reach" test after all. Again, because reaching tasks have never been part of the standard assessment of adult handedness, we cannot resolve this matter at this time. What does seem clear is that task analysis of the kind applied to handedness (e.g., Annett, 1985; Provins, 1956; Provins et al., 1982) could be profitably applied to the analysis of visually-guided reaching, with appropriate controls for sex and handedness.

References

Annett, M. (1970). The growth of manual preference and speed. *British Journal of Psychology, 61*, 545-558.

Annett, M. (1981). The genetics of handedness. *Trends in Neurosciences, 3*, 256-258.

Annett, M. (1985). *Left, right, hand and brain: The right shift theory*. London: Lawrence Erlbaum Associates.

Baldwin, J.M. (1890). Origin of right or lefthandedess. *Science, 16*, 247-248.

Baldwin, J.M. (1895). *Mental development in the child and the race*. New York: Macmillan.

Beck, C.H.M., & Barton, R.L. (1972). Deviation and laterality of hand preference in monkeys. *Cortex, 8*, 339-363.

Bishop, D.V.M. (1980). Handedness, clumsiness and cognitive ability. *Developmental Medicine and Child Neurology, 22*, 569-579.

Borod, J.C., Caron, H.S., & Koff, E. (1984). Left-handers and right-handers compared on performance and preference measures of lateral dominance. *British Journal of Psychology, 75*, 177-186.

Bradshaw, J.L., & Nettleton, N.C. (1981). The nature of hemispheric specialization in man (target article with commentaries). *The Behavioral and Brain Sciences, 4*, 51-91.

Bryden, M.P. (1977). Measuring handedness with questionnaires. *Neuropsychologia, 15*, 617-624.

Bryden, M.P. (1982). *Laterality: Functional asymmetry in the intact brain*. New York: Academic Press.

Carlson, D.F., & Harris, L.J. (1985). Development of the infant's hand preference for visually directed reaching: Preliminary report of a longitudinal study. *Infant Mental Health Journal, 6*, 158-172.

Cohen, G. (1973). Hemispheric differences in serial versus parallel processing. *Journal of Experimental Psychology, 97*, 349-356.

Cornwell, K.S., Harris, L.J., & Fitzgerald, H.I. (1991). Task effects in the development of hand preference in 9-, 13-, and 20-month-old infant girls. *Developmental Neuropsychology, 7*, 19-34.

Cronholm, J.N., Grodsky, M., & Behar, I. (1963). Situational factors in the lateral preferences of rhesus monkeys. *Journal of Genetic Psychology, 103*, 155-175.

Crovitz, H.F., & Zener, K. (1962). A group test for assessing hand and eye dominance. *American Journal of Psychology, 75*, 271-276.

Finch, G. (1941). Chimpanzee handedness. *Science, 94*, 117-118.

Franz, S.I. (1913). Observations on the preferential use of the right and left hands by monkeys. *Journal of Animal Behavior, 3*, 140-144.

Guiard, Y., Diaz, G., & Beaubaton, D. (1983). Left-hand advantage in right-handers for spatial constant error: preliminary evidence in a unimanual ballistic movement. *Neuropsychologia, 21*, 111-115.

Harris, L.J. (1980). Which hand is the 'eye' of the blind? A new look at an old question. In J. Herron (Ed.), *Neuropsychology of left-handedness* (pp. 303-329). New York: Academic Press.

Harris, L.J. (1983). Laterality of function in the infant: Historical and contemporary trends in theory and research. In G.Young, S.J.Segalowitz, C.M.Corter, & S.E.Trehub (Eds.), *Manual specialization and the developing brain* (pp. 177-247), New York: Academic Press.

Harris, L.J. (1986). James Mark Baldwin on the origins of right- and left-handedness: The story of an experiment that mattered. In A.B. Smuts & J.W. Hagen (Eds.), History of Research in Child Development. *Monographs of the Society for Research in Child Development* (pp. 44-64). Special Issue in Celebration of the Fiftieth Anniversary of the Society, Serial No. 211, *50*, Nos. 4-5.

Hawn, P.R., & Harris, L.J. (1983). Hand differences in grasp duration and reaching in two- and five-month-old infants. In G. Young, S.J. Segalowitz, C.M. Corter, & S.E. Trehub (Eds.), *Manual specialization and the developing brain* (pp. 331-348). New York: Academic Press.

Healey, J.M., Liederman, J., & Geschwind, N.(1986). Handedness is not a unidimensional trait. *Cortex, 22*, 33-53.

Hollis, W.A. (1875). Lopsided generations. *Journal of Anatomy and Physiology, 9*, 263-271.

Kimura, D. (1982). Left-hemisphere control of oral and brachial movements and their relation to communication. *Philosophical Transactions of the Royal Society of London, B298*, 135-149.

Kimura, D. (1983a). Sex differences in cerebral organization for speech and praxic function. *Canadian Journal of Psychology, 37*, 19-35.

Kimura, D., & Harshman, R.A. (1984). Sex differences in brain organization for verbal and non-verbal functions. In G.J. De Vries, J.P.C. De Bruin, H.B.M. Uylings, & M.A. Corner (Eds.), *Sex differences in the brain: The relation between structure and function: Vol. 61. Progress in brain research* (pp. 423-441). Amsterdam: Elsevier Science.

Lehman, R.A.W. (1970). Hand preference and cerebral predominance in 24 rhesus moneys. *Journal of the Neurological Sciences, 10*, 185-192.

Levy, J. (1974). Psychobiological implications of bilateral asymmetry. In S.J. Dimond & J.G. Beaumont (Eds.), *Hemisphere function in the human brain* (pp. 121-183). New York: Wiley.

Liederman, J. (1983). Mechanisms underlying instability in the development of hand preference. In G. Young, S.J. Segalowitz, C.M. Corter, & S.E. Trehub (Eds.), *Manual specialization and the developing brain* (pp. 71-92). New York: Academic Press.

MacNeilage, P.F., Studdert-Kennedy, M.G., & Lindblom, B. (1987). Primate handedness reconsidered (target article with commentaries). *The Behavioral and Brain Sciences. 10*, 247-303.

McKeever, W.F., & Van Deventer, A.D. (1975). Familial sinistrality and degree of left handedness. *British Journal of Psychology, 68*, 469-471.

Michel, G.F., & Harkins, D.A. (1985). Postural and lateral asymmetries in the ontogeny of handedness during infancy. *Developmental Psychobiology, 19*, 247-258.

Michel, G.F., Ovrut, M.R., & Harkins, D.A. (1986). Hand-use preference for reaching and object manipulation in 6- to 13-month-old infants. *Journal of Genetic, Social, and General Psychology, 111*, 407-428.

Moore, T. (1967). Language and intelligence: A longitudinal study of the first eight years. Part I: Patterns of development in boys and girls. *Human Development, 10*, 88-106.

Ogle, W. (1871). On dextral pre-eminence. *Transactions of the Royal Medical and Chirurgical Society of London, 54*, 279-301.

Oldfield, R.C. (1971). The assessment and analysis of handedness: The Edinburgh Inventory. *Neuropsychologia, 9*, 97-111.

Peters, M., & Durding, B. (1979). Left-handers and right-handers compared on a motor task. *Journal of Motor Behavior, 11*, 103-111.

Provins, K.A. (1956). Handedness and skill. *Quarterly Journal of Experimental Psycholgy, 8*, 79-95.

Provins, K.A., Milner, A., & Kerr, P. (1982). Asymmetry of manual preference and performance. *Perceptual and Motor Skills, 54*, 179-194.

Saling, M. (1983). Familial handedness, prenatal environmental adversity, and neonatal lateral organization. In G. Young, S.J. Segalowitz, C.M. Corter, & S.E. Trehub (Eds.), *Manual specialization and the developing brain* (pp. 275-284). New York: Academic Press.

Shucard, J.L., Shucard, D.W., Cummins, K.R., & Campos, J.J. (1981). Auditory evoked potentials and sex related differences in brain development. *Brain and Language, 13*, 91-102.

Tanner, J.M. (1978). *Fetus into man: Physical growth from conception to maturity.* Cambridge, Mass.: Harvard University Press.

Tapley, S.M., & Bryden, M.P. (1985). A group test for the assessment of performance between the hands. *Neuropsychologia, 23*, 215-221.

Teng, E.L., Lee, P-H., Yang, K-S., & Chang, P.C. (1979). Lateral preferences for hand, foot and eye and their lack of association with 4143 Chinese. *Neuropsychologia, 17*, 41-48.

Todor, J.I., & Cisneros, J. (1985). Accommodation to increased accuracy demands by the right and left hands. *Journal of Motor Behavior, 17*, 355-372.

Wells, G. (1979). Variation in child language. In P. Fletcher & M. Garman (Eds.), *Language acquisition: Studies in first language development* (pp. 377-395). Cambridge: Cambridge University Press.

Witelson, S.F. (1985). On hemispheric specialization and cerebral plasticity: Mark II. In C.T. Best (Ed.), *Hemispheric function and collaboration in the child* (pp. 33-85). Orlando, Florida: Academic Press.

Young, G., Segalowitz, S.J., Corter, C.M., & Trehub, S.E. (Eds.). (1983). *Manual specialization and the developing brain.* New York: Academic Press.

14
Rotational Behavior in Children and Adults

Stanley D. Glick

It is now well accepted that the two hemispheres of the human brain are different, or asymmetric, with regard to several functions. In most individuals the left hemisphere is more specialized for language processing and the right hemisphere for spatial ability and emotional expression. Because these attributes have generally been ascribed to the cortex, there has been very little interest in the possibility of there being functional asymmetries in subcortical regions. However, asymmetry in one subcortical structure, the corpus striatum, has been extensively investigated in rats, and recent studies have extended this work to humans. The present review summarizes this latter work and its potential for helping to understand the behavioral pathology and symptomatic treatment of particular neurological and psychiatric disorders.

Nigrostriatal Asymmetry and Circling in Rats

The propensity of animals to turn in circles was the starting point of research on cerebral lateralization in this laboratory. Although a relationship between turning and striatal function had been in the literature for more than a century (e.g., Wilson, 1914), it was not until about 1970 that sustained interest in this area developed. The discovery of L-DOPA's efficacy in treating the symptoms of Parkinsonism (Cotzias, Van Woert, & Shiffer, 1967) led to a search for animal models that could be used to evaluate new and potentially more effective agents for this disease. Such a model was provided when Ungerstedt (1971) demonstrated that rats with unilateral lesions of the nigrostriatal pathways would turn in circles (rotate) in response to a variety of dopaminergic drugs (e.g., d-amphetamine, L-DOPA, apomorphine). Soon afterwards, in this laboratory, it was discovered that such drugs would also induce rotation in normal rats (Jerussi & Glick, 1976a). Although normal rats rotated at lower rates than did lesioned rats, the rotation in normal as well as in lesioned rats was consistent in direction and correlated in magnitude upon repeated testing. It was further shown that normal untreated rats would rotate spontaneously at night (the more active half of their circadian cycle) and that this rotation was usually in the same direction as that induced by amphetamine during the day (Glick & Cox, 1978). Since the rotational behavior in lesioned rats was attributed to a functional asymmetry resulting from the unilateral loss of dopaminergic neurons, it was hypothesized that this behavior in normal rats was due to an endogenous asymmetry of these same neurons. Subsequent studies showed that normal rats had hemispheric differences in striatal dopamine levels (Zimmerberg, Glick, & Jerussi, 1974), dopamine metabolites (Jerussi, Glick, & Johnson, 1977), dopamine release (Glick, Carlson, Baird, Maisonneuve, & Bullock, 1988; Robinson, Becker, & Camp, 1983), dopamine uptake (Shapiro, Glick, & Hough, 1986), and dopamine receptors (Drew, Lyon, Titeler, & Glick, 1986; Glick, Lyon, Hinds, Sowek, & Titeler, 1988); most of these neurochemical differences were related to or correlated with rats' nocturnal or amphetamine-induced rotational behavior.

Dopamine Asymmetry and Rotation in Humans

A report that neuroleptic-induced dyskinesias are lateralized in humans (Waziri, 1980) initially suggested that an asymmetry in basal ganglia dopamine, similar to that observed in rats, would be identified in humans. Indeed, a reanalysis of postmortem data (Rossor, Garrett & Iverson, 1980) revealed that the human brain had neurochemical asymmetries in several structures and neurotransmitter systems (Glick, Ross & Hough, 1982). Most relevant to the foregoing animal work was the finding of a left-biased dopamine asymmetry in human globus pallidus. Pallidal dopamine appeared to be higher on the side contralateral to hand preferences -- analogous, perhaps, to striatal dopamine levels being higher on the side contralateral to side preferences in rats (Zimmerberg et al., 1974). The similar dopamine asymmetries in rats and humans indicated that a direct functional comparison between species might be possible if rotational behavior occurred and were measurable in humans. Accordingly, a device to monitor rotation in humans was designed and a prototype was built (Greenstein & Glick, 1981). The prototype was constructed from a compass, a Casio calculator, and an electronic interface. The needle position of the compass was monitored by four photoelectric sensors. As the subject moved in his/her environment, the sensors moved with the subject while the compass needle maintained a "fix" on magnetic North. This resulted in the needle moving over the sensors, breaking the light beams. The four sensors divided space into quadrants and allowed movements to be identified as either quarter (90°) or full (360°) turns. The calculator was programmed to store data continuously and, by pressing a key, to display successively left quarter turns, right quarter turns, left full turns and right full turns. If a subject entered four quadrants sequentially in the same direction (i.e., moves 360°), a full turn (as well as four quarter turns) was counted for the respective direction. If, however, a subject entered three quadrants sequentially in the same direction but then entered a quadrant from the opposite direction of the previous move, a new direction was begun and only quarter turns were counted. The output was thus determined by precisely the same logic employed in the analogous apparatus commonly used for rodents (Greenstein & Glick, 1975).

Early pilot studies with the prototype apparatus strongly suggested that "normal" persons rotated preferentially to the left or right during the course of an ordinary day. Although these tests, conducted on laboratory personnel, were not "blind" (i.e., subjects were aware of the kinds of movement being recorded), repeated tests on the same subjects showed that an individual's preferred direction of rotation was consistent from one day to the next. Subsequently, in collaboration with Dr. S. Leighton (Biomedical Engineering and Instrumentation Branch, National Institutes of Health) and Dr. H. S. Bracha (then at the Neuropsychiatry Branch, National Institute of Mental Health; now at the University of Arkansas for Medical Sciences), a more refined rotometer unit was designed and built -- it is now a small, lightweight, rechargeable device which the subject wears in a belt-mounted calculator case (the prototype was also worn in this way). The device consists of two functional components: the position sensor and the electronic processing circuit. As in the prototype, the position sensor entails a compass transducer system that uses magnetic North as an external reference. The essence of the electronic processing circuit is a microchip programmed with algorithms for generating left and right full (360°) turns as well as left and right quarter (90°) turns. This new rotometer, now available commercially from Beecher Co. (Silver Spring, Maryland), has already been utilized in several studies.

Sex and Side Differences in Adult Rotation

The first extensive study (Bracha, Seitz, Otemaa, & Glick, 1987) of rotation in normal humans involved students, physicians, and hospital employees at the University of California, San Diego; there were 74 males and 61 females ranging in age from 15-59 years. Each subject wore the rotometer for approximately eight hours during the course of a routine workday and was completely unaware of the type of measurements obtained by the device. Sixty-four subjects (31 males and 33 females) were tested a second time, about six weeks later. Hand, foot, and eye dominance were determined by using a 13-item laterality scale from the Halstead-Reitan battery (Reitan,

Table 14.1 Rotational Indices (Mean ± s.d.) in Adults and Children

	Net Rotations	% Preference
Adults*		
Males (N=74)	12.5 ± 13.0	59.3 ± 7.4
Females (N=61)	19.5 ± 17.9	62.1 ± 8.4
Children		
Males (N=26)	33.8 ± 37.4	56.5 ± 4.8
Females (N=30)	19.5 ± 25.6	55.6 ± 4.6

*Significant difference between sexes for both measures ($p<.02$ and $p<.05$, respectively, t-tests).

1979). Net rotation and percent preference were used as indices of rotational asymmetry: net rotation refers to the absolute difference between full turns in the preferred and non-preferred directions; percent preference refers to the percentage of full turns in the preferred direction. Extra quarter turns (XQT), a measure of non-lateralized activity (Greenstein & Glick, 1975), refers to the result of subtracting four times the number of full turns from the number of quarter turns.

Both measures of rotation were significantly greater in females than in males as seen in Table 14.1. This was attributable to a specific lateralized effect: whereas females made significantly more full turns in their preferred direction than did males (51.6 ± 3.3 vs. 38.8 ± 2.5, p <.01), there was no significant sex difference in non-preferred full turns (32.1 ± 2.6 vs. 26.3 ± 1.8); the sexes also did not differ in non-lateralized (total XQT) activity (896.5 ± 112.2 vs. 691.6 ± 92.3).

There were sex-dependent relationships between rotational behavior and the pattern of hand, foot, and eye dominance. Male subjects were found to be consistently left-hemisphere dominant (i.e., right-sided with respect to hand, foot, and eye dominance), had predominantly right-turning biases (28 right vs. 13 left) and, as a group, made significantly more full turns to the right than to the left (36.9 ± 3.7 vs. 31.0 ± 2.4); mixed dominance male subjects had predominantly left turning biases (19 left vs. 7 right) and made significantly more full turns to the left than to the right (34.1 ± 3.6 vs. 26.2 ± 2.7). In contrast, left-hemisphere dominant female subjects had predominantly left-turning biases (34 left vs. 18 right) and made significantly more full turns to the left than to the right (45.2 ± 4.0 vs. 37.0 ± 3.0); the few mixed dominance females had no overall group bias (5 left vs. 4 right; left/right full turns: 47.0 ± 8.7 vs. 44.7 ± 10.7).

Male turning preferences were generally in the same direction as hand, foot, and eye preferences whereas female turning preferences were generally opposite to the direction of hand, foot, and eye preferences. Furthermore, in males, the concordance with turning direction was clearest and significant (p<.02) only for eye preference whereas in females, the discordance with turning direction was clearest for hand preference and significant only for hand and foot preferences ($p<.02$ and $p<.05$, respectively).

Overall, when tested on two occasions, subjects showed consistent preferences ($p<.001$): 48 of 64 subjects rotated in the same direction from one test to the next, and both net rotations and percent preference were significantly correlated. These relationships were stronger when only left-hemisphere dominant subjects were considered, in as much as mixed dominance subjects evidenced no such consistency (e.g., 40 of 50 left-hemisphere dominant subjects had consistent directions of rotation compared to 8 of 14 mixed dominance subjects).

Rotational Behavior of Children

A study (Gospe, Mora & Glick, 1990) of prepubertal rotation involved 56 children between the ages of six and nine; there were 30 girls (mean age = 7.88 years) and 26 boys (mean age = 7.88 years). Subjects were recruited through local pediatric practices in the Albany area. Similar to the procedure in adults, each child's hand, foot, and eye preferences were determined using a modified laterality scale of the Halstead-Reitan battery (Reitan, 1979). Subjects were all evaluated in their homes. Each child wore the rotometer for four hours; activity was not restricted and all subjects were completely unaware of the purpose of the study and of the type of information being recorded by the device.

Overall, males and females did not differ significantly with respect to any of the measures of rotation (see Table 16.1) and activity (males vs. females: total full turns = 235.5 $\pm$ 169.6 vs. 183.0 $\pm$ 159.6; total XQT= 4943 $\pm$ 3006 vs. 3525 $\pm$ 2923). However, left-hemisphere dominant males were significantly stronger rotators than left-hemisphere dominant females (net rotations = 46.7 $\pm$ 46.3 vs. 11.4 $\pm$ 8.4). The most striking result was the occurrence of a left turning bias in both sexes: 24 of 30 girls (80%) and 20 of 26 boys (77%) rotated preferentially to the left. Moreover, in both sexes, net rotation and percentage preference were significantly higher in left-turning than in right-turning children. These data are depicted in Table 14.2.

Associations between direction of turning bias and hand, foot, or eye preference were apparent. In both sexes, turning preferences were generally in the opposite direction as hand, foot, and eye preferences, although this discordance was significant only for hand and foot preferences ($p<.05$ and $p<.01$, respectively for males; and $p<.02$ and $p<.05$, respectively, for females).

Table 14.2. Rotational Indices (Mean ± s.d.) in Left-turning vs. Right-turning Children*

	Net Rotations	% Preference
Left-turning		
Males (N=20)	40.3 ± 39.7	57.6 ± 4.8
Females (N=24)	23.0 ± 27.6	56.4 ± 4.8
Right-turning		
Males (N=6)	12.2 ± 16.5	52.7 ± 2.5
Females (N=6)	5.5 ± 4.7	52.2 ± 1.9

*Significant difference between directions for both measures ($p<.03$ and $p<.002$, respectively, ANOVAs).

Adults vs. Children: Similarities and Differences

In 1885 Ernst Mach wrote that "human beings and animals that have lost their direction move, almost without exception, nearly in circle." In 1966, it was noted by Howard and Templeton that people lost in fogs find that they walk in complete circles. The studies described above indicate that being lost, in fogs or otherwise, is not a prerequisite for spontaneous circling and that human beings, similar to many animal species including rats (Glick, Jerussi, & Zimmerberg, 1977; Glick, Shapiro, Drew, Hinds, & Carlson, 1986), mice (Glick, Zimmerberg, & Greenstein, 1976), gerbils (Schonfeld & Glick, 1980), hamsters (Jerussi & Glick, 1976b), dogs (Glick, 1976; Nymark, 1972), cats (Glick, Weaver, & Meibach, 1981), and monkeys (Glick & Cox, 1979), exhibit preferentially directed circling during the course of normal or routine activities. The precise functional significance of lateralized rotation in humans is not yet known as there are no data directly bearing on this issue. Studies in rats (Glick, Zimmerberg, & Jerussi, 1977; Zimmerberg, Strumpf, & Glick, 1978) have suggested that rotational behavior results from a persistent lateral or orienting bias — this bias appears to function as a proprioceptive reference (analogous, perhaps, to a biological compass) such that alterations of proprioceptive feedback from the biased side enable the organism to differentiate between left and right movement.

As in adult rats (e.g., Glick & Shapiro, 1985; Glick et al., 1986; Robinson et al., 1985), a sex difference in the magnitude of rotation was apparent in adult humans: females rotated more than males. In children, the sex difference appeared to be reversed, at least for left-hemisphere dominant subjects. The latter appeared to be attributable to a substantial age-related change in males as compared to females; even though the test periods were shorter, net rotations of male children averaged more than twice those of male adults whereas female children and adults had nearly equal values. Analogous developmental trends appear to occur in rats, with sex differences absent or reversed in juvenile rats as compared to adults (Hyde & Jerussi, 1983).

The vast majority of children preferred to rotate to the left, and this left turning bias was significant for both left-hemisphere dominant boys and left-hemisphere dominant girls. Left-hemisphere dominant women also had a left population bias, whereas left-hemisphere dominant men had a right population bias. In children of both sexes, hand and foot preferences were significantly associated with the opposite direction of rotation. Similar relationships were significant for adult females, whereas only an association of eye dominance with the same direction of rotation was significant in adult males. Several trends indicate, therefore, that developmental changes in rotational asymmetry are more prominent in males than in females. The mechanisms of hand, foot, and eye dominance are probably only very loosely related to the mechanism of turning in as much as such associations are sex and age dependent.

Adult subjects tested on two occasions exhibited consistent directions and magnitudes of turning. This is a very important finding because it strongly implies that turning biases are characteristic of the subjects themselves rather than being coincidently related to environmental events. Left-hemisphere dominant subjects had markedly more consistent biases than mixed dominance subjects. This result suggests that different behavioral asymmetries which, as defined here, are more poorly coordinated in mixed dominance subjects, are also more variably expressed across time in these subjects.

Altered Rotation in Neurological and Psychiatric Disorders

Rotational behavior has been measured in three studies of patients having neuropsychiatric disorders. In the first and perhaps most obvious study to conduct, five patients with hemi-Parkinson's disease were tested (Bracha, Shults, Glick & Kleinman, 1987). Although, the patients were medicated with levodopa/carbidopa, all exhibted marked unilateral Parkinsonian symptoms (Hoehn and Yahr scale, Type 1); the symptoms were predominantly right-sided in four of the subjects and predominantly left-sided in the fifth subject. All patients were found to exhibit spontaneous rotation toward the hemisphere that presumably had less dopaminergic activity (i.e., left-sided rotation in the four patients having right-sided symptoms and right-sided rotation in the one patient having left-sided symptoms). Considered as a group, the strength of rotation in the hemi-Parkinson's patients was significantly greater than the strength of rotation in a normal control group. This result strongly resembled the results of numerous studies with animals having unilateral lesions of the nigrostriatal pathways (Glick, Jerussi, & Fleisher, 1976; Pycock, 1980).

In 1973, Glick and Greenstein reported that administration of amphetamine to rats having unilateral ablations of frontal cortex induced ipsilateral rotation soon (1-7 days) after surgery and contralateral rotation at later times (15-30 days). The early change was attributed to a loss of a corticostriatal input and the later change to a compensatory striatal response; that is, it was hypothesized that the dopaminergic nigrostriatal system is modulated by the frontal cortex and that unilateral removal of such cortex produces denervation supersensitivity within the striatum. Bracha, Lyden, and Khansarinia (1989) recently reported clinical data bearing a strong resemblance to the animal findings. Outpatients having old (greater than six months) frontal or inferior-parietalcortical strokes exhibited marked contralateral rotational preferences (i.e., turning directed away from the side of the lesion). The authors inferred that there was "a delayed emergence of lasting ipsilateral striatal dopaminergic hyperactivity after anterior-lateral cortical insult in humans (1989, p. 273)."

Preliminary results have indicated that rotational behavior may also be abnormal in psychiatric disorders. Bracha (1987) reported that each of ten male unmedicated schizophrenic patients exhibited a left-sided rotational preference. He speculated that either there may be an asymmetric striatal involvement in schizophrenia or that the altered rotation may be secondary to a "frontal hemispatial neglect" syndrome in schizophrenia. Further studies will be necessary to substantiate this finding as well as to determine whether changes in rotational behavior can be used to monitor the efficacy of drug treatment.

Although the ultimate utility of rotational measurement in humans is far from being known, it is noteworthy that a behavioral asymmetry first identified in animals also occurs, with very similar characteristics, in humans. At least one kind of brain asymmetry appears to have evolved as part of a common mechanism underlying directional and/or spatial functions (Glick et al., 1977). The observations reviewed here suggest that concurrent animal and human studies will continue to reveal the importance of neural lateralization in both normal and pathological behavior.

Acknowledgements

The author is grateful for the numerous collaborative relationships that made possible much of the work reviewed here. In particular, I would like to thank Mr. S. Greenstein and Drs. S. Bracha, S. Leighton, and S. Gospe. Studies in animals were supported by NINCDS grant NS-14812 and NIDA grants DA-01044 and DA-03817.

References

Bracha, H. S. (1987). Asymmetrical rotational (circling) behavior, a dopamine-related asymmetry: preliminary findings in unmedicated and never-medicated schizophrenic patients. *Biological Psychiatry, 22*, 995-1003.

Bracha, H. S., Lyden, P. D., & Khansarinia, S. (1989). Delayed emergence of striatal dopaminergic hyperactivity after antero-lateral ischemic cortical lesions in humans;evidence from turning behavior. *Biological Psychiatry, 25*, 265-274.

Bracha, H. S., Seitz, D. J., Otemaa, J., & Glick, S. D. (1987). Rotational movement (circling) in normal humans: Sex difference and relationship to hand, foot and eye preference. *Brain Research, 411*, 231-235.

Bracha, H. S., Shults, C., Glick, S. D., & Kleinman, J. E. (1987). Spontaneous asymmetric circling behavior in hemi-Parkinsonism; a human equivalent of the lesioned-circling rodent behavior. *Life Sciences, 40*, 1127-1130.

Cotzias, G. C., Van Woert, M. H., & Shiffer, L. M. (1967). Aromatic amino acids and modification of Parkinsonism. *New England Journal of Medicine, 276*, 374-379.

Drew, K. L., Lyon, R. A., Titeler, M., & Glick, S. D. (1986). Asymmetry in D-2 binding in female rat striata. *Brain Research, 363*, 192-195.

Glick, S. D. (1976). Unpublished data on eight mongrel dogs.

Glick, S. D., Carlson, J. N., Baird. J. L., Maisonneuve, I. M., & Bullock, A. E. (1988). Basal and amphetamine-induced asymmetries in striatal dopamine release and metabolism: Bilateral in vivo microdialysis in normal rats. *Brain Research, 473*, 161-164.

Glick, S. D., & Cox, R. D. (1978). Nocturnal rotation in normal rats: correlation with amphetamine-induced rotation and effects of nigrostriatal lesions. *Brain Research, 150*, 149-161.

Glick, S. D., & Cox, R. D. (1979). Unpublished data on four rhesus monkeys.

Glick, S. D., & Greenstein, S. (1973). Possible modulating influence of frontal cortex on nigro-striatal function. *British Journal of Pharmacology, 49*, 316-321.

Glick, S. D., Jerussi, T. P., & Fleisher, L. N. (1976). Turning in circles: The neuropharmacology of rotation. *Life Sciences, 18*, 889-896.

Glick, S. D., Jerussi, T. P., & Zimmerberg, B. (1977). Behavioral and neuropharmacological correlates of nigrostriatal asymmetry in rats. In S. Harnad, J. Jaynes, L. Goldstein, R. W. Doty, & G. Krauthamer (Eds.), *Lateralization in the nervous system* (pp. 213-249). New York: Academic Press.

Glick, S. D., Lyon, R. A., Hinds, P. A., Sowek, C., & Titeler, M. (1988). Correlated asymmetries in striatal D_1 and D_2 binding: relationship to apomorphine-induced rotation. *Brain Research, 455*, 43-48.

Glick, S. D., Ross, D. A., & Hough, L. B. (1982). Lateral asymmetry of neurotransmitters in human brain. *Brain Research, 234*, 53-63.

Glick, S. D., & Shapiro, R. M. (1985). Functional and neurochemical mechanisms of cerebral lateralization in rats. In S. D. Glick (Ed.), *Cerebral lateralization in nonhuman species* (pp.157-183). Orlando: Academic Press.

Glick, S. D., Shapiro, R. M., Drew, K. L., Hinds, P. A., & Carlson, J. N. (1986). Differences in spontaneous and amphetamine-induced rotational behavior, and in sensitization to amphetamine, among Sprague-Dawley derived rats from different sources. *Physiology and Behavior, 38*, 67-70.

Glick, S. D., Weaver, L. M., & Meibach, R. C. (1981). Amphetamine-induced rotation in normal cats. *Brain Research, 208*, 227-229.

Glick, S. D., Zimmerberg, B., & Greenstein, S. (1976). Individual differences among mice in normal and amphetamine-enhanced locomotor activity: relationship to behavioral indices of striatal asymmetry. *Brain Research, 105*, 362-364.

Glick, S. D., Zimmerberg, B., & Jerussi, T. P. (1977). Adaptive significance of laterality in the rat. *Annals of the New York Academy of Sciences, 299*, 180-185.

Gospe, S. M., Mora, B. J., & Glick, S. D. (1990). Measurement of spontaneous rotational movement (circling) in normal children. *Journal of Child Neurology, 5*, 31-34.

Greenstein, S., & Glick, S. D. (1975). Improved automated apparatus for recording rotation in rats or mice. *Pharmacology, Biochemistry, and Behavior, 3*, 507-510.

Greenstein, S., & Glick, S. D. (1981). Asymmetrical behavior monitor. Invention Description (unpublished), Mount Sinai School of Medicine.

Howard, I. P., & Templeton, W. B. (1966). *Human spatial orientation*. New York: Wiley.

Hyde, J. F., & Jerussi, T. P. (1983). Sexual dimorphism in rats with respect to locomotor activity and circling behavior. *Pharmacology, Biochemistry and Behavior, 18*, 725-729.

Jerussi, T. P., & Glick, S. D. (1976a). Drug-induced rotation in rats without lesions: behavioral and neurochemical indices of a normal asymmetry in nigrostriatal function. *Psychopharmacology, 47*, 249-260.

Jerussi, T. P., & Glick, S. D. (1976b). Unpublished data on 24 hamsters.

Jerussi, T. P., Glick, S. D., & Johnson, C. L. (1977). Reciprocity of pre- and post-synaptic mechanisms involved in rotation as revealed by dopamine metabolism and adenylate cyclase stimulation. *Brain Research, 129*, 385-388.

Mach, E. (1885). *The analysis of sensations*. New York: Dover (translated and reprinted, 1959).

Nymark, M. (1972). Apomorphine provoked stereotypy in the dog. *Psychopharmacologia, 26*, 361-368.

Pycock, C. J. (1980). Turning behaviour in animals. *Neuroscience, 5*, 461-514.

Reitan, R. M. (1979). *Manual for administration of neuropsychological test batteries for adults and children*. Tucson, AZ: Reitan Neuropsychology Laboratories.

Robinson, T. E., Becker, J., & Camp, D. M. (1983). Sex differences in behavioral and brain asymmetries. In M. S. Myslobodsky (Ed.), *Hemisyndromes* (pp.91-128). Orlando: Academic Press.

Robinson, T. E., Becker, J., Camp, D. M., & Mansour, A. (1985). Variation in the pattern of behavioral and brain asymmetries due to sex differences. In S. D. Glick(Ed.), *Cerebral lateralization in nonhuman species* (pp.185-231). Orlando: Academic Press.

Rossor, M., Garrett, N., & Iversen, L. (1980). No evidence for lateral asymmetry of neurotransmitters in post-mortem human brain. *Journal of Neurochemistry, 35*, 743-745.

Schonfeld, A. R., & Glick, S. D. (1980). Circadian studies of the spontaneous rotational behavior of the gerbil. *Physiology and Behavior, 25*, 337-340.

Shapiro, R. M., Glick, S. D., & Hough, L. B. (1986). Striatal dopamine uptake asymmetries and rotational behavior in unlesioned rats: Revising the model? *Psychopharmacology, 89*, 25-30.

Ungerstedt, U. (1971). Postsynaptic supersensitivity after 6-hydroxydopamine induced degeneration of the nigrostriatal dopamine system. *Acta Physiologica Scandinavia Supplement, 367*, 69-93.

Waziri, R. (1980). Lateralization of neuroleptic-induced dyskinesia indicates pharmacologic asymmetry in the brain. *Psychopharmacology, 68*, 51-53.

Wilson, S. A. K. (1914). An experimental approach into the anatomy and physiology of the corpus striatum. *Brain, 36*, 425-427.

Zimmerberg, B., Glick, S. D., & Jerussi, T. P. (1974). Neurochemical correlate of a spatial preference in rats. *Science, 185*, 623-625.

Zimmerberg, B., Strumpf, A. J., & Glick, S. D. (1978) Cerebral asymmetry and left-right discrimination. *Brain Research, 140*, 194-196.

15 Implications of Primate Functional Asymmetries for the Evolution of Cerebral Hemispheric Specializations

Peter F. MacNeilage

"Ape-like they are, and simian.
Not like normal men and wimian".
Ogden Nash

Primate Functional Asymmetries: What Took Us So Long?

This book is a milestone in human attempts to understand ourselves and our world. It is the first book entirely devoted to the question of behavioral asymmetries in *all* primates. The existence of the book (for which the editors deserve our gratitude) is an indirect result of the accumulation of enough instances of lateralized behavior in nonhuman primates at the population level to demand the conclusion that this phenomenon exists, although there will, no doubt, continue to be tenacious resistance to this conclusion. In my opinion, this conclusion has profound implications for human evolution and the evolution of behavior in general. It comes in the context of the general acceptance, at least by the initiated, of the existence of functional asymmetries in other nonhuman mammals (Glick, 1985), and thus, in some ways, provides a missing link in an attempt to understand the evolution of hemispheric specialization in humans. In addition, left sided control of the song of some birds was discovered many years ago (See Arnold & Bottjer, 1985, for a review.) I wish to echo the sentiments of Geschwind (1985), who concluded that the recognition of asymmetry in the nonhuman nervous system in general "is likely to lead to major conceptual advances in fields as widely disparate as human evolution, linguistics, psychology and psychiatry. If these suppositions are correct, it will also have major philosophical repercussions." (1985, p. 248). Geschwind would have liked this book.

The implications of nonhuman asymmetries for human behavior are profound because they suggest that the presently accepted conclusion that hominid-specific lateralized brain specializations are keys to our uniqueness is wrong, and therefore needs to be replaced. The problem is most serious for handedness, which has hitherto been one of the main cornerstones of arguments for human uniqueness. For example, we find the following statements in the entries for Handedness and Hand in the *Oxford Companion to the Mind* (Gregory, 1987):

> HANDEDNESS is a characteristic peculiar to man. Only in man do individuals have a consistently preferred (and more skillful) hand for manual actions, and only man is predominantly right-handed.
> HAND. The anatomist Wood Jones (1879-1954) proposed that 'man's place in nature is largely writ upon the hand'. ...So mind may depend as much upon the hands as it depends on the senses for gaining knowledge and so developing. (1987, p. 300)

The implications of the existence of functional asymmetries in *all* primates are also profound for the study of nonhuman primates because they require a new view of their behavior and brain function as well, and because they suggest the possibility of nonhuman primate models for the study of human asymmetry (Ward, 1991).

How different will our thinking about functional asymmetries and their significance have to be in the future? Current thinking about functional asymmetries in vertebrates is nicely illustrated by two entries in the *Encyclopedia of Neuroscience* (Adelman, 1987). One, a two page entry entitled "Brain Asymmetry, Animal" (Shapiro, Drew, & Glick, 1987) deals exclusively with the "nonhuman animal brain". The other, a four page entry entitled "Brain Asymmetry, Functional Aspects" (Witelson, 1987) departs partially from the topic of human asymmetries in only one sentence:

> Functional asymmetry exists not only in adults, but also in young children and infants and at least to some extent in some nonhuman species. (1987, p. 156)

Nonhuman primates were not mentioned in either entry!

Why has the realization that nonhuman primates have functional asymmetries taken so long to occur? It has been known for thousands of years that humans have a functional asymmetry in the form of a characteristic right-handedness. And it has been known since Broca's classic paper of 1861 that language is predominantly lateralized to the same cerebral hemisphere that controls the right hand — the left hemisphere. And there has long been a basic acceptance of Darwin's view that all evolution is a matter of "descent with modification" (Darwin, 1859, p.420) and that "man" bears "the indelible stamp of his lowly origins" (Darwin, 1871, p.597). And it must be conceded that the Darwinian theory of evolution by natural selection is without competitors as a general theory of biology, encompassing all life forms. As Dawkins (1986) and Crick (1988) have recently noted, Darwin has given us what Crick has called "The secret of life" (1988, p. 25). Why then are we just now getting around to confirming the obvious implication, following from the facts of human asymmetry and evolution, that precursors to human functional asymmetries are likely to exist in our closest relatives?

The answer is that anthropocentrism is, and always has been, the dominant influence when humans come to contemplate their place in nature. For example, as Cartmill, Pilbeam, and Issac (1986) point out in a recent review of the field of physical anthropology since Darwin, humans have taken their task to be the documentation of the ways in which humans are special.

> In accepting this persistently pre-Darwinian definition of their problem, scientists who study evolution have saddled themselves with the paradoxical job of explaining how causes operating throughout nature have in the case of *Homo sapiens* produced an effect that is radically unlike anything else in nature." (1986, p. 410)

In other words, we have not faced up to the consequences of Darwinian theory with its cornerstones of descent with modification and the lowly origins of humans. To put it in an extreme form, very few people have become standard bearers for Darwin's contention that "he who would understand baboon would do more toward metaphysics than Locke" (deBeer, 1959).

Under the protective umbrella of the anthropocentric perspective it is possible to evade the issue of primate precursors to human functional asymmetries in several ways. First, with the aid of the functional eclecticism of much contemporary cognitive neuropsychology, with its emphasis on computer metaphors, it is possible to focus almost entirely on the first of the evolutionary biologist's three questions (Dobzhansky, 1965), the question of *mechanism* — "How does it work?" for human asymmetries. This leaves unadressed the other two questions with more phylogenetic implications, the question of *function* — "What does it do for the organism?", and the question of *origin* — "How did it get that way?". With respect to the question of other primates, most contemporary neuropsychologists would probably guess that "Prosimians" is part of the name for some football team. As Geschwind (1985) has pointed out, the main cause of the neglect of nonhuman asymmetries, in particular, is probably the overwhelming impression that language, perhaps the most radically lateralized function in humans, has no obvious parallel in other primates. In addition, as Geschwind also pointed out, the evidence in the past on whether other primates have a systematic handedness bias has been predominantly negative. As Harris (this volume) points out in a valuable historical review, an early period (1860-1910) with conflicting findings based mostly on inadequate evidence was succeeded by a period of experimental studies

(1910-1940) with mostly negative results. The subsequent continuation of this latter trend emboldened Warren (1980), whose views were until very recently widely cited without dissent, to conclude that "Monkeys are not little men in brown fur suits whose cerebral dominance is revealed by their handedness." (1980, p. 355). Thus anthropocentrism had reached the stage where any thought that hand preferences of other primates could be related to human hand preferences could be regarded as naive anthropomorphism.

But I would argue that in work on the evolution of handedness, anthropocentrism has had the result that the criteria for human performance has been set very high, and the possibility that other primates have traversed some stages on the route to human handedness has received scant attention. The result, as Calvin (1987) has pointed out, was "a hypothesis about primate handedness, which, because of the way the question was posed initially, had a tendency to converge on the null hypothesis when the data were sparse." (1987, p. 267). For example, Warren (1987) has concluded that "there is no compelling reason to regard "handedness" in monkeys as either analogous or homologous to human handedness" on the grounds that "From infancy, humans manifest stable hand preferences that are consistent in diverse situations and in multiple manual activities. Monkeys do not." (1987, p. 289).

These criteria both overestimate the consistency of handedness in human infants and adults, and rule out the possibility that some similarities between humans and other primates, together with some differences, could be grounds for either analogy or homology. The view that humans exhibit a totally consistent hand preference for everything has led to tests on monkeys that have, for the most part, explored only a limited range of manual possibilities that were easy to investigate — tasks such as "Simple Reaching", "Complex Reaching", and various tasks in which reaching is accompanied by or preceded by another relatively straightforward operation; e.g. tasks described as "Expose and Reach", "Stabilize and Reach", and "Manipulate and Reach". Such tasks are not part of handedness evaluations in humans. Here, culturally induced tasks, some of which have taken years to learn, such as writing, threading a needle, hammering, using scissors, etc, are used. The likelihood that, in the absence of cultural induction, preferences might be present but weaker in other primates has not been emphasized. Furthermore, Harris and Carlson (1988) have shown that simple reach tasks, the ones most often used in primate handedness studies, do not elicit anything close to uniform hand use in humans. For example, in infants tested during the first year, the maximum percentage of right hand reaches to objects in the midline was about 60%. And for adults, "percentages only slightly higher than that shown by the infants were observed." (1988, p. 727) (66-72% of dominant hand use). In addition, a bias towards use of the hand nearest the object when the object was not placed in the midline was observed in the human groups just as it has been in monkeys. Thus, contrary to Warren's assertion, a direct comparison between monkeys and humans showed that they are very similar in this particular respect.

An even worse situation exists with respect to the question of language evolution. Almost total emphasis has been put on the evolution of the highest level of language — the grammatical level. Who in this area has not taken a trip with Chomsky through some of the more complex sentences (a little like studying protein molecules for clues to the origins of basic molecular structure) to arrive at a conclusion such as "it is quite senseless to raise the problem of the evolution of human language from more primitive systems of communication that appear at lower levels of intellectual capacity." (Chomsky, 1973, p. 67). The evolution of the lower, phonological level, which to some extent must have preceded grammatical evolution (Studdert-Kennedy, 1990) has hardly been considered, beyond the conclusion that the two-tube vocal tract accorded us the potential for it (Lieberman, 1984). Such a conclusion leaves untouched the functional question of how this capacity was utilized. It certainly seems that no nonhuman primate has either a naturally occurring or a learned grammar, or even a phonology (MacNeilage, 1987), but is that sufficient to conclude that language had no significant precursors in either hemisphere? Chomsky goes a good deal further on this question than anyone who has claimed the de novo evolution of handedness in humans. He leads a now-sizeable group in asserting that language *could not have evolved by natural selection*. (For a discussion of this remarkable claim and a NeoDarwinian antidote to it, see Pinker and Bloom, 1990.)

As there is very little in this book on language evolution, I will not dwell on this topic. (For a systematic approach to the evolution of speech see Lindblom, MacNeilage, and Studdert-Kennedy, forthcoming). However, I would suggest the metaphor that grammar is to nonhuman primate communication as space flight is to propeller-driven flight. And perhaps phonology is to nonhuman primate communication as supersonic planes are to propeller planes. But I would argue that certain design features understood and implemented by Wilbur and Orville Wright remain fundamental to the more advanced devices (control? navigation?). Evidence that monkeys have a left hemisphere specialization for vocal communication (discussed later, and reviewed in Morris et al., this volume) suggests that prehominid communicative asymmetries may have been fundamental to the evolution of language. Beyond this I would only add that there are grounds for evolutionary continuity in the origins of both phonology and syntax. The syllable, the universal superordinate unit of phonological organization, might have had a precursor in the lipsmack, arguably the most frequent single communicative gesture in many other higher primates that occupy terrestrial habitats (Redican, 1975; MacNeilage & Davis, 1990). As to grammar, the tendency towards localization of syntactic capabilities in anterior cortex and lexical capabilities in posterior cortex (Nass & Gazzaniga, 1987), perhaps the most fundamental fact about the neurology of grammar, must have been influenced by capabilities of these regions of the brain that developed before hominid evolution.

The dangers of anthropomorphism in the study of other primates are certainly real. One of the appealing things about other higher primates as seen in zoos, and on funny postcards, calenders, etc, often with little pieces of language attributed to them, is that they are so like humans in certain respects while at the same time strikingly different. We must heed Lloyd Morgan's canon: "in no case may we interpret an action as the outcome of the exercise of a higher psychical faculty, if it can be interpreted as the outcome of the exercise of one which stands lower in psychological scale." (cited in Boring, 1950, p. 474). Note, however, that this canon is a two-edged sword, so to speak. But only the ban on anthropo*morphism* has been implemented. We have no separate canon warning us of the dangers of anthropocentrism, although it has always been a good deal more rampant than anthropomorphism ever was. Science must tread the thin line between excessive anthropocentrism and excessive anthropomorphism, and my opinion is that we lean precipitously in the direction of anthropocentrism when it comes to the question of the evolution of functional asymmetries. A most obvious example of our biases involves common usage of phrases such as "humans and animals", "humans and primates", or "humans and apes", without the use of the adjective "other" to qualify the last noun. We have just had a revolution that led to the placement of the word "human" in such phrases, rather than "man". Unfortunately, unlike human females, other animals can not fight such a battle. We should. In the title of this paper, the word "primate" refers to *all* primates.

One of the dangers of both anthropomorphism and anthropocentrism, against which we are warned by Preilowski (this volume), is that excessive focus will be placed on human characteristics, either as taxonomic modes or unattainable touchstones, at the potential cost of the understanding we might gain by considering each primate species in its own right. An extreme example of anthropomorphism is Lieberman's assertion that "The evolution of the biological bases of human syntax probably has its beginning in the neural mechanisms that structure the behavior of simple animals like mollusks." (1984, pp. 69-70). On the anthropocentric side is Corballis' observation that the finding of several tendencies towards left-handedness in primate populations "scarcely inspires confidence." (1987, p. 269).

A more subtle approach to the defense of a discontinuity stance regarding nonhuman-human functional relations than Warren's denial that any relation exists, is to note the existence of some evidence of asymmetries in other primates, but to give it no further consideration on the grounds that the evidence is weak (e.g. Corballis, 1987; 1989). Corballis even suggested that the evidence might be absent because all the positive findings, both left and right, might have been "influenced by biases in the immediate experimental environment." (1987, p. 270). True, the evidence of functional asymmetries in other primates is weak compared with that in humans (though it is getting stronger all the time). But why is it weak? It cannot be denied that one reason is that the asymmetries themselves are for the most part, though not entirely, (see Ward, Milliken, and Stafford, this volume, for several exceptions), inherently weaker and less consistent in their direction in other primates than in humans. But is that sufficient reason to accord them no role in theories of the origin of human asymmetries?

Another reason that evidence for nonhuman primate functional asymmetries is weak is that, in comparison with the human studies, virtually no work has been done on the question of asymmetries in other primates. By the most generous estimate there have been less than 100 systematic studies of hand preferences and communicative laterality in nonhuman primates. In contrast, McManus (1986) estimated that an overall total of 5000 studies of laterality, including handedness, had been published by 1985, virtually all of which were presumably done on human subjects. This imbalance by what must by now be a factor of well over 50:1 in attention to humans versus other primates obviously raises the question of whether nonhuman primates have been adequately investigated with respect to these questions. However, despite the imbalance in attention which is common to all types of comparison between humans and other primates, Ettlinger has concluded from a review of neuropsychological studies of nonhuman primates in four areas in which a functional discontinuity had been claimed, including cerebral lateralization, that " the discontinuities very clearly evident in 1963 have either been resolved or have become blurred." (1984, p. 685). With respect to cerebral lateralization, he concluded that the likelihood of a homology between humans and monkeys "has increased enormously." (p. 691). Later on in this chapter I will summarize evidence from over 30 studies of nonhuman primate handedness suggesting that population-level asymmetries in hand preference may exist in many species.

In principle, another reason for neglect of the question of functional asymmetries in other primates could be that we had a theory of the de novo evolution of functional asymmetries in hominids that was so satisfactory that it gave us no good reason for pressing our scientific attention on other primates. However, this is far from being the case. In the first sentence of a chapter entitled "Cerebral lateralization and specialization in human nervous system" in the new *Handbook of Physiology*, an organ of the American Physiological Society, Nass and Gazzaniga express the conventional discontinuity view of the evolution of human functional asymmetries, but also note the theoretical problem which arises from the view:

> A hallmark of human brain function is cerebral lateralization and specialization. Nowhere else in the phyla is there anything approaching the kind of localization seen in the human, thereby leaving the emergence of this organizational feature of human brain function begging for explanation. (1987, p. 701).

In fact, there has never been a theory of the origin of functional asymmetries in hominids that has been worked out in any detail or gained the sustained attention of a sizeable number of followers (See Corballis, 1983, for a review). The most common notion is that handedness first evolved with hominid tool construction and use, and that language then either evolved directly, via sign language (Hewes, 1973), or indirectly, via spoken language. Neither the necessity of a single type of lateralization for tools in particular, nor the mode of transfer of talent from the hand to the mouth has been adequately established (See Greenfield, 1991, for a courageous effort to address the latter question.) What has evolved has usually been described in terms of metafunctional capacities such as analysis versus synthesis, or serial versus parallel processing, capacities that could not have directly arisen from natural selection, because what is directly selected for are adaptive functions, not metafunctions. Consequently, even if these whole-hemisphere labels satisfactorily characterize the present status of the human brain, which they do not, there would still be a question of how they got that way. Indeed, depending on one's understanding of the word "theory", one might even agree with Allen's (1983) conclusion that "We have no theory" in this area. Again, Cartmill et al. put their finger on the problem:

> Unfortunately, as David Hume noted, a century before Darwin, science cannot say anything very satisfactory about the cause of any genuinely singular phenomenon. Unless we can replicate the supposed cause, or dispel the effect's singularity by taking it apart into replicable pairs of cause and effect, we can never show empirically that any proposed cause of a unique event was really adequate to do the job. (1986, p. 410)

Incidentally, it would not be beyond the bounds of orthodox Neodarwinian theory for a function without significant observable precursors to be present only in hominids, given the passage of at least 5 million years since the existence of an ancestor common to us and our nearest living (observable) relatives, the chimpanzees. Darwin himself often emphasized the possibility that gaps in the historical record could give the impression of de novo evolution. However, the fact that manual and vocal function constitute the two most complex action systems of humans and that both these functions and the organs responsible for them have an evolutionary history stretching back for hundreds of millions of years makes it seem unlikely that the entire basis for the asymmetries associated with them first evolved in hominids. If we are to understand the entire basis for the evolution of lateralized human specializations we must lessen our preoccupation with what is most special about humans. And we must watch for the anthropocentric tendency to trivialize possible precursors in other animals simply because they do not have the full bloom of modern human capabilities. From a Darwinian perspective lowly origins are to be expected.

There is an irony here. Humans are almost desparately interested in understanding themselves, and consequently, their origins. Each new find of a fragment of bone or a tooth from an ancestor of ours, provided it is a hominid, hits the front pages, and occasionally even motivates the establishment of a new research institute. Meanwhile, in dozens of neighboring zoos, hundreds of thoroughly bored *living* descendents of a common, scarcely pre-hominid, ancestor while away their time, taking a remote second place, in viewer interest, to the giant panda or the koala bear. It takes years and years of work by Jane Goodall, Dian Fossey, or Birute Galdikas to nudge us into some thoughts about how we might relate to our closest living relatives. Why? Because our *stories* of human evolution are usually of "the evolutionary transformation of some kind of primitive primate ancestor into a civilized human being." (Lewin, 1987, p. 32). The main events are "the shift from the trees to the ground — terrestriality; the change of posture from walking on four legs to balancing on two — bipedalism; the expansion of the brain, with flowering of intelligence and language — encephalization; and the emergence of technology, morals and society — civilization." (Lewin, 1987, pp. 32-33). The stories tend to conform to the myth of the hero, overcoming staggering odds to achieve humanity (Landau, 1984). For these stories, the ancestor needs to be adjudged as "primitive" (see the earlier quote from Chomsky) and to lack even the first of the series of liberating adaptations, but is otherwise of little interest. The irony is that these kinds of attempts at self-understanding necessarily deprive us of the self-understanding we could otherwise gain from more study of our prehominid heritage. Perhaps the existence, in embryonic form, of a new discipline of evolutionary psychology (see Bower, 1991 for an introduction), which so far seems to be doing a reasonable job of substituting Neodarwinian theory for myth, bodes well for our future understanding of this topic.

In summary, the evidence for functional asymmetries in other primates is relatively weak, partly because the asymmetries themselves are typically weak and nonuniform in the population, but also partly because the research effort to uncover them has been weak, owing to the somewhat self-defeating anthropocentric bias governing the scientific effort of humans to establish their place in nature by means of the myth of the hero. But there are reasons to believe that de novo evolution of human functional asymmetries is unlikely to be the whole story, and no satisfactory theory of de novo evolution of asymmetries in hominids presently exists. Consequently, it would seem useful, as a point of departure, to attempt to put together a coherent scenario for the evolution of functional asymmetries in *all* primates from whatever evidence there is of functional asymmetries in *other* primates. The following is a summary of such an attempt, which will provide a context for discussion of a number of other issues.(For more detailed treatments see MacNeilage,1987; 1991a; MacNeilage,Studdert-Kennedy, & Lindblom, 1987.)

Evolution of Primate Functional Asymmetries: A Postural Origins Theory

In an initial search of the primate handedness literature (MacNeilage et al., 1987), we found that among about 45 studies there were about a dozen statistically significant results and trends, and they formed a pattern which was only partly human-like. There were some right hand preferences, one in particular for demanding manipulative tasks (Beck and Barton, 1972). But there were also some left hand preferences, most of them confined to the act of manual prehension. We made the suggestion on the basis of these findings that other primates may indeed have population-level handedness biases. Our report, in the peer review journal *Behavioral and Brain Sciences*, was greeted, for the most part, with polite incredulity, perhaps best illustrated by the title of Ettlinger's (1987) peer commentary: "Primate Handedness: How nice if it were really so."

However, there was a small group of people who took the question seriously, including some who were prompted into it by our review, and in the next four years about 20 additional positive findings have been reported, many from better-designed studies. (For summaries of positive findings circa, 1991, see MacNeilage, 1991a or MacNeilage et al., 1991.) Now, only a short time after the first comprehensive review of the topic, the existence of handedness in other primates is beyond doubt. From these 30 or so positive findings, I summarize what I believe to be the main trends. In very general terms there may have been an inverse relation between the relative importance of left-handedness and right-handedness across the primate order (MacNeilage, 1991b). There have been several studies of the modern descendants of the earliest primates, the prosimians, most of them done by Ward and her colleagues (see Ward, Milliken, & Stafford, this volume), that have been almost unanimous in finding group trends towards left-hand preferences for acts of manual prehension. These animals, with few exceptions, do not have separate control of individual fingers and therefore do little that could be called manipulative. They also do little in the way of bimanual coordination. One important finding to which we will return later was that in two studies, hand preference was found to be posture-dependent (Sanford, Guin, & Ward, 1984: Forsythe, Milliken, Stafford, & Ward, 1988; See also Ward et al., this volume).

The left hand prehension preference is present but weaker in monkeys, (see Watanabe & Kawai, this volume, for an example) and has not been observed as a group effect for simple reaching in great apes. However, strong left-hand preferences have been found in monkeys and apes in tasks that make strong visuospatial demands, such as the plexiglass screen adjustment task of Vauclair and Fagot (this volume) and the goldfish catching task of King and Landau (this volume).

As I mentioned, there is negligible evidence of right-hand preferences in prosimians. But in monkeys and apes, predominant right-hand preferences have been reported for a number of tasks, including tasks involving manipulation, such as the joystick-controlled computer game described by Morris, Hopkins, Gilmore, and Washburn (this volume), and an auditory discrimination task with a manual key-press response for which all 30 monkeys tested preferred their right hand — not exactly a weak trend (Kuhl, 1988).

One study of 10 stumptailed macaques (Beck & Barton, 1972) is of particular importance because it showed that the left-hand preference for manual prehension and the right-hand manipulative preference can both be present in the same animals. Shafer (this volume) was the first to report a significant right-hand preference for all classes of act in a nonhuman primate population of 47 gorillas. The report of right limb locomotor initiation preferences in four species of apes by Heestand (1986) which, in our analysis (MacNeilage, Studdert-Kennedy, & Lindblom, 1988), was found to be correlated with right-hand preferences, suggested that postural asymmetries are involved in limb choice in higher primates as well as prosimians. Finally, a right-hand preference for the initiation of chest-beating in gorillas (Schaller, 1963) may be significant for communicative as well as manual evolution. Data on morphological asymmetries support the behavioral data. Falk, Pyne, Helmkamp, and De Rousseau (1988) have found an upper limb bone asymmetry favoring the right side in a large sample of macaques — a phenomenon well-known in humans (Ruff & Jones, 1981).

Thus we have a picture that suggests that the first primate handedness was left-handedness, and the type of right-handedness predominant in humans evolved later. How can we give an evolutionary explanation for this pattern. In our review, we suggested that the first evolutionary step may have been the evolution of a left hand-right hemisphere visuospatiomotor specialization for unimanual predation, and that the human right hemisphere visuospatial specialization may be in part a residue of this development. It is known from paleontological studies that the earliest primates were small, unimanual predators, and that most of them were vertical clingers and leapers (Fleagle, 1988). The modern prosimians that have been shown to have the strongest left-hand preferences — bushbabies — are also unimanual predators and vertical clingers and leapers. The ancestors of these animals apparently made an extremely momentous adaptation to life in the trees. Although having a heritage of about 1/2 a billion years of bilaterally symmetrical midline predation involving the mouth, they switched to the hand as the organ of predation, presumably because it was harder to get the mouth on prey in the small-branch niche these animals favored. I would emphasize the extreme degree of specialization involved here. Some modern prosimians can catch insects and small animals, sometimes even on the wing, with a movement faster than the eye can see (Charles-Dominique, 1984).

The postural demands of such an adaptation are obvious. For the left hand to be free in vertical clinging, the other three limbs are needed to cling to the limb to provide postural support, with the right side of the body bearing the brunt. This fact, and the findings mentioned earlier of postural dependence in amount of left-hand preference in prosimians, led us to suggest that complementary to the left hand-right hemisphere predation specialization in prosimians there may have evolved a right side-left hemisphere postural support specialization.

In higher primates, the data suggest that the left hand predation specialization may have become less important, to the point of not being normally visible in humans (though visible, I believe, in some *experimental* situations — MacNeilage et al., 1991) while right-handedness evolved. The question, of course, is why would not left-handedness, having arisen, continue to be preferred? In higher primates, vertical clinging and leaping drops out as an adaptation leaving the forelimbs less involved in postural support. Higher primates can sit. Secondly, many species became more omnivorous than the typical lower primate, which involved more versatile food acquisition behaviors — including invasive foraging, in which force is exerted on a food object, and food processing. A number of higher primates evolve a human-like opposable thumb, and became capable of highly versatile hand-internal movement control. We suggested that the right side of the body, being physically stronger and perhaps capable of finer on-line movement control adjustments because of its postural heritage, may have been better preadapted than the left, with its primarily ballistic capability, for both the force application and the manipulation required for the increasing omnivory of many higher primates. Thus we suggest a Postural Origins theory of right-handedness, preceding hominid evolution.

Where does the evolution of language fit into all this? A very important datum in this respect, mentioned earlier, is that there is very good though not plentiful evidence that monkeys may have a left hemisphere specialization for vocal communication. Two studies (discussed also in Morris et al., this volume) complement each other in showing this. A piece of background information here is that in Dichotic Listening tests of humans, in which two different syllables are presented simultaneously, one to each ear, most right-handers correctly report more stimuli coming in their right ear. As auditory transmission from ear to brain is considered to be predominently crossed, this right ear advantage is considered to indicate the presence of a left hemisphere specialization for speech. Peterson and his colleagues (1978) showed that if Japanese macaque monkeys are required to discriminate between different calls of members of their own species, they perform better when the stimuli are presented to the right ear (with noise in the other ear) than in the opposite condition. In a complementary study, Heffner and Heffner (1984) showed that all five monkeys with lesions in the left hemisphere homolog of Wernicke's area in humans (superior posterior temporal lobe) showed a performance deficit on this discrimination task, while none of the five animals with a lesion in the right hemisphere homolog of Wernicke's area showed deficits.

These results, taken with the handedness results, suggest that, contrary to the standard anthropocentric view, neither right-handedness nor a left hemisphere specialization for vocal communication first evolved in hominids. Thus, if we wish to argue that these two specializations are linked we need to postulate developments which predated hominid evolution. Our suggestion is as follows. Much communication in the earliest prosimians probably involved the assumption of various sterotyped whole-body postures. Even vocal communication involves a sizeable proportion of body musculature. Given the obvious postural constraints on the performance of such communicative gestures in a habitat in which the available postural support varies in three-dimensional space, with every movement that the animal makes, it would not be surprising if a lateralized postural control mechanism became useful for communicative purposes.

This suggestion is of course speculative. There is no evidence for a link between a postural specialization and a communicative specialization in other primates. But I believe there is some evidence in humans. Left-handers are a subgroup that allows the possibility of evaluating the relation between postural and language specializations. About 50% of left-handers are right-footed. Footedness obviously involves a whole-body postural asymmetry. One can't use a single foot for a task while standing up without most of the rest of the body serving a postural support role. It was this subgroup of left-handed right-footers that was primarily responsible for a very surprising finding reported in 1980 by Searleman who was studying the relations between various functional asymmetries in a very large number of subjects. He reported that footedness was more closely related to language laterality, as measured by the dichotic listening test, than was handedness. In fact he found that relatively more of his left-handed right-footers had a right ear advantage than did his group of right-handers (39 out of 40), and that their average advantage was also larger. A student working in my laboratory (Maki, 1990) has recently replicated these findings of Searleman. She also used a factor analysis to show that on one factor, which had high loadings on hand acts that involved whole-body postural support, the left-handed right-footers were more often right-handed than left-handed, thus strengthening further the evidence for a postural link with the left hemisphere language specialization. This result is highly troublesome for the standard anthropocentric theory, for which the link between handedness and language is due to some common fine skill factor, often considered to be associated with sequencing. Left-handed right-footers have a closer link between hand preference for tasks involving whole body posture, and language laterality, than between hand preference for fine skill tasks — including writing — and language laterality. A next step is to study the 5% of right-handers who are left-footed to see whether language laterality goes with posture or fine skill in that group as well.

This postural origins theory is controversial, as is typically the case with new theories. (See for example Annett, 1991; Cicchetti, 1987; Dennenberg, 1988; Vauclair & Fagot, 1987; Guiard, 1987; Hauser et al., 1991: Jerison, 1987; McManus, 1987; Steklis & Marchant, 1987; Warren, 1987.) But regardless of the fate of this theory, the fact of the matter is that nonhuman primates do have handedness and we will need a theory that provides an explanation of this. This theory will need to give an account of the relation between handedness in hominids and other primates. There is presently no alternative conception to the postural origins theory on the matter of the evolution of human asymmetries which does not have serious problems, including the Right-Shift theory of Annett (1985). (See MacNeilage et al., 1991 for a discussion of problems of the Right-Shift theory.) And there is presently no other theory of any sort which attempts to deal with the evolution of functional asymmetries across the primate order. The alternative, suggested by Hauser et al. (1991), of eschewing theory because it is premature reflects a lack of realization of the value of theories as heuristic tools. As Remy de Gourmont said: "Two elements are needed to form the truth — a fact and an abstraction." The relatively coherent account given by the postural origins theory of the evolution of functional asymmetries across the primate order has already served, and we believe will continue to serve, as a useful point of departure for further consideration of these questions. However, there are also a number of other theoretical issues, including some not directly bearing on the postural origins theory, which deserve emphasis at present. In the concluding section of this paper I will consider these theoretical issues.

Theoretical Issues for the Future

Perhaps the most important development in recent research on nonhuman primate asymmetries has been the greater weight given to ethological questions — questions regarding the understanding of complex naturally occurring behavior. This contrasts favorably with an earlier epoch dominated by the question of what would be an easily administered independent variable which might reveal a hand preference in a caged animal. It is to be expected that hand use asymmetries in other primates will eventually be best understood in the context of the patterns of naturally occurring behavior characteristic of each species. The work of Ward and her colleagues (this volume) has been a model of the way in which this perspective might be pursued. In the process they have done a great deal to establish Prosimians as a taxon to be reckoned with.

Adoption of an ethological perspective, important as it is, is not inconsistent with experimental approaches to the understanding of functional asymmetries. However, a watchword here should be "challenge the animals". The work by King and Landau (this volume) on goldfish catching is a particularly good example of this, as is work by a number of other contributors (Fragaszy & Adams-Curtis; Morris et al.; Preilowski; Vauclair & Fagot; all in this volume). Particularly important for the future would seem to be the kind of high-technology, user-friendly testing facility of Morris et al. (this volume) though it requires friendly users; this group is to be congratulated for their foresight and perseverence in the development of these elegant facilities. More emphasis on the use of complex manipulation tasks of the kind discussed by Fragaszy and Adams-Curtis (this volume) should also be rewarding. Shafer (this volume) observed that manipulation was the only act category in which all 47 gorillas showed a hand preference. And, after all, this is the type of activity that is usually thought to have led to the origin of right-handedness in hominids. A dual focus on preference and ability and how they relate to each other will also be of value. In this regard, the distinction made by Vauclair and Fagot (this volume; see also Fagot & Vauclair, 1991) between handedness and manual specialization, though possessing some logical problems, in my opinion should prove salutory. Two questions that have been almost totally neglected have been the relation between hand preference and bimanual coordination (though see Trevarthen, 1978), and the possibility that force application might be associated with hand preference. Fragaszy and Adams-Curtis (this volume) concluded from their study of crab-eating macaques that "bimanual tasks are more likely to show lateral differences than unimanual tasks, and at the same time be less vulnerable to spurious results with limited sample sizes." Lutz-Maki and MacNeilage (1991) recently confirmed, in chimpanzees, the prediction from the postural origins theory that there should be a right hand-side preference for tasks involving time-extended force application.

One important implication of the work of Ward and her colleagues is that vertical posture in particular may exert selection pressure towards asymmetric hand use (see also Olson et al., 1990). Ward et al. (this volume) have pointed out that the strongest preferences in all primates tend to be found in humans and prosimians, and among prosimians there is a tendency for species which favor a *vertical* clinging posture to have stronger hand preferences. Thus, we may have the interesting situation that both left- and right-hand preferences seem to have been selected in animals which favor vertical posture, but in different taxa, and for different types of act.

This possibility is to some extent at variance with the common assumption that it was primarily the adoption of a *terrestrial* habitat by hominids which provided the main selection pressure for handedness. Obviously, selection pressures on asymmetrical hand use in vertical postures and in terrestrial habitats were not mutually exclusive in early hominids — they had both attributes in common. However, much emphasis has been put on the possibility that the evolution of tool construction and use was the primary event that led to right-handedness in humans, and much of this kind of behavior does not require vertical posture. What has been particularly emphasized in conceptions of evolution of handedness in humans is the development of manipulative skill and the elaborate sequences of acts that go with it. From this standpoint the signal failure of Fragaszy and her colleagues to find population-level handedness trends in capuchin monkeys, a taxon known for their manipulative skill, is of interest (e.g., Fragaszy & Mitchell, 1990). Although only small numbers of animals have been studied, the lack of any consistent trends seems to suggest that these animals do not typically favor the left or the right hand. This result suggests that perhaps the terrestrial habitat *is* important. It may be that versatile invasive foraging in an arboreal habitat, often done from a vertical posture, which is characteristic of these animals, selects against favoritism of one hand, because that may lead to a disadvantage in food acquisition under some particular conditions — conditions in which the animal can only position itself to obtain food from its non-favored side (Fragaszy & Mitchell, 1990). But if this is so, then we might expect stronger handedness trends than have so far been demonstrated, to be found in nonhuman primates which are invasive foragers with a predominantly terrestrial habitat — many monkeys and baboons.

With the accumulation of larger bodies of information about asymmetries in other primates, some questions which were difficult to evaluate before are becoming more tractable. Sex differences is one of these questions. Ward et al.(this volume) have found an intriguing tendency for prosimian females to be like human females in being slightly more often right-handed. In the light of the increasing evidence (e.g. Kimura, 1987) that there are profound sex differences in brain organization, including brain asymmetries, in humans, phylogenetic implications of these differences become of greater interest. Age differences are also a topic of interest. A number of studies have reported that hand preferences become stronger in older non-human primates just as they do in humans. There is less agreement on the presence of age-related differences in *direction* of hand preference. Ward et al. (this volume) report another interesting similarity between prosimian and human findings, a tendency for an increase in right-hand preference with age. Intensive longitudinal case studies of the kind reported by Fagot (this volume) should become an increasingly important approach to this question. This work may be stimulated in part by the first clear demonstration of a very early ontogenetic parallel in direction of hand preference between humans and other primates. Hopkins and Bard (this volume) report a counterpart, in neonatal chimpanzees, to the human right hand thumb-sucking preference, recently shown to be present from the fourth month of fetal age (Hepper, Shahidullah, & White, 1990). Another recently established parallel between human and nonhuman manual asymmetries, relevant to ontogeny, is the finding that female gorillas, like humans, prefer to hold their offspring on their left side (Manning & Chamberlain, 1990). Presumably this reflects a homology in hemispheric specialization between humans and gorillas.

The right cerebral hemisphere has often been accorded a special status in emotional behavior in humans (Nass & Gazzaniga, 1987). That it might play a similar role in nonhuman primates is suggested by the recent finding that under conditions of threat, monkey mothers strongly prefer to retrieve their offspring with their left hand (Hatta & Koike, 1991). In this connection, it should be noted that Ward et al.(this volume) point to the possibility of an important effect of arousal in hand choice. This possibility has, to my knowledge, received no attention in human handedness studies, as the anthropocentric perspective encourages us to regard human hand use (and, of course, language), as purely cognitive affairs, unsullied by affect. However, in this context, the opinion of Bullock (1984) perhaps deserves some attention: "A significant likelihood that can be examined only in unusually well-known species is that many crucial behaviors correlated with brain adaptations are those used only rarely, during life crises, not the everyday behavior we are familiar with." (1984, p. 477). His examples of life crises include "escape from predators."

As long as single groups of nonhuman primates continue to prefer one hand for one type of task and the other hand for the other (Beck & Barton, 1972; King & Landau, this volume), the question of classification of tasks in terms of their demands will remain with us. We have been (justifiably) criticized by Vauclair and Fagot (this volume) for concluding that while a task of adjusting clear plexiglass screens (on which baboons and gorillas showed a left-hand preference, Fagot & Vauclair, 1988a, b) primarily involved visual-spatial analysis, a task involving a joystick-controlled cursor hitting a visual target (on which monkeys and chimpanzees showed a right-hand preference) primarily involved manipulation (Hopkins, Washburn, & Rumbaugh, 1989; Morris et al., this volume). We believe that it is the demand for a particular time/space relation in joystick movement in the tracking task, a demand lacking in the screen adjustment task, that invokes a manipulatory specialization in the former task. But we can't presently prove it. What is desirable, though not easy, is to have the same animals do both types of tasks. Then if the hand preference dissociation remains, vary the demands of the tasks in order to find out what factors are influencing hand choice. It is probably premature to expect that we could, at present, make an a priori classification of tasks in terms of what they demand of the animal, as suggested by Hauser et al. (1991), especially as the same task might often make different demands on animals of different species. Methods must instead be developed hand-in-hand with theories.

We have been criticized for ignoring or underestimating the importance of the many negative results that have been obtained in the search for population hand preference biases in nonhuman primates, including most of the earliest studies. One critical reviewer of a research proposal concluded that even if we were to get positive results on tests of the Postural Origins theory, there would still be confusion in the area due to the negative results. This is a little like suggesting that we should not search for gold because most people have been unsuccessful. Instead, as we have argued elsewhere (MacNeilage et al., 1987), most of these negative results are not inconsistent with the contention that the phenomenon of population-level handedness biases exists. There have been very few cases in which virtually identical studies alternatively result in positive and negative findings, a phenomenon which, if rife, would be worrisome. My contention, obviously shared by some contributors to this volume (Morris et al.; Vauclair & Fagot; Ward et al.) though apparently not by others (Lehmann; Preilowski) is that there are now enough finds to tell us that there is gold in this territory. The question therefore becomes not whether but where.

In what sense can it be claimed that the studies of nonhuman primates discussed in this volume provide a missing link between humans and other animals. I have reviewed a good deal of evidence for this claim. However there is still a lot missing. There are virtually no neuropsychological studies on brain-behavior relations in the great apes, our nearest relatives, even though the behavioral studies of language-trained chimpanzees summarized by Morris et al. strongly suggest a human-like distibution of functional specializations in both hemispheres. This basically leaves the neurally oriented advocate of evolutionary continuity between humans and their ancestors playing against a stacked deck (not that I would advocate otherwise in this particular case). Neuropsychological studies of hemispheric specializations in monkeys, our more distant relatives (See Morris et al. for a partial summary. See also MacNeilage et al., 1987 for discussion of limitations of these studies) gives mixed results on the issue of evolutionary continuity - a not altogether surprising outcome. While the human right hemisphere specialization for face recognition seems to have a parallel in monkeys, there is evidence that the left hemisphere has an edge in some types of spatial capabilities while the human left hemisphere does not (See Hamilton & Vermeire, 1988 for a brief summary).

Some would suggest that the link from other mammals to humans via other primates can be assumed, even in the absence of substantial data on other primates. For example, Dennenberg (1989) makes the case that there is evidence in rats of right hemisphere emotional and spatial specializations, and of other nonprimate lateralities remarkably similar to what is known about behavioral lateralization in humans. Glick (this volume) reports that preferred direction of rotation in humans is an asymmetry "with very similar characteristics" to the one he first observed in rats. Glick has presented evidence (see also Glick & Shapiro, 1985) that these tendencies in humans and rats are associated with very similar neurochemical asymmetries in the corpus striatum. In the case of both turning preferences and emotional control asymmetries, findings in rats seem to have important clinical implications for humans (Glick, this volume; Robinson, 1985). It is not always easy to make straightforward links between human asymmetries and asymmetries in non-primate mammals. Problems include differences in dependent variables being studied, and non-trivial strain differences in subpopulations of rats. However, there is more than enough evidence that attempts to make the connection, greatly assisted by more widespread use of invasive procedures than studies of other primates, will continue to be worthwhile. And the importance of evidence from nonhuman primates in the chain of reasoning will be obvious. What should not be overlooked is that hand use in other primates offers a tremendously rich noninvasive source of information, not only about functional asymmetries but about the evolution of complex action systems in general.

Coda

It is to be hoped that the reader of this book, however casual, will find that the area of primate functional asymmetries has now become a viable research area in its own right, one with enormous promise for the future. Of course, people are conservative. It is natural, and in one case it has even been considered profitable (Annett, 1991) for those who are basically interested only in human behavior to resist the demanding step of crossing interdisciplinary boundaries. However, I believe that such a step comes with the territory. We are never going to adequately understand human functional asymmetries by studying one primate.

Acknowledgement

My thanks to Bjorn Lindblom for his comments on the manuscript.

References

Adelman, G. (Ed.) (1987). *The encyclopedia of neuroscience*. Boston: Birkhauser.

Allen, M. (1983). Models of hemispheric specialization. *Psychological Bulletin, 73*, 73-104.

Annett, M. (1985). *Left,right, hand and brain: The right shift theory*. Hillsdale, N.J.: Erlbaum.

Annett, M. (1991). Predicting from the Right Shift Theory. *Behavioral and Brain Sciences, 14*, 338-341.

Arnold, A.P., & Bottjer, S.W. (1985). Cerebral lateralization in birds. In S.D. Glick (Ed.) *Cerebral lateralization in nonhuman species* (pp. 11-39). Orlando: Academic Press.

Beck, C.H.M., & Barton, R.L. (1972). Deviation and laterality of hand preference in monkeys. *Cortex, 8*, 339-363.

Boring, E.G. (1950). *A history of experimental psychology (2nd ed)*. New York: Appleton Century Crofts.

Bower B. (1991). Darwin's Minds: Psychologists probe the descent of the human psyche. *Science News, 140*, 232-234.

Broca, P. (1861). Remarks on the seat of the faculty of articulate language, followed by an observation of aphemia. In G. von Bonin (Ed.) *The cerebral cortex* (pp 49-72). Springfield, IL: C. C. Thomas, 1960.

Bullock, T.H. (1984). Comparative Neuroscience holds promise for quiet revolutions. *Science, 225*, 473-478.

Calvin, W.H. (1987). On evolutionary expectations of symmetry and toolmaking. *Behavioral and Brain Sciences, 10*, 267-268.

Cartmill, M., Pilbeam, D., & Issac, G. (1986). One hundred years of paleoanthropology. *American Scientist, 74*, 410-20.

Charles-Dominique, P. (1984). Bushbabies, lorises and pottos. In D. Hamilton (Ed.), *The Encyclopedia of Mammals* (pp. 332-339). New York: Facts on File Publications.

Chomsky, N. (1973). *Language and mind*. New York: Harcourt Brace Jovanovich.

Cicchetti, D.V. (1987). On viewing the evidence for primate handedness: some biostatistical considerations. *Behavioral and Brain Sciences, 10*, 268-269.

Corballis, M.C. (1983). *Human laterality*. New York: Academic Press.

Corballis, M.C. (1987). Straw monkeys. *Behavior and Brain Sciences, 10*, 269-70.

Corballis, M.C. (1989). Laterality and Human Evolution. *Psychological Review, 96*, 492-505.

Crick, F. (1988). *What mad pursuit*. New York: Basic Books.

Darwin, C. (1859). *On the origin of species by means of natural selection or the preservation of favored races in the struggle for life*. London: John Murray.

Darwin, C. (1871). *The descent of man*. New York: Great Books, Encyclopaedia Britanica, 1952.

Dawkins, R. (1986). *The blind watchmaker*. New York: Norton.

deBeer, G. R. (1959). Darwin's journal. *Bulletin of the British Museum [Natural History] Historical Series 2*, No. 1, 1-21.

Dennenberg, V. (1988). Handedness hangups and species snobbery. *Behavioral and Brain Sciences, 11*, 721-722.

Dennenberg, V. (1989). Laterality in animals: Brain and behavioral asymmetries and the role of early experience. In D.L. Molfese, & S.J. Segalowitz (Eds.), *Brain lateralization in children: Developmental implications* (pp. 59-72). New York: Guilford.

Dobzhansky, T. (1965). Mendelism, Darwinism and Evolutionism. *Proceedings of the American Philosophical Society, 109* 205-215.

Ettlinger, G. (1984). Humans, apes, and monkeys: The changing neuropsychological viewpoint. *Neuropsychologia, 22*, 685-696.
Ettlinger, G. (1987). Primate handedness: How nice if it were really so. *Behavioral and Brain Sciences, 10*, 271-273.
Fagot, J., & Vauclair, J. (1988a). Handedness and manual specialization in the baboon. *Neuropsychologia, 26*, 795-804.
Fagot, J., & Vauclair, J. (1988b). Handedness and bimanual coordination in the lowland gorilla. *Brain, Behavior and Evolution, 32*, 89-95.
Fagot, J., & Vauclair, J. (1991). Manual laterality in nonhuman primates: A distinction between handedness and manual specialization. *Psychological Bulletin, 109*, 76-89.
Falk, D., Pyne,L., Helmkamp, C., & De Rousseau. C.J. (1988). Skeletal asymmetry in forelimb of *Macaca mulatta.American Journal of Physical Anthropology, 77*, 1-6.
Fleagle, J.C. (1988). *Primate adaptation and evolution.* New York: Academic Press.
Forsythe, C., Milliken, G.W., Stafford, D.K., & Ward, J.P. (1988). Posturally related variations in the hand preferences of the ruffed lemur (*Varecia variegata variegata*). *Journal of Comparative Psychology, 102*, 248-250.
Fragaszy, D.M., & Mitchell, S.R. (1990). Hand preference and performance on unimanual and bimanual tasks in capuchin monkeys (*Cebus apella*). *Journal of Comparative Psychology, 104*, 275-282.
Geschwind, N. (1985). Implications for evolution, genetics and clinical syndromes. In S.D. Glick (Ed.), *Cerebral lateralization in nonhuman species* (pp.267-278). Orlando: Academic Press.
Glick, S.D. (Ed.) (1985). *Cerebral lateralization in nonhuman species.* Orlando: Academic Press.
Glick, S.D., & Shapiro R.M. (1985). Functional and neurochemical mechanisms of cerebral lateralization in rats. In S.D. Glick (Ed.), *Cerebral lateralization in nonhuman species* (pp. 157-183). Orlando: Academic Press.
Greenfield, P.M. (1991). Language, tools and brain: The ontogeny and phylogeny of hierarchically organized sequential behavior. *Behavioral and Brain Sciences, 14*, 531-596.
Gregory, R.L. (Ed.) (1987). *The Oxford companion to the mind.* Oxford: Oxford University Press.
Guiard, Y. (1987). Precursors to what? Theory is lacking for handedness in humans. *Behavioral and Brain Sciences, 10*, 276-277.
Hamilton, C.R., & Vermeire, B.A. (1988). Cognition, not handedness is lateralized in monkeys. *Behavioral and Brain Sciences, 11*, 723-725.
Harris, L.J., & Carlson, D.F. (1988). Hand preference for visually guided reaching in human infants and adults. *Behavioral and Brain Sciences, 11*, 726-727.
Hatta, T., & Koike, M. (1991). Left hand preference in frightened mother monkeys in taking up their babies. *Neuropsychologia, 29*, 207-209.
Hauser, M., Perry, P., Manson, J.H., Ball, H., Williams, M., Pearson, E., & Berard, J. (1991). It's all in the hands of the beholder. *Behavioral and Brain Sciences, 14*, 342-344.
Heestand, J.E. (1986). *Behavioral lateralization in four species of apes.* Unpublished Ph.D. dissertation, University of Washington, Seattle, WA.
Heffner, H.E., & Heffner, R.S. (1984). Temporal lobe lesions and the perception of species-specific vocalizations by macaques. *Science, 226*, 75-76.

Hepper, P.G., Shahidullah, S., & White, R. (1990). Origins of fetal handedness. *Nature, 347*, 431.

Hewes, G.W. (1973). Primate communication and the gestural origin of language. *Current Anthropology, 14*, 5-24.

Hopkins, W.D., Washburn, D.A., & Rumbaugh, D.M. (1989). A note on hand use in the manipulation of joysticks by two rhesus monkeys and three chimpanzees (*Pan troglodytes*). *Journal of Comparative Psychology, 103*, 91-94.

Jerison, H.J. (1987). Which hand lost its cunning? *Behavioral and Brain Sciences, 10*, 278-279.

Kimura, D. (1987). Are men's and women's brains really different? *Canadian Psychology, 28*, 133-147.

Kuhl, P.K. (1988). On handedness in primates and human infants. *Behavioral and Brain Sciences, 11*, 739-741.

Landau, B. (1984). Human evolution as narrative. *American Scientist, 72*, 262-268.

Lewin, R. (1987). *Bones of contention*. New York: Simon & Shuster.

Lieberman, P. (1984). *The biology and evolution of language*. Cambridge, Mass.: Harvard University Press.

Lindblom, B., MacNeilage, P.F., & Studdert-Kennedy, M.G. (1991). Evolution of Speech (In preparation).

Lutz-Maki, S.G., & MacNeilage, P.F. (1991). Right side preference for force application in the chimpanzee. *American Journal of Primatology, 24*, 117.

MacNeilage, P.F. (1987). The evolution of hemispheric specialization for manual function and language. In S. Wise (Ed.), *Higher brain functions: recent explorations of the brain's emergent properties* (pp.285-308). New York: Wiley.

MacNeilage P.F. (1991a). The postural origins theory of neurobiological asymmetries in primates. In N. Krasnegor, D. Rumbaugh, M.G. Studdert-Kennedy & R. Schiefelbusch (Eds.), *The biological foundations of language development* (pp.165-188). Hillsdale, N.J.: Erlbaum.

MacNeilage, P.F. (1991b). Patterns of handedness across the primate order. In A. Ehara, T. Kimura, O. Takenaka, and M. Iwamonto (Eds.), *Primatology today* (pp.275-278). Amsterdam: Elsevier Scientific Publications.

MacNeilage, P.F., & Davis, B.L. (1990). Acquisition of speech: the achievement of segmental independence. In W.J. Hardcastle & A. Marchal (Eds.), *Speech production and speech modelling* (pp. 55-68). Dordrecht: Kluwer.

MacNeilage, P.F., Studdert-Kennedy, M.G., & Lindblom, B. (1987). Primate handedness reconsidered. *Behavioral and Brain Sciences, 10*, 247-263.

MacNeilage, P.F., Studdert-Kennedy, M.G., & Lindblom, B. (1988). Primate handedness: A foot in the door. *Behavioral and Brain Sciences, 11*, 748-758.

MacNeilage, P.F., Studdert-Kennedy, M.G. & Lindblom, B. (1991). The other theory, the other hand and the other attitude. *Behavioral and Brain Sciences, 14*, 344-349.

Maki, S. (1990). *An experimental approach to the postural origins theory of neurobiological asymmetries in primates*. Unpublished Ph.D. Dissertation, University of Texas at Austin, Austin, TX.

Manning, J.T., & Chamberlain, A.T. (1990). Left sided cradling preferences in great apes. *Animal behavior, 39*, 1225-1227.

McManus, I.C. (1986). Review of 'Left Side, right side' by Alan Beaton. *British Journal of Psychology, 77*, 419-421.

McManus, I.C. (1987). On the one hand, on the other hand: Statistical fallacies in laterality research. *Behavioral and Brain Sciences, 10*, 282-4.

Nass, R., & Gazzaniga, M. (1987). Cerebral lateralization and specialization in human nervous system. In Plum F. (Ed.), *Handbook of Physiology: Section 1. The Nervous System* (pp. 701-762). Bethesda: American Physiological Society.

Olson, D.A., Ellis, J.E., & Nadler, R.D. (1990). Hand preferences in captive gorillas, orang-utans and gibbons. *American Journal of Primatology, 20*, 83-94.

Petersen, M., Beecher, B., Zoloth, S., Moody, D., & Stebbins, W. (1978). Neural lateralization of species-specific vocalizations by Japanese macaques (*Macaca fuscata*). *Science, 202*, 324-327.

Pinker, S., & Bloom, P. (1990). Natural language and natural selection. *Behavioral and Brain Sciences, 13*, 707-784.

Redican, W.K. (1975). Facial expressions in nonhuman primates. In L.A. Rosenblum (Ed.), *Primate behavior: Developments in field and laboratory research, Vol. 4* (pp. 103-194). New York: Academic Press.

Robinson, R.G. (1985). Lateralized behavior and neurochemical consequences of unilateral brain injury in rats. In S. Glick (Ed.), *Cerebral lateralization in nonhuman species* (pp. 135-156). New York: Academic Press.

Ruff, C.B., & Jones, H.H. (1981). Bilateral asymmetry in cortical bone of the humerus and tibia - sex and age factors. *Human Biology, 53*, 69-86.

Sanford, C.G., Guin, K., & Ward, J.P. (1984). Posture and laterality in the bushbaby. *Brain, Behavior and Evolution, 25*, 217-224.

Schaller, G.B. (1963). *The mountain gorilla*. Chicago: Chicago University Press.

Searleman, A. (1980). Subject variables and cerebral organization for language. *Cortex, 16*, 239-254.

Shapiro, R.M., Drew, K.L., & Glick, S.D. (1987). Brain asymmetry, animal. In G. Adelman (Ed.), *Encyclopedia of neuroscience* (pp. 150-152). Boston: Birkhauser.

Steklis, H.D., & Marchant, L.F. (1987). Primate handedness: reaching and grasping for straws? *Behavioral and Brain Sciences, 10*, 284-286.

Studdert-Kennedy, M.G. (1990). This view of language. *Behavioral and Brain Sciences, 13*, 758-759.

Trevarthen, C. (1978). Manipulative strategies of baboons and the origins of cerebral asymmetry. In M. Kinsbourne (Ed.), *Asymmetrical function of the brain* (pp. 329-389). Cambridge: Cambridge University Press.

Ward, J.P. (1991). Prosimians as animal models in the study of neural lateralization. In F.L. Kitterle (Ed.), *Cerebral lateralization: Theory and research: The Toledo symposium* (pp.1-17). Hillsdale N.J.: Erlbaum.

Warren, J.M. (1980). Handedness and Laterality in Humans and Other Animals. *Physiological Psychology, 8*, 351-359.

Warren, J.M. (1987). Primate Handedness: Inadequate analysis, invalid conclusions. *Behavioral and Brain Sciences, 10*, 288-289.

Witelson, S. (1987). Brain asymmetry, functional aspects. In G. Adelman (Ed.), *The Encyclopedia of neuroscience* (pp. 152-156). Birkhauser: Boston.

Vauclair, J., & Fagot, J. (1987). Visually guided reaching in adult baboons. *Behavioral and Brain Sciences, 10,* 287.

Author Index

A

B

C

D

Z

Subject Index